Introduction to Ordinary Differential Equations

Zane C. Motteler
Gonzaga University

Prindle, Weber & Schmidt
Boston, Massachusetts

To my parents Roy and Betty Motteler
They ringed me round with glad cries of freedom

Prindle, Weber & Schmidt, Inc.
53 State Street, Boston, Mass. 02109

Library of Congress Catalog Card Number: 72-186170

SBN 87150-136-8

Printed in the United States of America

Preface

This book is intended for a first course in differential equations. For many students this will be their first upper division course after a standard calculus sequence. I assume that this "standard" calculus course includes an introduction to infinite series, including Taylor and Maclaurin series, as well as an introductory treatment of vectors and linear algebra. In order that this text be as self-contained as possible, I have added an appendix which summarizes those topics less likely to have been treated in detail in an elementary calculus course: certain manipulations with integrals, results on infinite series, and a brief introduction to complex numbers.

Much of this book is based on a one semester course which I have taught many times at Gonzaga University. At this relatively small institution a typical class contains a mixture of students with varying majors (engineering, physical sciences, mathematics), and widely differing career goals (elementary and secondary teacher, programmer, technician, engineer, scientist, even professor). For such a mixed bag of students, experience has taught me that a course with a practical orientation, toward solving rather than theorizing about differential equations, is the best. To me, this approach is equally valid at the large university as well. Sophomores and juniors are still learning what Newton did and why. It is for this reason that this book emphasizes the practical techniques of solution and includes numerous examples of physical applications. Without the latter, I have found that even a course in methods of solution which, to a mathematician, is quite adequately mechanical and applied, nevertheless seems quite theoretical to the typical student at this level. Naturally I do mention existence and uniqueness theorems where appropriate, and I even prove a few of the simpler ones. But a presentation of the rigorous abstract theory of differential equations is not the purpose of this text.

Some may question what appears to be the premature introduction of partial differential equations and of the separation of variables technique. Unfortunately, there seems to be no better way of motivating the study of series solutions and their properties than by the anticipation of this subject for a future course. From physics we obtain a clear indication of the type of solution we must choose, and consultation of a model in this way is always a fruitful course to pursue in applied mathematics.

A special word of gratitude goes to two colleagues, Professor Stanley Lukawecki of Clemson University and Professor Foster Brooks of Kent State University, both of whom painstakingly reviewed several drafts of this book and offered numerous valuable suggestions that contributed significantly to its present form.

Finally a word of thanks to a number of individuals who helped beyond measure: to my four former students, and always friends, Jeff Jones, Mark Castner, Mary Ann Henry (now Stevens), and Betty Fessler (now Braet), who worked almost all of the problems; to a patient and long-suffering editor, Mr. John Martindale; and to my wife Marilynn, who typed the various drafts and whose support kept me going when discouragement and fatigue could have induced me to quit.

Zane C. Motteler
Spokane, Washington

Contents

Appendix

1

Definitions of Terms; First-Order Equations

The motivation for, and immediate consequence of, Newton's development of the calculus was the formulation and solution of a set of differential equations partially describing the dynamics of planetary motion and explaining Kepler's three observational laws. Indeed, the calculus has as its first stimulus and as its ultimate goal to set up and to solve differential equations, the solutions of which describe observed physical facts with some degree of accuracy. This is not always apparent from the way the elementary calculus is usually taught, since we so often motivate the development of the derivative and integral by means of geometrical examples and of physical applications which are shorn of all but the most trifling reference to relatively trivial differential equations. Only toward the end of the traditional course does one begin to study differential equations, but at least by this time all the tools are at hand.

Differential equations can be studied from at least two points of view: (1) the purely mathematical, in which differential equations are studied as abstract mathematical entities about which one proves theorems; and (2) the utilitarian, where closer attention is paid to those equations which arise from physical problems, and where intuition is used more than rigor to justify the results. Our approach to the subject will be somewhat more the latter than the former.

Differential equations arise in the study of natural phenomena in which change is apparent as time passes: the motion of bodies, chemical reactions, the flow of heat, and so on. As our study proceeds, we shall refer to many of these examples for the heuristic insight which they give into the mathematical process. First, however, we shall need to establish our notation and terminology.

1.1 Definitions

Let $F(u_1, u_2, \ldots, u_{n+2})$ be a given single-valued function of the $n + 2$ variables $u_1, u_2, \ldots, u_{n+2}$. We shall usually assume that F is defined and (at least) continuous for all real values of the variables (in this text we shall only briefly treat the case in which the variables assume complex values). Then an expression of the form

$$F(x, y, y', y'', \ldots, y^{(n)}) = 0 \tag{1.1}$$

is called an *ordinary differential equation* in which x is the independent variable and $y = y(x)$ is an unknown function of x which is to be determined. The notations $y', y'', \ldots, y^{(n)}$ represent the first n derivatives of y with respect to x. An *initial value problem* consists of the differential equation plus n so-called *initial conditions*

$$y(a) = c_1, y'(a) = c_2, \ldots, y^{(n-1)}(a) = c_n. \tag{1.2}$$

Although we have chosen to denote the variables by x and y, the reader should be aware that any other two letters could just as well have been used.

An example of an initial value problem is the following, which will be derived in Section 1.3:

$$\dot{m}(t) + \kappa m(t) = 0, \quad m(0) = m_0;$$

this problem describes the ongoing process of radioactive decay in a piece of radioactive material with initial mass m_0. The solution $m(t)$ represents the amount of mass of the original material still remaining after t units of time.

The reader should note that the dot notation ($\dot{m}$ as above, $\dot{x}$, $\ddot{x}$, etc.) is often used to represent derivatives when the independent variable represents time. In fact, all the following notations are equivalent:

$$x'(t), \dot{x}(t), \frac{dx}{dt}, D_t x \quad \text{(or just } Dx\text{)}.$$

The prime ($'$), dot ($\dot{}$), and D notations are used for brevity whenever there is no confusion as to the identity of the independent variable; when confusion is likely to arise, we shall use d/dt or D_t, d/dx or D_x, etc.

In section 1.6 we shall study another initial value problem

$$\dot{x}_2 + \frac{1}{k}x_2 = \frac{1}{k}\left(\frac{1}{2}a_1 t^2 - 11\right), \quad x_2(0) = -11,$$

the solution of which describes the motion of an automobile second in line at a traffic light, under certain specified conditions.

A function $y = f(x)$ is a *solution* of (1.1) if it is n times differentiable and if the substitution of $f^{(k)}(x)$ for $y^{(k)}$ $(k = 0, 1, 2, \ldots, n)$ in (1.1) reduces it to an identity in x. It is important to note that, in the process of solving a differential equation, we may obtain the solution function either in the explicit form $y = f(x)$ or in an implicit form such as $g(x, y) = 0$. For example, the solution of the differential equation $x + yy' = 0$ can be represented in implicit form as $x^2 + y^2 = c$, or in explicit form as $y = \sqrt{c - x^2}$ and $y = -\sqrt{c - x^2}$, where c is an arbitrary constant. The second form of the solution is more clumsy in this case, since it is necessary to give two formulas in order to encompass the whole solution.

A *solution of the initial value problem* (1.1), (1.2) is a function $y = y(x)$ which is a solution of the differential equation (1.1) in the sense defined above, and which satisfies the initial conditions (1.2). For example, either $y = \sqrt{1 - x^2}$ or $x^2 + y^2 = 1$ represents a solution of the initial value problem $x + yy' = 0$, $y(0) = 1$. The goal of this course is to learn how to obtain or to approximate the solutions of ordinary differential equations and initial value problems. We shall devote a portion of our time to the approximation of solutions, not only because the widespread use of electronic digital computers makes this a desirable technique, but also because it is very difficult or even impossible to obtain explicit or implicit formulas for the solutions of most differential equations.

The *order* of a differential equation is the order of the highest derivative present; thus, the order of Equation (1.1) is n, and the order of the equation in the previous paragraph is 1. The *degree* of a differential equation (when defined) is the power to which the highest derivative is raised. The equation

$$(y'')^2 + 3x^2y = 5$$

is of second degree, while the degree of

$$y'' \sin y'' = 4xy - e^x$$

is not defined. Degree is defined only for polynomial functions, while the sine function is transcendental and not a polynomial at all. In fact, as we recall from calculus, the Maclaurin expansion for the sine,

$$\sin u = u - \frac{u^3}{3!} + \frac{u^5}{5!} - \frac{u^7}{7!} + \cdots$$

contains no highest power of u.

As a further example, note that the equation

$$(y'')^2 + (y')^3 - y = 0$$

is of *second* degree, because the highest derivative, y'', is raised to the second power. This equation is *not* third degree, since y' is not the highest derivative present. The degrees of lower derivatives are irrelevant in determining the degree of an equation.

A *linear ordinary differential equation* is of the form

$$a_n(x)y^{(n)} + a_{n-1}(x)y^{(n-1)} + \cdots + a_1(x)y' + a_0(x)y = f(x), \tag{1.3}$$

where $a_0(x), \ldots, a_n(x)$ and $f(x)$ are given functions of x. Note that y and all its derivatives occur to the first power, multiplied only by functions of the independent variable x. Equation (1.3) is called *homogeneous* if $f(x) \equiv 0$, otherwise *inhomogeneous*. Much of our attention in this book will be directed at linear equations; however, our preliminary study of first-order equations and our later chapter on numerical techniques will point out some of the problems in dealing with nonlinear equations.

Problems

In the following problems, determine the order of the equation, the degree (if defined), whether or not the equation is linear, and, if it is linear, whether or not it is homogeneous. Note that a linear equation is always of first degree. State reasons for your answers.

1. $\sqrt{1 + (y'')^2} + y' = x^2 + 2.$
2. $\sqrt{1 + x^2 y'''} + (\sin x)\, y' + \cos x = 0.$
3. $y^3(y^{(4)})^2 + 2yy'' + y^2 - 1 = 0.$
4. $5y'' + 6y' - 7y = 0.$
5. $(x^2 - y^2)y' + 3xy = 0.$
6. $17xe^{-x}y - 5(\sin x)\, y' = 1.$
7. $3x(\sin x)yy' + \frac{3}{2}y^2 \sin x + \frac{3}{2}xy^2 \cos x = 0.$
8. $yy''' + y'' + y' + 15xy = \csc x.$
9. $5xyy'' + (y')^3 + 17xe^y = 0.$
10. $3y^{(5)} + \sin y''' - 8y' = 1 - x^2.$
11. $(y''')^3 y'' y' = ye^x.$
12. $\cos(y''y') - ye^x = \cos(xy).$
13. $y' + 1 = (y')^2 - e^x.$
14. $x^2(\sin x)\, y'' + y'/(1 + x^2) = e^{-x} \cos x.$
15. $(y''')^4 y + 67x(y')^3 + 2xy^2 = 0.$

16. $(x^2 - 1)\,y'' + 2xy' - 7xy = 0.$
17. $5y^{(4)} + 7y''' - 12y'' + y = x^2 - 3x + 2.$
18. $2 \ln\ y' - \sec y = xy^2 \cos x.$

In the following problems, determine whether or not the given function is a solution of the given equation or initial value problem. Note that a function can be a solution of a differential equation and not be a solution of an initial value problem involving that equation.

19. $e^x + 2e^{-x};\, y'' - y = 0.$
20. $\frac{3}{2}xy^2 \sin x;\, 3x(\sin x)yy' + \frac{3}{2}y^2 \sin x + \frac{3}{2}xy^2 \cos x = 0.$
21. $e^{2x};\, y'' - 4y = 0,\, y(0) = 1,\, y'(0) = 1.$
22. $y \cos x - 1 = 0;\, y' \cos x - y \sin x = 0,\, y(0) = 2.$
23. $e^{-2x};\, y''' + y'' - y' - 2y = 0.$
24. $3e^x - e^{-x};\, y''' + y'' - y' - 2y = 0.$
25. $5 \cos x;\, y''' - 2y'' + y' - 2 = 0,\, y(0) = 5.\, y'(0) = 0,\, y''(0) = -5.$
26. $x^{-1};\, xy'' + 3xy' + y = 0.$
27. $-\frac{1}{2}x \cos x + 3 \sin x;\, y'' + y = \sin x.$
28. $x^2 + 2x + 1;\, (x^2 - 1)y'' + 2y' - 2y = 0.$
29. $4x^2y = \sin x;\, 4x^2y' - 8xy = \cos x.$
30. $3x^2 + 1;\, \frac{1}{6}y'' - \frac{1}{6}xy' + y = 1.$
31. $ax + b;\, y'' = 0.$
32. $3x + 2;\, y'' = 0;\, y(0) = 3,\, y'(0) = 2.$
33. $5 \cos x - 1;\, y^{(5)} - y^{(4)} + 5y''' - 5y'' + 4y' - 4y = 4.$
34. $3 \sin 2x + e^x - 1;\, y^{(5)} - y^{(4)} + 5y''' - 5y'' + 4y' - 4y = 4.$
35. $e^{-x} - 1;\, y^{(5)} - y^{(4)} - 4y''' - 5y'' + 4y' - 4y = 4.$

1.2 The Simplest Differential Equations

Naturally, the simplest of all differential equations is

$$y' = f(x). \tag{1.4}$$

Why do we say that this is the simplest? Not only because it is linear and first order, with y' on one side and a function of x alone on the other, but also because every student who has studied the integral calculus can instantly write down its solution:

$$y(x) = \int f(x)dx + c, \tag{1.5}$$

where c is an arbitrary constant and where $\int f(x)dx$ represents the indefinite integral, or antiderivative, of $f(x)$. Clearly the value of y given by (1.5) is an explicit solution of equation (1.4), as can be verified by substituting it into the equation. We also see that expression (1.5) gives *an infinite number of solutions* of the differential equation, because c can assume any constant value whatsoever. If we combine equation (1.4) with an initial condition

$$y(a) = b, \tag{1.6}$$

then the solution of the initial value problem (1.4), (1.6) is clearly

$$y(x) = \int_a^x f(t)dt + b. \tag{1.7}$$

We use the dummy variable t under the integral sign to avoid confusion with the variable upper limit x. Again, substitution of (1.7) into (1.4), (1.6) verifies that it is indeed a solution.

Once we have exhibited a solution of a differential equation, the question of *existence* of a solution has been answered. The only remaining question is that of *uniqueness*: Is (1.7) the only possible solution of the initial value problem (1.4), (1.6)? The answer is yes, for suppose that (1.4), (1.6) has two solutions, $y_1(x)$ and $y_2(x)$, i.e.,

$$\begin{aligned} y_1' &= f(x), \quad y_1(a) = b; \\ y_2' &= f(x), \quad y_2(a) = b. \end{aligned}$$

Then by subtracting the second line from the first, we obtain

$$y_1' - y_2' = (y_1 - y_2)' = 0, \quad y_1(a) - y_2(a) = (y_1 - y_2)(a) = 0;$$

that is, the function $y_1 - y_2$ has a zero derivative and goes through the point $(a, 0)$. The only continuous differentiable function having these two properties is the identically zero function, i.e., $y_1 - y_2 \equiv 0$, implying that $y_1 \equiv y_2$. Hence the two "different" solutions of (1.4), (1.6) are in fact identical, so the problem has only one solution, as claimed. [The reader should recall the important distinction between equality ($=$) and identity ($\equiv$) when applied to functions. To say $f(x) = 0$ merely implies that $f(x)$ is zero for *certain* values of x, as in the equation $x^2 - 4 = 0$; to say $f(x) \equiv 0$ means that $f(x)$ vanishes for *every possible value* of x.]

It should be noted here that the previous paragraph is rather naive, in that $f(x)$ must obviously be such that it is integrable between a and x so that a solution exists in the first place. Also, we have said that the only solution of the initial value problem $y' = 0$, $y(a) = 0$ is the function $y(x) \equiv 0$. How do we know this? In mathematics it is never safe to take a seemingly simple fact for granted. Indeed, this statement must be proved, and is left to the reader in the problems at the end of this section.

We actually have the following:

Theorem

The initial value problem

$$y' = f(x), \quad y(a) = b,$$

where f is integrable for values of x near a, has a unique solution that is given by

$$y = \int_a^x f(t)dt + b.$$

The most general ordinary differential equation of the first order and degree is of the form

$$y' = f(x, y),$$

where the right side depends on y as well as x. We have seen above that one can state and prove an existence and uniqueness theorem for this equation when y is missing; it is no longer a trivial problem, however, when y is present. Nevertheless, we can state an existence and uniqueness theorem here and give an indication of the proof. First we shall need to define one term:

Definition

Suppose $f(x, y)$ is defined for all x in an interval I and all y in an interval J. Then f is said to satisfy a *Lipschitz condition in y on $I \times J$** if there exists a constant N such that

$$|f(x, y_1) - f(x, y_2)| \leqq N|y_1 - y_2|$$

for all y_1 and y_2 in J and for all x in I. (Figure 1.1.)

The restriction on f imposed by a Lipschitz condition is not an especially strong one. For instance, every function having a bounded partial derivative with respect to y satisfies a Lipschitz condition; for, by the mean value theorem, for each x, y_1, and y_2 there exists an η between y_1 and y_2 such that

$$f(x, y_1) - f(x, y_2) = f_y(x, \eta)(y_1 - y_2),$$

* Recall that $I \times J$, called the *Cartesian product of I and J*, is simply the set of all ordered pairs (x, y) with x in I and y in J.

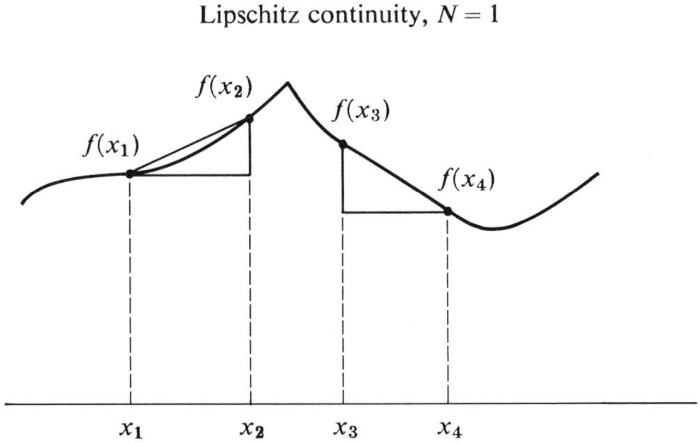

Figure 1.1

$$|f(x_2) - f(x_1)| \leq |x_2 - x_1|$$
$$|f(x_4) - f(x_3)| \leq |x_4 - x_3|$$

where f_y is the partial derivative of f with respect to y. Because $|f_y(x, \eta)|$ is bounded by a constant (say, M) we have

$$|f(x, y_1) - f(x, y_2)| \leqq M|y_1 - y_2|.$$

Our existence and uniqueness theorem is as follows:

Theorem 1.1

Let $f(x, y)$ be continuous and bounded for all (x, y) in $I \times J$ and satisfy a Lipschitz condition in y there. Let (a, b) be a point in $I \times J$. Then there exists a constant c and one and only one continuously differentiable function $y = y(x)$ on $|x - a| < c$ satisfying the initial value problem $y' = f(x, y)$. $y(a) = b$.

We shall not go into the details of the existence proof since it is rather lengthy and would take us too far afield from our primary task of learning practical methods of solution. However, it might be instructive to outline the proof.

We define a sequence of functions $\{y_n(x)\}$ by means of the equations

$$y_0(x) = b, \quad y_1(x) = b + \int_a^x f(t, y_0(t))\, dt, \ldots,$$

$$y_{n+1}(x) = b + \int_a^x f(t, y_n(t))\, dt, \ldots.$$

We must then prove that the sequence $\{y_n\}$ converges uniformly to a function y which satisfies

$$y(x) = b + \int_a^x f(t, y(t))\, dt$$

and is hence a solution of the initial value problem. The Lipschitz condition on f is crucial in this proof, as it is in the uniqueness. Unlike many existence proofs in mathematics, this one (often called "the method of successive approximations") is constructive in nature and gives a quantitative estimate of the actual solution. The method is due to E. Picard.*

The uniqueness proof is much easier than the existence proof; for this reason, and because it is much shorter, we shall carry it out in detail. As always, a uniqueness proof begins with the assumption that there are two solutions of the problem $y_1(x)$ and $y_2(x)$, and it is necessary to prove that they are identical. Since both functions are solutions, we have

$$y_1(x) = b + \int_a^x f(t, y_1(t))\, dt,$$

$$y_2(x) = b + \int_a^x f(t, y_2(t))\, dt.$$

Subtracting the second equation from the first, we obtain

$$y_1(x) - y_2(x) = \int_a^x [f(t, y_1(t)) - f(t, y_2(t))]\, dt.$$

Recalling that $|\int g(t)\, dt| \leqq |\int |g(t)| dt|$ for any integrable function g, we take absolute values of both sides and apply this inequality to obtain

$$|y_1(x) - y_2(x)| \leqq \left|\int_a^x |f(t, y_1(t)) - f(t, y_2(t))|\, dt\right|$$

and, by the Lipschitz condition on f, this yields

$$|y_1(x) - y_2(x)| \leqq N \left|\int_a^x |y_1(t) - y_2(t)|\, dt\right|.$$

* For a complete proof, see, for example, M. Tenenbaum and H. Pollard, *Ordinary Differential Equations* (Harper & Row, New York, 1963), p. 734.

The integral on the right is bounded by the length of the interval times the maximum of the integrand over that integral, so

$$|y_1(x) - y_2(x)| \leqq N\,|x - a| \max |y_1(t) - y_2(t)|,$$

where the maximum is taken over all t between a and x. We can now use this with the preceding inequality to obtain

$$\begin{aligned} |y_1(x) - y_2(x)| &\leqq N \left| \int_a^x N|\,t - a\,|\max|\,y_1(t) - y_2(t)\,|dt \right| \\ &= N^2 \frac{(x-a)^2}{2} \max |y_1(t) - y_2(t)|. \end{aligned}$$

Using the latter result in the same inequality again we obtain

$$\begin{aligned} |y_1(x) - y_2(x)| &\leqq N \left| \int_a^x N^2 \frac{(t-a)^2}{2} \max|\,y_1(t) - y_2(t)\,|dt \right| \\ &= N^3 \frac{|x-a|^3}{3!} \max |y_1(t) - y_2(t)|. \end{aligned}$$

It is a simple matter to show by induction that if we repeat the process we will have

$$|y_1(x) - y_2(x)| \leqq N^n \frac{|x-a|^n}{n!} \max |y_1(t) - y_2(t)|$$

for all integer values of n. $N^n|x - a|^n/n!$ is simply the nth term in power series for $\exp(N|x - a|)$,* and since this series converges for all values of x, its nth term goes to zero as n goes to infinity. Furthermore, because y_1 and y_2 are continuous on $|x - a| \leqq c$, it follows that $\max |y_1(t) - y_2(t)|$ is a finite constant, and so

$$|y_1(x) - y_2(x)| \leq \lim_{n \to \infty} \frac{N^n|x-a|^n}{n!} \max |y_1(t) - y_2(t)| = 0.$$

Since x is arbitrary, y_1 and y_2 must be equal for all values of x and are hence identical; therefore the solution is unique.

* We shall often use the notation $\exp(\cdots)$ in place of $e^{(\cdots)}$, especially when the quantity in parentheses is somewhat involved.

Some time back we saw that it is possible to solve the differential equation $y' = f(x)$ with the expression $y(x) = \int f(x)\,dx + c$, containing one arbitrary constant: This expression is called the general solution of the differential equation. The expression

$$y(x) = \int_a^x f(t)\,dt + b,$$

the solution of the initial value problem, is called a *particular solution* of the differential equation, because it is a solution in which a particular range of integration and a particular value of the constant c have been chosen.

The goal of the study of first-order equations is to reduce them to the form (1.4), or something equally simple, in order to be able to integrate and solve. While it is seldom possible to put an equation of the first order into a form quite as simple as (1.4), one can often put a first-order equation into the form

$$y'f(y) = g(x) \tag{1.8}$$

in which, at least, the left side contains only y and its derivative, and the right side is a function of x alone. A first-order equation that can be put into form (1.8) is said to have *variables separable*. This equation can always be solved provided that $g(x)$ is integrable with respect to x and $f(y)$ is integrable with respect to y (by the chain rule), and its solution is

$$\int f(y)\,dy = \int g(x)\,dx + c.$$

Of course, there is always a chance, once the variables have been separated, that we can recognize $y'f(y)$ as the derivative of some other function of y, say $h(y)$, so that $y'f(y) = h'(y)$, where the prime denotes differentiation with respect to x. In this case (1.8) becomes

$$h'(y) = g(x), \tag{1.9}$$

for which the general solution is

$$h(y) = \int g(x)\,dx + c;$$

we recognize this expression as being an *implicit* equation defining the solution, which we may or may not be able to solve for y as an explicit function of x.

An example of a first-order equation with variables separable is the following:

$$y(x - 2) + 4(x - 1)y' = 0.$$

This can be written

$$4(x - 1)y' = -y(x - 2),$$

so that

$$\frac{4y'}{y} = -\frac{(x-2)}{(x-1)} = -\frac{(x-1)-1}{(x-1)} = -\left(1 - \frac{1}{x-1}\right) = \frac{1}{x-1} - 1$$

is the result after separation of variables. Integrating, we obtain

$$4 \ln y = \ln y^4 = \ln(x - 1) - x + c, \tag{1.10}$$

which upon exponentiation gives

$$e^{\ln y^4} = y^4 = e^{\ln(x-1)-x+c} = e^{\ln(x-1)}e^{-x}e^c = k(x - 1)e^{-x}, \tag{1.11}$$

where $k = e^c$.* Either form, (1.10) or (1.11), is acceptable as a solution of the given differential equation. Note, however, that in the first solution, y is defined only for $x > 1$, and in the second, for $x \geqq 1$. Neither expression gives a solution which is valid for all values of x, but there is no reason to expect that it should; Theorem 1.1 guarantees validity only in an interval. Note also that both expressions are implicit equations for solutions of the problem and that in $4 \ln y$, y cannot be negative, whereas in $\ln y^4$ or y^4 it can be either positive or negative. Thus when we pass algebraically from one form of the solution to another, not only the domain of validity of the solution, but also the range of values of y, may change. Finally, an implicit solution of a differential equation may, as in the case of y^4, define y as a multiply valued function of x.

Naturally most first-order differential equations do not have variables separable. However, it is sometimes possible to make a change of variables so that the variables in the new equation can then be separated and the equation solved. For example, this can be done for equations of the form $F(x, y, y') = 0$ which are of first order and degree when F has the property that $F(\alpha x, \alpha y, y') = \alpha^k F(x, y, y')$ for some integer k; in other words, the

* In performing calculations like these, the reader should recall the basic law of exponents $e^{a+b} = e^a e^b$, and that by definition of the natural logarithm, $e^{\ln u} = u$ and $\ln e^u = u$, for any u whatsoever for which $\ln u$ has meaning.

substitution of αx and αy for x and y does not change the differential equation. A function F with the above property is said to be *homogeneous in x and y of degree k*. The differential equation $F(x, y, y') = 0$ is called *homogeneous*; note that this use of the term "homogeneous" differs from the usage defined in the previous section regarding linear equations. No confusion should arise from this fact, since the two usages will always occur in different contexts.

For example, $xy' - y = 0$ is homogeneous of degree one because $(\alpha x)y' - \alpha y = \alpha(xy' - y)$. Similarly, the reader can verify that the equation

$$2xyy' \cos(y/x)^2 - [2y^2 \cos(y/x)^2 + x^2 \sin(y/x)^2] = 0$$

is homogeneous of degree two. This second equation, because of the repeated occurrence of the expression (y/x), suggests that a possible method of solution might be to use the substitution $v(x) = y(x)/x$. In fact, this method does work, although we have to restrict x to nonzero values unless it turns out after y has been found that y/x has a definite limit as x approaches zero.

Let us apply this substitution in our first equation, $xy' - y = 0$. Then, since $y(x) = xv(x)$, we have $y' = v + xv'$, and

$$xy' - y = x(v + xv') - xv - xv' - 0.$$

Therefore (assuming $x \neq 0$), $v' = 0$ and $v = c$, leading to the general solution $y = cx$. We see that in this case the solution is defined for $x = 0$.

The second of the above equations is more difficult. We have first

$$2x\, vx(v + xv') \cos v^2 - (2v^2x^2 \cos v^2 + x^2 \sin v^2) = 0,$$

which upon simplification yields

$$2x^3vv' \cos v^2 - x^2 \sin v^2 = 0;$$

with the variables separated, we obtain

$$2vv' \frac{\cos v^2}{\sin v^2} = \frac{1}{x}, \qquad \text{provided } x \neq 0.$$

To integrate the left side, we make the substitution $u = v^2$ in the integral

$$\int 2vv' \frac{\cos v^2}{\sin v^2}\, dx = \int 2v \frac{\cos v^2}{\sin v^2}\, dv,$$

and, because $du = 2v\,dv$ and $\cos u/\sin u = \cot u$, the integral becomes simply

$$\int \cot u\,du = \ln \sin u = \ln \sin v^2,$$

so that the solution of the differential equation is $\ln \sin v^2 = \ln x + k$, which upon exponentiating and replacing v by y/x, becomes $\sin(y/x)^2 = cx$, where $c = e^k$.

There are other substitutions which can also be made to reduce certain first-order equations to ones with variables separable; we shall consider two additional ones here.

1. When the same expression involving x and y occurs more than once in an equation, call it by a new name. Thus, in

$$(3x + 4y - 1) + (3x + 4y + 2)y' = 0,$$

make the substitution $v = 3x + 4y$. Then $v' = 3 + 4y'$, and we have

$$(v - 1) + (v + 2)(\tfrac{1}{4}v' - \tfrac{3}{4}) = 0,$$

which gives

$$\frac{1}{4}v' = \frac{3}{4} - \frac{v-1}{v+2} = \frac{-v+10}{4(v+2)}.$$

The variables are now separable, and we have

$$-1 = v'\frac{(v+2)}{(v-10)} = v'\frac{(v-10+12)}{v-10} = v'\left(1 + \frac{12}{v-10}\right),$$

which has the solution

$$v + 12\ln(v - 10) = -x + c$$

or, in terms of the original variables x and y,

$$4y + 12\ln(3x + 4y - 10) = -4x + c.$$

2. When the equation is of the form

$$(ax + by + c)y' + (dx + ey + f) = 0,$$

then replace x by $\tilde{x} + x_0$ and y by $\tilde{y} + y_0$, where (x_0, y_0) is the solution of the pair of linear equations

$$ax + by + c = 0, \quad dx + ey + f = 0,$$

provided that these two equations have a single solution $(ae - bd \neq 0)$. This has the effect of removing the constant terms c and f; geometrically, it amounts to a translation of the origin to the point (x_0, y_0). The equation then becomes homogeneous of degree one and can be solved by making the substitution $\tilde{y} = v\tilde{x}$. For example, consider

$$(x + 3y - 4) + (3x - y - 2)y' = 0.$$

The solution of the pair of equations

$$x + 3y - 4 = 0, \quad 3x - y - 2 = 0$$

is $(1, 1)$. Thus, replacing x by $\tilde{x} + 1$ and y by $\tilde{y} + 1$, we obtain

$$(\tilde{x} + 3\tilde{y}) + (3\tilde{x} - \tilde{y})\tilde{y}' = 0.$$

Now, making the substitution $\tilde{y} = v\tilde{x}$, $\tilde{y}' = v + \tilde{x}v'$, we have

$$(\tilde{x} + 3v\tilde{x}) + (3\tilde{x} - v\tilde{x})(v + \tilde{x}v') = 0,$$

which upon simplification becomes

$$\tilde{x}v'(3 - v) = v^2 - 6v - 1.$$

After separating variables and completing the square, we have

$$-\frac{-v'(v - 3)}{(v - 3)^2 - 10} = \frac{1}{\tilde{x}}, \qquad \tilde{x} \neq 0,$$

which yields

$$-\tfrac{1}{2}\ln[(v - 3)^2 - 10] = \ln \tilde{x} + k;$$

or, on multiplying through by -2,

$$\ln[(v - 3)^2 - 10] = -2\ln \tilde{x} + k = \ln \tilde{x}^{-2} + k = \ln(1/\tilde{x}^2) + k.$$

Exponentiating both sides, we obtain

$$(v-3)^2 - 10 = v^2 - 6v - 1 = \frac{c}{\tilde{x}^2}, \qquad \text{where } c = e^k.$$

Thus we have

$$(\tilde{y}/\tilde{x})^2 - 6(\tilde{y}/\tilde{x}) - 1 = c/\tilde{x}^2,$$

or

$$\tilde{y}^2 - 6\tilde{x}\tilde{y} - \tilde{x}^2 = c.$$

To get back to the original variables, we recall that $\tilde{x} = x - 1$ and $\tilde{y} = y - 1$. Thus after some further manipulations we obtain the final solution

$$y^2 - x^2 - 6xy + 4y + 8x - 6 = c,$$

or, by absorbing the 6 into the arbitrary constant,

$$y^2 - x^2 - 6xy + 4y + 8x = d, \qquad \text{where } c + 6 = \text{d}.$$

Problems

Find the general solutions of the following equations:

1. $y(1 + x)y' = (1 + y^2)x$.
2. $3y - x^2y' = 0$.
3. $y^3 + x^4y' = 0$.
4. $\cos\theta\rho' - \rho = 0$. ($\rho' = d\rho/d\theta$.)
5. $\operatorname{sech} z = t^2z'$.
6. $x^2yy' - (1 + x^2) = 0$.
7. $1 - x - y + xy = y'$.
8. $e^{x^2}y' + (y^2 - 1)x = 0$.
9. $(x \sin x - \cos x)/y = 4y'$.
10. $(2 + y)y' + xy + 2y - x - 2 = 0$.
11. $e^{x^2}y' = e^x(y + 1)(2x - 1)$.
12. $xy + (1 - x^2)y' = 0$.
13. $xy^2 + (1 + x^2)(1 + y)y' = 0$.
14. $x \sin y = (\sec x)y'$.
15. $(1 - x^2)(1 + y^2)y' = 3$.

Solve the following equations by making the appropriate change of dependent variable as discussed in the text.

16. $(3x + 2y - 2) + (3x + 2y + 2)y' = 0.$
17. $(x^3 + y^3) + 3xy^2y' = 0.$
18. $2x(x^2 - y) + (x^2 - y + 3)y' = 0.$
19. $(x + y) + (x - y)y' = 0.$
20. $(x + y - 1) + (x - y + 1)y' = 0.$
21. $(6y + x - 2)y' + (-6x + y + 7) = 0.$
22. $2xy + y'(x^2 - y^2) = 0.$
23. $(4y - 5x + 3)y' + (6x - 5y + 8) = 0.$
24. $(5x^4 - 6x^3y + 21x^2y^2 - 5y^4)y' + (20x^3y - 9x^2y^2 + 14xy^3) = 0.$
25. $(-18y + 7x - 6)y' + (8x + 7y - 2) = 0.$
26. Use the mean value theorem to show that the only continuously differentiable solution of $y' = 0$ is $y \equiv$ constant.

1.3 Two Applications

Certain chemical elements are inherently unstable in that the nuclei of their atoms spontaneously discharge particles and/or rays, and by so doing become atoms of a new element. Such unstable elements are called *radioactive*, and the process by which they are transmuted into new elements is called *radioactive decay*. The rate at which atoms are breaking down can be determined experimentally by detecting and counting the ejected particles with the use of such instruments as Geiger counters. A curious fact emerges from this measurement: The rate of radioactive decay of an element is directly proportional to the amount of mass present. Thus, for instance, a two-gram mass will show twice as many decays in a given time interval as a one-gram mass. If we represent the mass of the element at time t by $m(t)$, then the rate of change of mass is given by $\dot{m}(t)$, and the experimental law stated above is given by

$$\dot{m}(t) \propto m(t).$$

The constant of proportionality $-\kappa$ (where κ is positive) can be determined experimentally, so that we can rewrite this as

$$\dot{m}(t) = -\kappa m(t);$$

the negative sign indicates that the amount of mass is decreasing. Assuming that an initial mass m_0 is present at time $t = 0$, we have the following initial value problem:

$$\dot{m} + \kappa m = 0, \quad m(0) = m_0. \tag{1.12}$$

One might ask the following two questions: (1) How much mass will remain after a given amount of time? (2) How much time must pass in order that the mass present be reduced to half the original amount? (This amount of time is called the *half-life* of the element.)

In deriving this initial value problem we have made a number of simplifying assumptions. We have assumed that $m(t)$ is a continuous function, when in fact the total mass is made up of a finite though extremely large number of atoms. For another, most radioactive elements decay into other radioactive elements, which also decay. We must assume that the measurements are capable of distinguishing between the decays of the original element and those of the radioactive products of its decay. Both of these assumptions are reasonable in practice, and a great deal of useful information can be obtained from solving the differential equation. For example, we can determine the half-lives of elements which have extraordinarily long half-lives, ones which are far too long to be determined by observation.

In Chapter 6 we shall see a somewhat more complicated problem involving biological organisms, in which the population of one (the "predator") declines in the absence of the other (the "prey"), following a law exactly like that of radioactive decay. On the other hand, the population of the prey, in the absence of the predator, grows in direct proportion to the number of organisms present. When both organisms are together in the same environment, a system of equations (in which the two populations are the unknowns) describes the rather complicated relationship between them.

To solve the equation of radioactive decay, we write the equation as

$$\dot{m} = -\kappa m$$

and divide both sides of the equation by m, assuming for the moment that $m(t) \neq 0$ for any t. We thus obtain

$$\dot{m}/m = -\kappa.$$

We can now see that the left side is simply the derivative with respect to t of the natural logarithm of m, so that the solution is

$$\ln m = -\kappa t + c,$$

c being an arbitrary constant. We can solve this equation for m as an explicit function of t by exponentiating, and we obtain

$$e^{\ln m} = m = e^{-\kappa t + c} = e^c e^{-\kappa t} = ke^{-\kappa t},$$

where k is a new arbitrary constant. Thus, since $m = k\exp(-\kappa t)$, we see that the mass of a radioactive element decreases exponentially with time. We also see that our assumption that $m(t) \neq 0$ for any t was justified. Utilizing the initial condition $m(0) = m_0$, we have

$$m_0 = ke^{-\kappa 0} = k,$$

so that the function

$$m(t) = m_0 e^{-\kappa t}$$

is the complete solution of the initial value problem of radioactive decay.

Our second application concerns one of Kepler's three laws of planetary motion. In order to prove the law mathematically, we first have to derive the equation of motion of a particle under the influence of a central force. In the long run the efforts will be worthwhile, since we shall return to this equation in future chapters as we develop the techniques to attack it.

We assume, then, that a point mass P is moving in the Cartesian plane and that the only force acting upon it is acting in the direction of the origin O and depends only on the distance from the origin. It will be most convenient to use polar coordinates $(r(t), \theta(t))$ to specify the position of the particle at time t, and we use unit vectors $\boldsymbol{U}$ and $\boldsymbol{V}$ pointing in the directions of increasing r and θ, respectively, to represent the basis at this point (Figure 1.2). Note

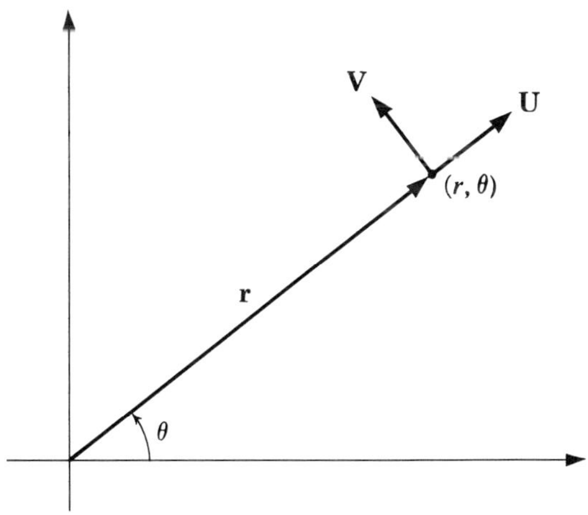

Figure 1.2

also that $\boldsymbol{U}$ and $\boldsymbol{V}$ vary with time—not in length, but in direction only—and that $\boldsymbol{U}$ and $\boldsymbol{V}$ do not change either in direction or in magnitude as r varies, since they both depend only on θ:

$$\boldsymbol{U} = \cos\theta\,\boldsymbol{i} + \sin\theta\,\boldsymbol{j},$$
$$\boldsymbol{V} = -\sin\theta\,\boldsymbol{i} + \cos\theta\,\boldsymbol{j},$$

where $\boldsymbol{i}$ and $\boldsymbol{j}$ are the customary unit vectors pointing in the direction of increasing x and y, respectively. Thus we have

$$\dot{\boldsymbol{U}} = (-\sin\theta\,\boldsymbol{i} + \cos\theta\,\boldsymbol{j})\dot{\theta} = \dot{\theta}\,\boldsymbol{V},$$
$$\dot{\boldsymbol{V}} = (-\cos\theta\,\boldsymbol{i} - \sin\theta\,\boldsymbol{j})\dot{\theta} = -\dot{\theta}\,\boldsymbol{U}.$$

(Because $\boldsymbol{i}$ and $\boldsymbol{j}$ do not change either in direction or magnitude from one point to another, their derivatives vanish.) Thus, since the position vector $\boldsymbol{r}$ of the point mass P is $\boldsymbol{r} = r\boldsymbol{U}$, the velocity vector is

$$\dot{\boldsymbol{r}} = \dot{r}\boldsymbol{U} + r\dot{\boldsymbol{U}} = \dot{r}\boldsymbol{U} + r\dot{\theta}\boldsymbol{V}.$$

The acceleration vector is then

$$\ddot{\boldsymbol{r}} = \ddot{r}\,\boldsymbol{U} + r\dot{\boldsymbol{U}} + \dot{r}\dot{\theta}\boldsymbol{V} + r\ddot{\theta}\boldsymbol{V} + r\dot{\theta}\dot{\boldsymbol{V}},$$

which, after simplification, gives

$$\ddot{\boldsymbol{r}} = (\ddot{r} - r\dot{\theta}^2)\boldsymbol{U} + (r\ddot{\theta} + 2\dot{r}\dot{\theta})\boldsymbol{V}.$$

Because of our hypothesis that the force $\boldsymbol{F}$ acting on P is in the direction of the origin and depends only on r, we have

$$\boldsymbol{F} = -f(r)\boldsymbol{U};$$

$f(r)$ is positive, and the negative sign represents the fact that it is an attractive force toward the origin, that is, opposite the direction of $\boldsymbol{U}$. According to Newton's second law, the rate of change of momentum is proportional to the force impressed, and is in the same direction as the force; mathematically, this is the familiar statement

$$\boldsymbol{F} = m\ddot{\boldsymbol{r}},$$

or, in our case,

$$-f(r)\boldsymbol{U} = m(\ddot{r} - r\dot{\theta}^2)\boldsymbol{U} + m(r\ddot{\theta} + 2\dot{r}\dot{\theta})\boldsymbol{V},$$

where m is the mass of the particle at P. We thus obtain a pair of differential equations describing the motion at P:

$$m(\ddot{r} - r\dot{\theta}^2) = -f(r), \quad m(r\ddot{\theta} + 2\dot{r}\dot{\theta}) = 0.$$

Let us examine the latter equation, which, because $m \neq 0$, says that $r\ddot{\theta} + 2\dot{r}\dot{\theta} = 0$. Upon multiplying through by r, we find that

$$0 = r^2\ddot{\theta} + 2r\dot{r}\dot{\theta} = \frac{d}{dt}(r^2\dot{\theta}) = 0.$$

The only function whose derivative vanishes identically is a constant, as we showed in the previous section, and hence we have

$$r^2\dot{\theta} = c. \tag{1.13}$$

This statement leads to Kepler's second law: the rate at which the radius vector from the origin to P sweeps out area is always constant (Figure 1.3); in other words, in equal time intervals, it sweeps out equal areas. Thus if r is large, the rate of change of the angle θ must be small, and conversely.

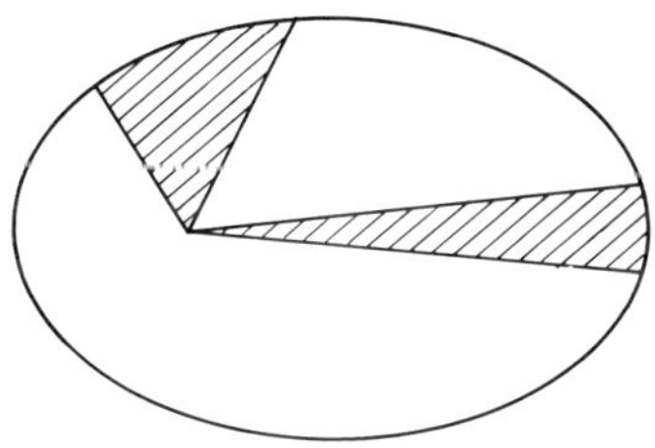

Figure 1.3

To see that (1.13) indeed implies the statement of Kepler's second law, we recall that the area as a function of time is given by

$$A(t) = \int_0^{\theta(t)} \int_0^{r(t)} r\,dr\,d\theta = \int_0^{\theta(t)} \tfrac{1}{2}r^2(t)d\theta = \int_0^t \tfrac{1}{2}r^2(t)\dot{\theta}(t)dt;$$

substituting c for $r^2\dot{\theta}$ from (1.13), we have $A(t) = \frac{1}{2}ct$. Thus, the area swept out by $\boldsymbol{r}$ between $t = t_1$ and $t = t_2$ is

$$A(t_2) - A(t_1) = \tfrac{1}{2}ct_2 - \tfrac{1}{2}ct_1 = \tfrac{1}{2}c(t_2 - t_1).$$

Since the area depends only on the *time interval*, we have proved our claim.

In future sections, after we have developed the techniques, we shall derive the first and third laws of Kepler. In order to do this we shall have to complete the solution of the equations of motion of the particle P, which is by no means a trivial process.

Problems

1. In a substance with radioactive decay constant κ, let T represent the half-life. Find a formula for T.

2. Below are given the decay constants for certain radioactive elements. Compute the half-lives of these elements.

a. Lithium-8: $0.81547\ \text{sec}^{-1}$.
b. Beryllium-12: $36.482\ \text{sec}^{-1}$.
c. Carbon-14: $1.2013 \times 10^{-4}\ \text{yr}^{-1}$.
d. Neon-24: $0.003817\ \text{sec}^{-1}$.
e. Aluminum-26: $9.3668 \times 10^{-7}\ \text{yr}^{-1}$.
f. Sulfur-35: $7.9948 \times 10^{-3}\ \text{day}^{-1}$.
g. Iron-59: $1.5403 \times 10^{-2}\ \text{day}^{-1}$.
h. Cobalt-60: $0.13152\ \text{yr}^{-1}$.
i. Strontium-90: $2.4755 \times 10^{-2}\ \text{yr}^{-1}$.
j. Lead-204: $4.951 \times 10^{-20}\ \text{yr}^{-1}$.
k. Radium-226: $4.2734 \times 10^{-4}\ \text{yr}^{-1}$.
l. Thorium-232: $4.9866 \times 10^{-11}\ \text{yr}^{-1}$.
m. Uranium-235: $9.7215 \times 10^{-10}\ \text{yr}^{-1}$.
n. Uranium-238: $1.5369 \times 10^{-10}\ \text{yr}^{-1}$.

3. Discuss the simplifying assumptions that would have to be made in order to conclude that the motion of the particle P is a reasonable mathematical model for the motion of a planet around its primary.

4. Complete the process of induction begun in the uniqueness theorem; i.e., show that the truth of

$$|y_1(x) - y_2(x)| \leqq \frac{N^n|x - a|^n}{n!} \max|y_1(t) - y_2(t)|$$

for $n = k$ implies its truth for $n = k + 1$.

1.4 Exact Equations

Equations with variables separable ultimately result from the *explicit* differentiation of expressions equating a function of x alone to a function of y alone. Exact equations, which we shall consider in this section, result from the *implicit* differentiation of a function of the two variables x and y. Indeed, if we have $f(x, y) = c$, then we know from the calculus that we can find the derivative of y as a function of both x and y by computing

$$\frac{\partial f}{\partial x} + \frac{\partial f}{\partial y} y' = 0, \tag{1.14}$$

and then solving for y'. This is obviously a differential equation of first order and degree for y, and its general solution is simply the original expression $f(x, y) = c$.

Any first-order first-degree equation of the form (1.14) is called *exact*. Our problem in this section will be to determine a testing procedure which will tell us when a general equation of the first order and degree

$$m(x, y) + n(x, y)y' = 0 \tag{1.15}$$

is exact; and when it is exact, to develop a method for finding the functional relationship $f(x, y) = c$ of which it is the total derivative.

Let us return to equation (1.14) in order to see what we can deduce from an equation already known to be exact. In this case we have $m(x, y) = \partial f/\partial x$, and $n(x, y) = \partial f/\partial y$. We immediately note that

$$\frac{\partial m}{\partial y} = \frac{\partial^2 f}{\partial y\,\partial x} \quad \text{and} \quad \frac{\partial n}{\partial x} = \frac{\partial^2 f}{\partial x\,\partial y},$$

provided, of course, that f possesses second derivatives. Now, provided that these second derivatives not only exist, but are continuous (which we shall always assume to be the case), we have

$$\frac{\partial^2 f}{\partial x\,\partial y} = \frac{\partial^2 f}{\partial y\,\partial x};$$

that is, the order in which the partial derivatives are taken is irrelevant. Consequently

$$\frac{\partial m}{\partial y} = \frac{\partial n}{\partial x}. \tag{1.16}$$

We have therefore deduced that (1.16) holds whenever equation (1.15) is known *a priori* to be exact. Since $\partial m/\partial y$ and $\partial n/\partial x$ can easily be calculated from (1.15) and checked for equality, we are impelled to ask next the converse question, whether the truth of formula (1.16) implies that equation (1.15) is exact, and, if so, how we might go about finding the solution $f(x, y) = c$ of (1.15).

Hence let us assume that we are given equation (1.15), and we have verified by explicit calculation that (1.16) is true. We then wish if possible to find a function $f(x, y)$ such that

$$m = \frac{\partial f}{\partial x}, \quad n = \frac{\partial f}{\partial y}. \tag{1.17}$$

Certainly, if there were an $f(x, y)$ such that the first of these equations held, then by integrating both sides with respect to x and holding y constant, we would have

$$f(x, y) = \int m(x, y)dx + c(y),$$

where the "constant" of integration c may depend on y, since y has been held constant in this reversal of the process of partial differentiation with respect to x. Then from the second equation in (1.17) we would have

$$\frac{\partial f}{\partial y} = n(x, y) = \frac{\partial}{\partial y}\int m(x, y)dx + c'(y) = \int \frac{\partial m}{\partial y}(x, y)dx + c'(y),$$

and from (1.16) we obtain

$$n(x, y) = \int \frac{\partial n}{\partial x}(x, y)dx + c'(y)$$

or

$$c'(y) = n(x, y) - \int \frac{\partial n}{\partial x}(x, y)dx. \tag{1.18}$$

The right side of this equation is a function of y alone, since its partial with respect to x vanishes:

$$\frac{\partial}{\partial x}\left(n - \int \frac{\partial n}{\partial x}dx\right) = \frac{\partial n}{\partial x} - \frac{\partial}{\partial x}\int \frac{\partial n}{\partial x}dx = \frac{\partial n}{\partial x} - \frac{\partial n}{\partial x} = 0.$$

Hence (1.18) can be solved for $c(y)$ simply by integrating with respect to y:

$$c(y) = \int\left[n(x, y) - \int \frac{\partial n}{\partial x}(x, y)dx\right]dy$$

and consequently

$$f(x, y) = \int m(x, y)dx + \int\left[n(x, y) - \int \frac{\partial n}{\partial x}(x, y)dx\right]dy = c, \tag{1.19}$$

where c is an arbitrary constant, is the general solution of (1.15), as can now be verified by substitution. Thus we have proved

Theorem 1.2
The equation (1.15) is exact if and only if $\partial m/\partial y = \partial n/\partial x$, and its general solution is given by (1.19).

In practice, although we could write down the solution of an exact equation directly by means of (1.19), it is usually simpler to proceed from the equation itself. Let us demonstrate the usual technique by solving the following example:

$$x^2y' + 2xy = 3x^2.$$

First we rewrite this in the form of (1.15),*

$$(2xy - 3x^2) + x^2y' = 0,$$

where $m(x, y) = 2xy - 3x^2$ and $n(x, y) = x^2$. We must now verify that this equation is exact, by calculating $\partial m/\partial y = 2x$, and $\partial n/\partial x = 2x$, which are indeed equal. If they were not, the equation would not be exact and it would be necessary to use some other technique to solve it. Next, we note that since

$$f(x, y) = \int m(x, y)dx + c_1(y) = \int n(x, y)dy + c_2(x),$$

in this case we have the two expressions

$$\begin{aligned} f(x, y) &= \int (2xy - 3x^2)dx + c_1(y) = x^2y - x^3 + c_1(y), \\ f(x, y) &= \int x^2dy \qquad\qquad + c_2(x) = x^2y + c_2(x). \end{aligned} \tag{1.20}$$

Both of these expressions must be equal; by comparing the two lines, we perceive that we must have $c_1(y) = 0, c_2(x) = -x^3$. Using either one of these in the appropriate one of the expressions (1.20), we obtain

$$f(x, y) = x^2y - x^3 = c,$$

which is the general solution of the given equation.

* Note that this step is not absolutely necessary, since, if the left side is an exact derivative, we will have solved the problem, since the right side is clearly the exact derivative of $x^3 + c$.

Let us run through a second example:

$$2y(x+1)(\sin y^2)y' = \cos y^2.*$$

In the standard form (1.15), this is

$$\cos y^2 - 2y(x+1)(\sin y^2)y' = 0,$$

so that

$$m(x, y) = \cos y^2, \qquad n(x, y) = -2y(x+1)\sin y^2.$$

(Always take into account the proper signs for m and n!) Then

$$\frac{\partial m}{\partial y} = (-\sin y^2)(2y), \qquad \frac{\partial n}{\partial x} = -2y \sin y^2,$$

which are equal; hence the equation is exact. So

$$f(x, y) = \int \cos y^2\, dx \qquad\qquad + c_1(y) = x \cdot \cos y^2 + c_1(y),$$

$$f(x, y) = \int (-2y)(x+1)\sin y^2\, dy + c_2(x) = (x+1)\cdot \cos y^2 + c_2(x).$$

Again comparing the two lines, which must be equal, we obtain that $c_1(y) = \cos y^2$, and $c_2(x) = 0$. Therefore

$$f(x, y) = (x+1)\cos y^2 = c$$

is the general solution.

Sometimes it is possible to turn a nonexact equation into an exact equation by multiplying through by some function $\mu(x, y)$. If such a function exists, it is called an *integrating factor*. When such a function μ exists, it may depend on x or on y or on both variables. It is possible to test a first-order first-degree equation to see whether it has an integrating factor depending on x alone (for instance); we derive this test by assuming that

$$m(x, y) + n(x, y)y' = 0$$

is not exact, but that

* Note that the variables are separable in this equation, so it could also be solved as in Section 1.2.

$$\mu(x)m(x, y) + \mu(x)n(x, y)y' = 0$$

is exact. Since it is, we must have

$$\frac{\partial}{\partial y}(\mu m) = \frac{\partial}{\partial x}(\mu n)$$

or

$$\mu\frac{\partial m}{\partial y} = \mu\frac{\partial n}{\partial x} + \mu' n.$$

Therefore we have

$$\mu' n = \mu\frac{\partial m}{\partial y} - \mu\frac{\partial n}{\partial x},$$

or, finally, provided $n \neq 0$, $\mu \neq 0$,

$$\frac{\mu'(x)}{\mu(x)} = \frac{\partial m(x, y)/\partial y - \partial n(x, y)/\partial x}{n(x, y)}.$$

The left side of this equation is simply the derivative of $\ln \mu(x)$; the right side must be a function of x alone in order for the equation to be meaningful. If it is, say

$$\mu'/\mu = F(x),$$

then

$$\ln \mu = \int F(x)dx,$$

or, finally,

$$\mu = \exp\left[\int F(x)dx\right]$$

We have omitted an arbitrary constant of integration in the calculation of μ, since we need only a particular integrating factor and not the most general one.

Naturally, it is not always true for every equation that the expression

$$\frac{\partial m/\partial y - \partial n/\partial x}{n}$$

is a function of x alone; but if it is, then the preceding technique shows how to find an integrating factor. For example, consider the equation

$$e^x \cos y - \sin y\, y' = 0.$$

This equation is not exact; but

$$\frac{\partial m/\partial y - \partial n/\partial x}{n} = \frac{-e^x \sin y - 0}{\sin y} = -e^x,$$

which does depend on x alone. Therefore $F(x) = -e^x$ and the integrating factor is $\mu = \exp[\int F(x)dx] = \exp(-e^x)$. Therefore the equation obtained by multiplying through by μ,

$$\exp(-e^x)e^x \cos y - \exp(-e^x)\sin y\, y' = 0$$

is exact, and it is the work of only a moment to see that its general solution is

$$\exp(-e^x)\cos y = c.$$

There is a similar test to see whether the integrating factor is a function of y alone; if we call the factor $v(y)$ this time, then v must satisfy

$$\frac{v'(y)}{v(y)} = \frac{\partial n/\partial x - \partial m/\partial y}{m},$$

where the right side must depend on y alone. The derivation of this formula is similar to that of the preceding one, and is left to the reader.

The linear equation of first order is a special case in which the integrating factor depends only on one variable; we shall study it in the next section.

In the general case, when the integrating factor μ depends on both x and y, it is not so easy to find it, and there are no hard and fast rules. One usually finds μ by inspection, by recognizing a fragment of the derivative of a more complicated expression which lacks a factor to make it an exact derivative. The process is best illustrated by working a couple of examples.

In the equation

$$2y + 5xy' = 0,$$

we notice that the exponent of x is one higher in the second term than in the first, while just the reverse is true of y. Hence in all likelihood this differential equation resulted from canceling a factor of the form $x^n y^m$ from a perfect derivative. Thus, let us see whether it is possible to choose n and m so that

$$x^n y^m(2y + 5xy') = 2x^n y^{m+1} + 5x^{n+1} y^m y' = 0 \qquad (1.21)$$

is exact. For this to be so, we must have

$$\frac{\partial}{\partial y}(2x^n y^{m+1}) = \frac{\partial}{\partial x}(5x^{n+1} y^m),$$

which implies that $2(m + 1)x^n y^m = 5(n + 1)x^n y^m$ or $2(m + 1) = 5(n + 1)$. Any number of values for n and m will do; the smallest positive ones are $n = 1$, $m = 4$. Upon making this substitution, equation (1.21) becomes exact and its solution is $x^2 y^5 = c$.

Another such equation is

$$y - xy' = y^2 \cos x.$$

We recognize that the left side is the numerator of the expression which we get upon differentiating x/y. Hence upon dividing both sides by y^2, we obtain

$$\frac{y - xy'}{y^2} = \cos x,$$

and the left side becomes the perfect derivative of x/y. Thus the solution is $x/y = \sin x + c$.

Problems

Determine which of the following equations is exact, and solve the exact ones by the method described in this section. If any of the nonexact equations can be solved by the methods of earlier sections, do so.

1. $(x^2 - y^2) + x(2x - 2yy') = 0$.
2. $xy' \sin(xy) + y \sin y - 2x \sin x^2 = 0$.
3. $(3x^2 y - y \cos x)y' + [x + 3xy^2 + (y^2/2)\sin x] = 0$.
4. $(x^2 - y^2)y' - (2xy - x^2) = 0$.
5. $y' = (\cos x \sinh y)y' - \sin x$.
6. $xy' - y = 0$.
7. $e^y y' = -(1 + \ln x)$.
8. $y' = -e^{-y}(1 + \ln x)$.
9. $(3y^2 + 6xy + 12x^3)y' + (3y^2 + 36x^2 y + 4x^3) = 0$.
10. $e^y \cos 2y + x(e^y \cos 2y - 2e^y \sin 2y)y' = 0$.
11. $2xy - y^2 \cos x + xy^2 \sin x + (x^2 - 2xy \cos x)y' = 0$.

12. $(3x^2y + y \sin xy) + (x^3 - x \sin xy)y' = 0.$
13. $(x^2 - y^2)y' + (2xy - x^2) = 0.$
14. $(3x + 2y) + (2x - 3y - 2)y' = 0.$
15. $(4x - 5y) + (5x - 2y + 1)y' = 0.$

In each of the following, find the integrating factor and solve.

16. $(y^2 + x)y' - y = 0.$
17. $xy' + y(1 + \ln yx) = 0.$
18. $y - 2xy' = 2xy^3.$
19. $xy'(y^2 - x) + xy = y^2.$
20. $x(y^2 - 1)y' + y(y + 1) = 0.$
21. $y \cos y + (\cos y - y \sin y)y' = 0.$
22. $3 \ln \sin y + x \cot y\, y' = 0.$
23. $2x(y^3 + 1) + 3y^2 \tan x^2\, y' = 0.$
24. $4x^3y + x^4(1 + y)y' = 0.$
25. $\ln y + \dfrac{(1 + x^2)\arctan x}{y} y' = 0.$
26. Derive the formula given in the text for an integrating factor which depends only on y.

1.5 Linear Equations; Bernoulli's Equation

During most of this course we shall occupy ourselves with linear equations of higher order; linear equations of the first order will be our model for this study. Consequently we shall study them in considerable detail in order to learn as much as possible about their properties, which will give us an idea of what to expect in the case of general linear equations. The general form of a linear equation of first order, we recall, is

$$a_1(x)y' + a_0(x)y = g(x). \tag{1.22}$$

Now since both y and y' appear to the first power and in separate terms from each other, it appears that there is some possibility that (1.22) may have come from differentiating an expression of the form

$$F(x)y = G(x); \tag{1.23}$$

indeed, if we differentiate this, we obtain

$$F(x)y' + F'(x)y = G'(x)$$

which is assuredly of the same form as (1.22). The difficulty is that (1.22) may not be exact. While it may have been obtained ultimately from an expression of the form (1.23), the cancellation of any common factors would have destroyed its exactness. For example,

$$e^x y - e^{-x} = c$$

differentiates into

$$e^x y' + e^x y + e^{-x} = 0,$$

which can be rewritten as

$$y' + y + e^{-2x} = 0,$$

which is no longer an exact equation even though it is strictly equivalent to the original equation.

Returning to (1.22), if we divide both sides through by $a_1(x)$ [assuming $a_1(x) \neq 0$], we have

$$y' + P(x)y = f(x), \tag{1.24}$$

where $P(x) = a_0(x)/a_1(x)$ and $f(x) = g(x)/a_1(x)$. We do this because it will be a bit easier to handle this equation if the coefficient of y' is 1. In the form (1.24), we see that the first-order linear equation is not exact unless $P(x) \equiv 0$, a very special circumstance indeed. But a true mathematician is never daunted by such a state of affairs; instead he resorts to his natural ingenuity. In this case, we recognize that if there is to be an integrating factor, it will be a function of x alone, say $v(x)$. Upon multiplying through by $v(x)$, then, we have

$$v(x)y' + v(x)P(x)y = v(x)f(x), \tag{1.25}$$

which must be exact, meaning that the left side is the derivative of y times some function of x. What must this $v(x)$ be in order to produce this phenomenon? From the previous section, we must have

$$\frac{\partial}{\partial x}[v(x)] = \frac{\partial}{\partial y}[v(x)P(x)y]. \tag{1.26}$$

Carrying out the differentiations, we obtain

$$v'(x) = v(x)P(x),$$

in which the variables are separable:

$$\frac{v'(x)}{v(x)} = P(x)$$

or

$$\ln v(x) = \int P(x)dx.$$

Then, exponentiating, we have*

$$v(x) = \exp\left[\int P(x)dx\right].$$

Putting this into (1.25), we obtain

$$\exp\left[\int P(x)dx\right]y' + \exp\left[\int P(x)dx\right]P(x)y = \exp\left[\int P(x)dx\right]f(x), \tag{1.27}$$

and since $[\int P(x)dx]' = P(x)$, the left side of (1.27) is simply

$$\left\{y \exp\left[\int P(x)dx\right]\right\}'.$$

We have thus succeeded in displaying the left side of (1.25) as the derivative of a product, so that upon integration we have

$$y \exp\left[\int P(x)dx\right] = \int \exp\left[\int P(x)dx\right]f(x)dx + c$$

or

$$y = \exp\left[-\int P(x)dx\right]\left\{\int \exp\left[\int P(x)dx\right]f(x)dx + c\right\}. \tag{1.28}$$

This is an actual formula for the explicit solution of a linear equation which has been put in the form (1.24). We could, of course, memorize (1.28) and use it to write down immediately the solution of any first-order linear equation.

* Note that this is simply a special case of the formula developed in the previous section for a more general first-order first-degree equation with an integrating factor depending only on x.

But it is often more convenient (and easier to remember) to put the equation in the form (1.24) by dividing through by the coefficient of y' (if it is not already 1), then computing $\int P(x)dx$ and exponentiating it to get the integrating factor $v(x)$, and, after multiplying the equation through by $v(x)$, recalling that the left side is simply $(yv)'$.

We shall illustrate this technique shortly with some examples. Before we do this, however, it will be instructive for our future study of linear equations of higher order to examine the formula (1.28) with some care. First of all, if equation (1.24) is homogeneous, i.e.,

$$y' + P(x)y = 0,$$

then the general solution of this associated homogeneous equation* is

$$y_h(x) = c/v(x).$$

(The subscript h stands for "homogeneous.") Secondly, the expression (1.28) is a solution of the inhomogeneous equation (1.24) for any value of c at all, and in particular for $c = 0$, so that

$$y_p(x) = \left[\int v(x)f(x)dx\right]/v(x)$$

is a *particular solution* of (1.24). (The subscript p, of course, stands for "particular.") Thus, we see by inspection that the *general solution* of the inhomogeneous equation is

$$y(x) = y_h(x) + y_p(x),$$

that is, *the general solution of the inhomogeneous linear equation of first order is the sum of the general solution of the associated homogeneous equation and a particular solution of the inhomogeneous equation.* This is a very important fact, as we shall see in the next chapter, because this same statement is true of linear equations of any order. By this means the problem of solving linear equations is reduced to two simpler ones: (1) Find the general solution of the associated homogeneous equation; (2) find a particular solution of the inhomogeneous equation.

Let us now apply what we have learned to solve the following linear equation:

$$xy' - 2y = x^4e^x.$$

* So-called because its left side is the same as that of $y' + P(x)y = f(x)$.

First, we put this equation into the form (1.24) by dividing through by the coefficient of y':

$$y' - \frac{2}{x} y = x^3 e^x. \tag{1.29}$$

We now note that $P(x) = -2/x$ (taking the sign into account), so that

$$\int P(x)dx = \int(-2/x)dx = -2 \ln x = \ln(x^{-2}).$$

Therefore the integrating factor $v(x)$ is $\exp[\ln(x^{-2})] = 1/x^2$. Multiplying (1.29) through by $v(x)$, we have

$$\frac{1}{x^2} y' - \frac{2}{x^3} y = xe^x.$$

The left side of this equation is obviously $(x^{-2}y)'$; integrating the right side by parts yields $(xe^x - e^x)$, so we obtain

$$x^{-2}y = (x - 1)e^x + c,$$

and hence

$$y = x^2(x - 1)e^x + cx^2$$

is the general solution.

Let us consider one further example:

$$xy' + y = xy \tan x + \sec x,$$

which, when placed in the form (1.24), is

$$y' + \left(\frac{1}{x} - \tan x\right) y = \frac{1}{x} \sec x.$$

Here $P(x) = (1/x - \tan x)$, so

$$\int P(x)dx = \int (1/x)dx - \int \tan x \, dx = \ln x + \ln \cos x = \ln(x \cos x),$$

and

$$v(x) = \exp\left[\int P(x)dx\right] = \exp(\ln x \cos x) = x \cos x.$$

Multiplying both sides by $x \cos x$ yields

$$x \cos x y' + (\cos x - x \sin x)y = 1.$$

Since the left side is $(yv)' = (yx \cos x)'$, we obtain after integration

$$xy \cos x = x + c$$

or

$$y = (1 + c/x)\sec x.$$

A differential equation of the form

$$y' + P(x)y = f(x)y^n$$

is called a *Bernoulli differential equation.* Unless $f(x) \equiv 0$ or $n = 0$ or 1, Bernoulli's equation is nonlinear because of the factor y^n on the right side, even though the left side has the same form as the left side of a linear equation. However, Bernoulli's equation can be transformed into a linear equation by means of a change of dependent variable. To see what this change should be, we first multiply through by y^{-n} $(n \neq 1)$ in order to get all the terms involving y on the left side. We then have

$$y^{-n}y' + P(x)y^{-n+1} = f(x).$$

The first term is essentially the derivative of y^{-n+1}; in fact,

$$\frac{1}{-n+1}(y^{-n+1})' + P(x)y^{-n+1} = f(x).$$

Thus, it appears reasonable to make the substitution $v = y^{-n+1}$; this yields

$$v' + (-n + 1)P(x)v = (-n + 1)f(x),$$

which is now a linear equation for v, and which we can therefore solve by the methods of this section.

For example, to solve

$$y' - \frac{1}{x}y = \frac{x}{y^3},$$

we multiply through by y^3, obtaining

$$y^3y' - \frac{1}{x}y^4 = x,$$

or

$$\frac{1}{4}(y^4)' - \frac{1}{x}y^4 = x,$$

and upon substituting $v(x) = y^4$, we have the linear equation for v

$$v' - \frac{4}{x}v = 4x.$$

The integrating factor is $\exp[-\int(4/x)dx] = \exp(-4\ln x) = x^{-4}$; thus we have

$$(x^{-4}v)' = 4x^{-3}.$$

Integrating this, we obtain

$$x^{-4}v = -2x^{-2} + c.$$

Thus, we finally have

$$v = y^4 = -2x^2 + cx^4.$$

Problems

Solve the following equations by the methods of this section if they are linear or Bernoulli type; use the methods of previous sections if not.

1. $y' - 2xy = -e^{x^2}\sin x$.
2. $x \sin x\, y' + (2\sin x + x\cos x)y = (x + 2)e^x$.
3. $x(x + 3)y' + 3(x + 2)y = (1 + \ln x)/x$.
4. $xy' - (1 + x)y = x^2e^{2x}$.
5. $2xy^3 + 3x^2y^2y' = \sin x + x\cos x$.

6. $\cos x\, y' - y \sin x = \sec^2 x.$
7. $yy' = e^{-y} x \ln x.$
8. $y' = e^y \ln x.$
9. $(x^2 + 1)^2(xy' + y) = -2x.$
10. $y \sin 4x + \sin^2 2x\, y' = x^2.$
11. $y' - y = e^{2x}.$
12. $4y' + y = 0.$
13. $4y' + y = \sin x.$
14. $4y' + y = x.$
15. $4y' + y = e^x.$
16. $4y' + xy = x \exp(-x^2/8).$
17. $2xy \ln y + x^2(\ln y + 1)y' = \cos x.$
18. $y' = \sin(x + y) - \cos x \sin y.$
19. $(\cosh x)y' + (\sinh x)y = e^x.$
20. $y + (x^2 + 1)(\arctan x)y' = 2x^3 + 2x.$
21. $4x^3y' + 3x^2y = -y^{-3} \sin x.$
22. $y^2 + 2xyy' = \frac{1}{x} - 1.$
23. $y - y' = e^{-x}y^4 \cos x.$
24. $y + y'(1 + x^2)\arctan x = (1 + x^2)y^{-2}.$
25. $4xy' - (1 + x)y = 2e^{-2x}y^5.$
26. $4x^3y + 5x^4y' = x^4 y \cot x.$
27. $4y' = 3y^{-3} \cos 3x.$
28. $3y' \cos x - y \sin x = 3x^2y^{-2}.$
29. $2y' + y = 3(e^x y\sqrt{1 - x^2})^{-1}.$
30. $y + 3xy' \ln x = 3x^3y^{-2}.$

1.6 Application of Linear Equations to Traffic and Radioactive Decay Problems

Let us first consider the problem of describing the motion of a car which is second in line at a stop light under the following conditions. At time $t = 0$, the light turns green and the first car accelerates uniformly with an acceleration of a_1, which continues for the period of time under consideration. We assume that the first car is ten feet long and that the front of the second car is one foot from the rear of the first. Then the question we pose is the following: what must the acceleration $a_2(t)$ of the second car be if it is to remain ten feet behind the first car for each ten miles per hour of its speed, plus the initial one foot? Admittedly this problem is oversimplified. For, first, we assumed that car 1 could instantly change his acceleration from 0 to a_1, and then keep it constant, whereas in reality there will be a continuous (if short) change from 0 to a_1, and then some slight variation from that value. Second, car 1 will obviously not continue to accelerate indefinitely, but will cease to accelerate when he has reached a certain speed, hopefully within the speed limit. We have also assumed that the driver of car 2 has perfect reactions and starts his car at the same instant as the driver of car 1. Any driver who has been idled in traffic knows how optimistic this assumption is.

Consequently, having made all these assumptions, we must ask how useful our solution will be. The answer cannot be very precise, just "not too bad." It will be a tentative first step in helping to solve the maddening problem of why it takes a line of cars so long to go through a signal. In the exercises we shall pursue more complicated versions of the problem, considering cars farther back in line, the effect of reaction time, and the effect caused by a car reaching its maximum speed.

To derive the equation describing the motion of the second car, let us represent by $x_1(t)$ the distance car 1 travels from the stop light and by $x_2(t)$ the corresponding distance for car 2. Then the speed and acceleration of car 1 are $\dot{x}_1$ and $\ddot{x}_1$, respectively, and of car 2, $\dot{x}_2$ and $\ddot{x}_2$. Before we derive an equation describing the motion of car 2, let us first look at car 1. We know that its acceleration from $t = 0$ on is a constant, a_1. We also know that at time $t = 0$, its distance from the light is zero and it is stopped, so its speed is also zero. In equation form, we can state this as follows:

$$\ddot{x}_1 = a, \quad x_1(0) = 0, \quad \dot{x}_1(0) = 0.$$

The first equation is the equation of motion of car 1, and the next two represent the initial conditions which tell us how the motion begins. This initial value problem is so simple that we can solve it by the methods of calculus; by integrating twice, we obtain

$$x = \tfrac{1}{2}a_1 t^2 + ct + d,$$

where c and d are arbitrary constants. Applying the initial conditions, we see that c and d must both vanish; hence the expression

$$x_1(t) = \tfrac{1}{2}a_1 t^2$$

describes the position of car 1 for all values of time within the range being considered.

We know, next, that the speeds of the two cars are related by the fact that their distance apart must be ten feet for every ten miles per hour which car 2 is going, plus one foot. In equations, this says that

$$x_1 - 10 - x_2 = k\dot{x}_2 + 1$$

where, if the units are feet and seconds, $k = 0.6818$ sec., to four decimal places. Solving this equation for the variable x_2, we obtain

$$\dot{x}_2 + \frac{1}{k}x_2 = \frac{1}{k}(x_1 - 11).$$

This is a first-order linear equation for $x_2(t)$. All the quantities on the right side are known. This equation, together with the initial condition stating that car 2 begins one foot behind car 1 (and hence 11 feet behind the signal) gives us the following initial value problem for car 2:

$$\dot{x}_2 + \frac{1}{k}x_2 = \frac{1}{k}\left(\frac{1}{2}a_1t^2 - 11\right), \quad x_2(0) = -11. \tag{1.30}$$

Why do we need only *one* initial condition for this equation? There are two reasons why this should be so: First, the equation is only of first order, so upon integration we obtain only one constant of integration to be evaluated; and second, the second initial equation $\dot{x}_2(0) = 0$ is already implied by the equation and the first initial condition [simply substitute $t = 0$ and $x_2(0) = -11$ into the equation itself].

To solve equation (1.30), note that it is already in the form (1.24). The integrating factor $v(t)$ is equal to

$$\exp\left[\int P(t)dt\right] = \exp\left[\int (1/k)dt\right] = \exp(t/k).$$

Thus after multiplying through by $v(t)$ the left side is simply the derivative of x_2v, and the right side becomes

$$\frac{1}{k}e^{t/k}(\tfrac{1}{2}a_1t^2 - 11),$$

which can be integrated by parts to yield

$$\tfrac{1}{2}a_1e^{t/k}(t^2 - 2kt + 2k^2 - 22/a_1) + c;$$

thus the general solution of the traffic problem is

$$x_2(t)e^{t/k} = \tfrac{1}{2}a_1(t^2 - 2kt + 2k^2 - 22/a_1)e^{t/k} + c$$

or

$$x_2(t) = \tfrac{1}{2}a_1(t^2 - 2kt + 2k^2 - 22/a_1) + ce^{-t/k}.$$

Using the initial condition $x_2(0) = -11$, we have

$$-11 = \tfrac{1}{2}a_1(2k^2 - 22/a_1) + c = -11 + a_1k^2 + c$$

so that $c = -a_1k^2$, and we have

$$x_2(t) = \tfrac{1}{2}a_1(t^2 - 2kt + 2k^2 - 22/a_1) - a_1k^2e^{-t/k}.$$

What we are actually seeking, of course, is the acceleration of car 2. To get this, we differentiate the above expression twice:

$$v_2(t) = \dot{x}_2(t) = a_1t - a_1k + a_1ke^{-t/k},$$

$$a_2(t) = \ddot{x}_2(t) = a_1 - a_1e^{-t/k} = a_1(1 - e^{-t/k}).$$

This expression answers our question completely; the acceleration of car 2 starts out at time zero equal to zero, then increases and approaches a_1 exponentially with increasing time (Figure 1.4). Using the value $k = 0.6818$ second, we see that a_2 approaches a_1 quite rapidly; after 0.6818 second, a_2 is already about 0.64 a_1.

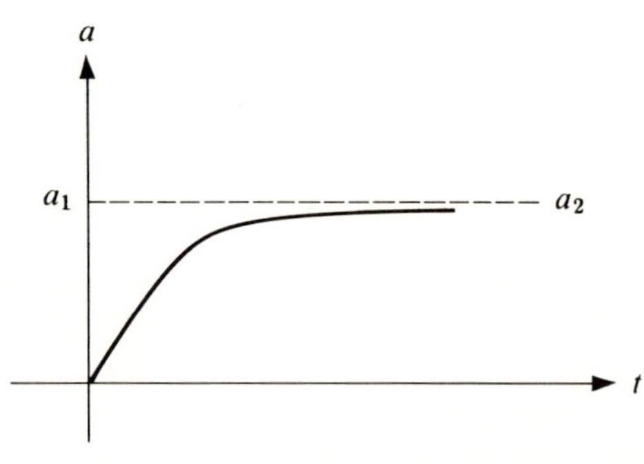

Figure 1.4

In the radioactive decay problem,

$$\dot{m} + \kappa m = 0, \quad m(0) = m_0,$$

we note that this equation is linear in m, so that we can solve it just as well by finding the integrating factor as we can by the method of separation of variables used earlier. Indeed, the integrating factor is

$$\exp\left(\int \kappa\, dt\right) = e^{\kappa t},$$

so that we have immediately

$$me^{\kappa t} = c,$$

or

$$m = ce^{-\kappa t},$$

and the initial condition gives $c = m_0$ as before, so that

$$m(t) = m_0 e^{-\kappa t}.$$

Problems

1. In the traffic problem, what would be the equation of motion and initial value problem for a third car in line, assuming as before that car 2 is ten feet long, car 3 is one foot behind it at $t = 0$, that car 3 begins moving at the same instant as car 2, and that car 3 maintains a distance of ten feet behind car 2 for each ten miles per hour of its speed, plus the initial one foot?

2. Suppose, in the traffic problem originally considered, that after a time T has elapsed, the acceleration of car 1 ceases and that car 1 proceeds at a uniform speed from then on. What will be the equation of motion and initial value problem for car 2 if all the other assumptions remain unchanged?

3. Study the traffic problem in the case where the driver of car 2 does not begin until a reaction time of τ seconds has elapsed. If car 2 can accelerate at a maximum rate of $A > a_1$, then how long will it take him to catch up to within ten feet of car 1 for each mile per hour of his speed, and how will his motion proceed from there?

4. Referring to problem 1, what is the acceleration of car 3?

5. What is the solution of the initial value problem in problem 2?

6. From the previous problem, compute the acceleration $a(t)$ for $t \geqq T$ and graph and describe its behavior. Is it necessary for car 2 to apply his brakes?

1.7 Second-Order Equations which can be Reduced to First-Order Equations

There are two types of second-order equations which can be reduced to first-order equations by means of a change of variables, and thus solved by the methods of this chapter. The first consists of equations of the form

$$F(y, y', y'') = 0$$

which, because x does not appear explicitly, is called the type with *independent variable missing*. The second consists of those equations of the form

$$F(x, y', y'') = 0$$

which have y missing and accordingly are called the type with *dependent variable missing*. We shall consider each of these types separately, the latter first because it is easier.

(1) *Dependent variable missing*. In this type we make the substitution $y' = p(x)$ and $y'' = p'(x)$, thus reducing it to the first-order equation

$$F(x, p, p') = 0.$$

The reason for this substitution is rather obvious; an equation with dependent variable missing is really a first-order equation for the unknown function y'. We solve this first-order equation for y', then integrate a second time to obtain y. For example, consider the equation

$$y'' + y' = x.$$

After substitution of p for y', we have

$$p' + p = x,$$

a first-order linear equation for p; its integrating factor is e^x, so

$$(e^x p)' = xe^x$$

and thus

$$e^x p = (x - 1)e^x + c_1.$$

Solving for p,

$$p = y' = (x - 1) + c_1 e^{-x}$$

and, upon integrating a second time, we obtain

$$y = \tfrac{1}{2}x^2 - x - c_1 e^{-x} + c_2.$$

Notice that the general solution contains two arbitrary constants because of the fact that the original equation was of second order, and solving it involved integrating twice.

A second example of this type is

$$xy'' = 2y'$$

which, upon substitution and separation of variables, becomes

$$\frac{p'}{p} = \frac{2}{x}.$$

Integrating, we obtain

$$\ln p = \ln y' = 2 \ln x + c_1 = \ln x^2 + c_1.$$

Exponentiating, we have

$$y' = k_1 x^2, \qquad \text{where } k_1 = e^{c_1},$$

and therefore

$$y = k_1 x^3/3 + k_2.$$

(2) *Independent variable missing.* In this case, we make up for the lack of an independent variable by acting as if y were independent. We do this by making the substitution $y' = p(y)$; then

$$y'' = \frac{dy'}{dx} = \frac{d}{dx}p(y) = \frac{d}{dy}p(y)\frac{dy}{dx} = p\frac{dp}{dy}.$$

Thus the differential equation becomes

$$F\left(y, p, p\frac{dp}{dy}\right) = 0,$$

which is again a first-order equation which can be solved by standard methods.

For example, let us consider

$$y'' - 2y' = 0.$$

After making the above substitution, we have

$$p\frac{dp}{dy} - 2p = 0.$$

Thus, if $p \neq 0$,

$$\frac{dp}{dy} = 2,$$

and hence

$$p = 2y + c_1.$$

Putting y' into this equation for p, we have

$$y' = 2y + c_1, \qquad \text{or} \qquad y' - 2y = c_1,$$

which is a first-order linear equation for y, with integrating factor e^{-2x}; thus

$$(e^{-2x}y)' = c_1 e^{-2x},$$

and hence

$$e^{-2x}y = (-c_1/2)e^{-2x} + c_2,$$

so that we finally have

$$y = k_1 + c_2 e^{2x}. \qquad (k_1 = -c_1/2.)$$

As a more difficult example of the same type, let us consider the following:

$$y'' - 3(y')^3 = 0.$$

After making our substitution, we have

$$yp\frac{dp}{dy} - 3p^3 = 0$$

or

$$\frac{1}{3p^2}\frac{dp}{dy} = \frac{1}{y}. \qquad (p \neq 0.)$$

Hence, upon integrating,

$$-1/3p = \ln c_1 y \qquad \text{(why?)}$$

so that

$$3y' \ln c_1 y = -1.$$

Integrating a second time, we find that

$$3y(\ln c_1 y - 1/c_1) = -x + c_2.$$

Problems

1. $y'' + 2y' = 3e^x$.
2. $y'' - 3y' = 2\cos x$.
3. $y'' - 1 - (y')^2 = 0$.
4. $xy'' - y' \ln y' = 0$.
5. $\ln[y''/(2x + 1)] + y' - x^2 - x = 0;\ y(0) = 0,\ y'(0) = 0$.
6. $y'' - y = 0$.
7. $y'' = 2yy'$.
8. $y''y^2 + y' = 0$.
9. $y'' = y'e^y$.
10. $y^3y'' - 1 = 0$.

1.8 Clairaut's Equation: a Non-First-Degree Equation with a Geometric Application

We have noticed that the general solution of a first-order ordinary differential equation involves a single arbitrary constant; it is always an expression of the form

$$f(x, y, c) = 0. \tag{1.31}$$

For each value of the constant c we get a particular solution of the equation. In other words, c is a parameter which, as it varies, gives us a whole family of solutions of the differential equation. Such a family is called a *one-parameter family of solutions*. Graphically, each member of the family (1.31) is a curve in the (x, y) plane.

Conversely, given a one-parameter family of curves as a set of equations (1.31), it is possible to derive a differential equation of which (1.31) is the one-parameter family of solutions, whenever it is possible to eliminate the arbitrary constant c between (1.31) and its total derivative

$$\frac{\partial f}{\partial x} + y' \frac{\partial f}{\partial y} = 0.$$

For example, consider the equations

$$x^2 y = ce^x.$$

The total derivative of this family is

$$2xy + x^2 y' = ce^x.$$

This differential equation is *not* the one we seek because it still contains the arbitrary constant c. We still need to eliminate c in order to obtain the differential equation of the family. From the first equation, $c = x^2 ye^{-x}$. Substituting in the second, we obtain

$$2xy + x^2 y' = x^2 y$$

or

$$xy' + 2y = xy$$

as the differential equation whose solution is the one-parameter family $x^2 y = ce^x$.

Another example of a one-parameter family of curves is the collection of straight lines

$$y = cx + f(c), \tag{1.32}$$

which are nonvertical with slope c, and the y intercept of which is a given function $f(c)$ of the slope. Differentiating this equation with respect to x, we obtain simply $y' = c$, and substituting this back, we obtain *Clairaut's differential equation*

$$y = xy' + f(y'). \tag{1.33}$$

The set of equations (1.32) represents a one-parameter family of solutions of Clairaut's equation (1.33). For instance, the Clairaut equation

$$y = xy' + \sin y',$$

which is nonlinear, not first degree, and cannot even readily be solved for y', has as a one-parameter family of solutions the collection of straight lines

$$y = cx + \sin c.$$

Because Clairaut's equation is in general nonlinear, and, moreover, like the example above, may not even be solvable for y', our existence and uniqueness results may not apply to it. Of course, a family of solutions does exist and is given by the formula (1.32); but it is possible that uniqueness may not hold. The Clairaut equation may have other solutions which cannot be obtained from (1.32) by giving a particular value to c. Such solutions, if they exist, are called *singular solutions*, and we shall be concerned with attempting to find methods to obtain them.

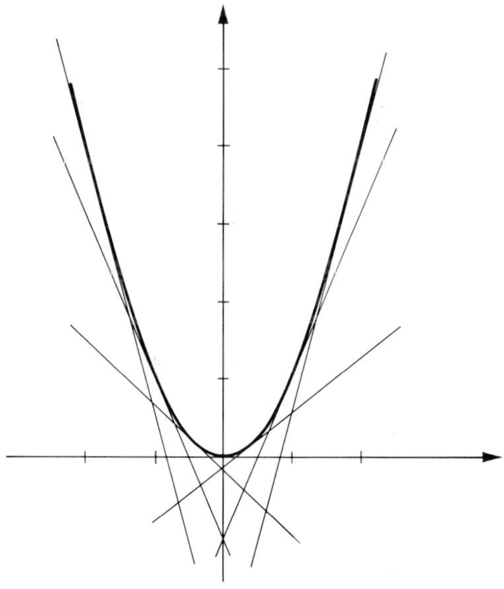

Figure 1.5

For example, let us consider the family of lines which are all tangent to the parabola $y = x^2$. (Figure 1.5.) Let us obtain their equations in the form

$$y = cx + f(c)$$

and determine the Clairaut equation which they all satisfy. If a tangent line goes through the point (x_0, x_0^2), then its slope is $c = 2x_0$. Hence its equation is

$$y - x_0^2 = 2x_0(x - x_0)$$

or

$$y = (2x_0)x - x_0^2.$$

Put into the form of solutions of Clairaut's equation, this reads

$$y = cx - \tfrac{1}{4}c^2.$$

Hence the Clairaut equation of this family is

$$y = xy' - \tfrac{1}{4}(y')^2. \tag{1.34}$$

Each straight line which is tangent to the parabola $y = x^2$ is a particular solution of this equation. Because these lines are tangent to the parabola, the parabola at each point has the same slope as a solution of the equation and thus we might begin to suspect that $y = x^2$ is itself a solution of (1.34). In fact, since $y = x^2$, we have $y' = 2x$, and thus

$$xy' - \tfrac{1}{4}(y')^2 = x(2x) - \tfrac{1}{4}(2x)^2 = 2x^2 - x^2 = x^2 = y.$$

Therefore equation (1.34) has not only the one-parameter family of solutions

$$y = cx - \tfrac{1}{4}c^2,$$

but also the singular solution

$$y = x^2,$$

which is singular precisely because it cannot be obtained from the family for any value of c. Uniqueness fails for this Clairaut equation because at any point on the parabola there are two solutions: the parabola itself, and the tangent line to the parabola.

Whenever a curve exists which is tangent to each member of a one-parameter family, such as the parabola in Figure 1.5, it is called the *envelope* of the family. Clearly because at each point the envelope has the same slope as a solution, it must also be a solution; however, as noted in the example above, it is a singular solution. Thus, our problem is as follows: given a one-parameter family of solutions of Clairaut's equation, how can we determine

(1) whether or not the family has an envelope, and (2) if so, what the envelope is?

Both questions can be answered by the following observations: first, each point on the envelope, if it exists, must satisfy the equation

$$y - cx - f(c) = 0 \tag{1.35}$$

for some particular value of c which depends on the point chosen. This can be considered as one of two parametric equations necessary to describe the envelope. One other is necessary in order for us to be able to eliminate c between the two and solve for the Cartesian equation of the curve. For instance, we can take the partial derivative of (1.35) with respect to the parameter c, obtaining

$$x + f'(c) = 0. \tag{1.36}$$

Why differentiate with respect to c, an erstwhile constant? Because, as remarked above, two different points on the envelope will in general correspond to two *different* values of c. Thus c, a particular value of which gives one solution of the equation when put into (1.35), becomes a parameter which *varies* from one point to the next as we traverse the singular solution or envelope. If c can be eliminated between (1.35) and (1.36), then the resulting expression will be the equation of the envelope, provided that it is a solution of the original differential equation.

For example, let us use this technique to find the envelope of the solutions of

$$y = xy' - \tfrac{1}{4}(y')^2,$$

which we already know to be $y = x^2$. We write the two equations

$$y - cx - \tfrac{1}{4}c^2 = 0, \quad x + \tfrac{1}{2}c = 0.$$

The second equation is the partial of the first with respect to c. It gives us $c = 2x$, and putting this into the first, we have $y = x^2$ as required.

As another example, to find the envelope of the one-parameter family of straight lines

$$x \cos \theta + y \sin \theta = 1,$$

where θ is a fixed constant, we differentiate with respect to θ and obtain

$$-x \sin \theta + y \cos \theta = 0,$$

which can be solved for θ:

$$\tan\theta = \frac{y}{x} \quad \text{or} \quad \theta = \arctan\frac{y}{x} = \arccos\frac{x}{\sqrt{x^2 + y^2}} = \arcsin\frac{y}{\sqrt{x^2 + y^2}}.$$

Putting this value for θ into the first equation, we obtain

$$\frac{x^2}{\sqrt{x^2 + y^2}} + \frac{y^2}{\sqrt{x^2 + y^2}} = 1,$$

or finally,

$$\sqrt{x^2 + y^2} = 1,$$

the unit circle.

To determine whether this is, indeed, the envelope of the family of straight lines given, we need to find the Clairaut equation of the family and see if the unit circle is a solution. Since

$$y = -x\cot\theta + \csc\theta,$$

placing $y' = -\cot\theta$ and $\csc\theta = \sqrt{1 + (y')^2}$, we have the Clairaut equation

$$y = xy' + \sqrt{1 + (y')^2}.$$

For the upper half of the unit circle, $y = \sqrt{1 - x^2}$, so $y' = -x(1 - x^2)^{-1/2}$ and

$$\begin{aligned} xy' + \sqrt{1 + (y')^2} &= -x^2(1 - x^2)^{-1/2} + \sqrt{1 + x^2(1 - x^2)^{-1}} \\ &= \frac{1 - x^2}{\sqrt{1 - x^2}} = \sqrt{1 - x^2} = y. \end{aligned}$$

Hence the upper half of the unit circle is part of the envelope; one can show similarly that the lower half is.

Problems

Solve the following equations and determine the envelope, if any.

1. $y = xy' - \ln y'$.
2. $y = xy' - 2(y'/3)^{3/2}$.
3. $y = xy' - y' \arccos y' + \sqrt{1 - (y')^2}$.
4. $y = xy' - (y')^2 + y'$.
5. $y = xy' + (y')^4$.

6. $y = xy' + (1/4y')$.

7. A curve is such that the sum of the intercepts of each of its tangent lines is constant. Find the equation of this curve if this constant is 1.

2

Derivations and Some Properties of Linear Equations

In this chapter we shall formulate some of the properties of linear equations and their solutions, with particular attention being fixed later in the chapter on equations with constant coefficients. Much of what we study will be primarily theoretical, but it will be highly important for practical applications which we shall take up in the succeeding chapters. Linear equations are extremely important, and we devote such a large amount of time to them for two main reasons: first, linear equations are the first (or most basic, or lowest degree) approximations to a mathematical description of most physical phenomena (as in the case of the linearized pendulum equation, for instance); and second, because of the nice properties which all linear equations share, they are relatively easy to deal with, and their solutions are not hard to find.

2.1 Derivation of some Linear Equations and Systems

Suppose that a body with mass m is suspended at the end of a spring. It is a well-known experimental fact (Hooke's law) that, if the mass is displaced from its equilibrium position, then the force generated by the spring to pull the mass back to equilibrium is directly proportional to the displacement from equilibrium, as long as this displacement is not so large as to alter the spring by bending it out of shape. This can be expressed by the equation $F = -kx$, where F is the force, x is the displacement, and k is the positive constant of proportionality. The negative sign occurs because the force is pulling the mass in the opposite direction from the displacement.

If we were to release the mass after displacing it, the mass would of course vibrate back and forth around the equilibrium position. In order to describe the motion of the mass, we shall need to derive the differential equation of motion. To do this, we must recall Newton's law of motion, which states that the force exerted by the spring on the mass is equal to the product of the mass and its acceleration, $F = ma$. Since $a = \ddot{x}$, where the displacement x is now a function of t, we have $F = m\ddot{x}$. Equating the two expressions for F which we now have, we obtain

$$m\ddot{x} = -kx,$$

the equation of motion of a mass m on a vibrating spring.

This equation was derived under the assumption that there were no forces on the mass other than the Hooke's law force $F = -kx$; but there may well be other forces, either externally imposed driving forces or frictional forces, either of which could strongly affect the motion of the mass. To take these forces into account, we shall use $f(t)$ to represent the (known) driving forces, and we shall assume that the frictional force is proportional to the velocity and equals $-r\dot{x}$, where r is a positive constant. The negative sign is used because the frictional force is always in the opposite direction from the velocity, i.e., it tends to retard the motion. Using these assumptions, we obtain

$$m\ddot{x} = -kx - r\dot{x} + f(t)$$

or

$$m\ddot{x} + r\dot{x} + kx = f(t).$$

What initial conditions do we need? Intuitively, since we have a second-order equation, we would expect a solution to contain two arbitrary constants, so that two initial conditions would be necessary to determine their values. Physically, we realize that not only can we give the mass an initial displacement from equilibrium, but also we can give it a push or a pull as we release it. Thus we can specify both the initial position $x(0)$ and the initial velocity $\dot{x}(0)$. These two items can be combined with the equation of motion to give the initial value problem:

$$\begin{gathered} m\ddot{x} + r\dot{x} + kx = f(t), \\ x(0) = x_0, \qquad \dot{x}(0) = v_0. \end{gathered} \tag{2.1}$$

How well does this system describe the motion of the mass on the vibrating spring? Again, not too badly, although there are some problems involved. We have assumed that it is possible to constrain the motion to one dimension, which is not always easy to do. We have assumed that the spring obeys

Hooke's law, which is an approximation, albeit a good one for small displacements. We have also assumed that friction is proportional to $\dot{x}$, when in fact it may depend on higher powers such as $(\dot{x})^2$ and $(\dot{x})^3$ as well. This dependence, though slight, is there, but if it were inserted into the equation it would cause an unpleasant complication by its nonlinearity.

A simple pendulum consists of a mass m (called the *bob*) fixed to the end of a rigid bar, one end of which is pinned to a support which constrains the pendulum to oscillate back and forth in a plane (Figure 2.1). Normally we assume that the bar is weightless, or at least that its mass is so small in comparison with m that it can be neglected in our calculations. We use l to represent the (fixed) length of the bar, and $\theta(t)$ to represent the angular displacement (in radians) of the pendulum from the vertical.

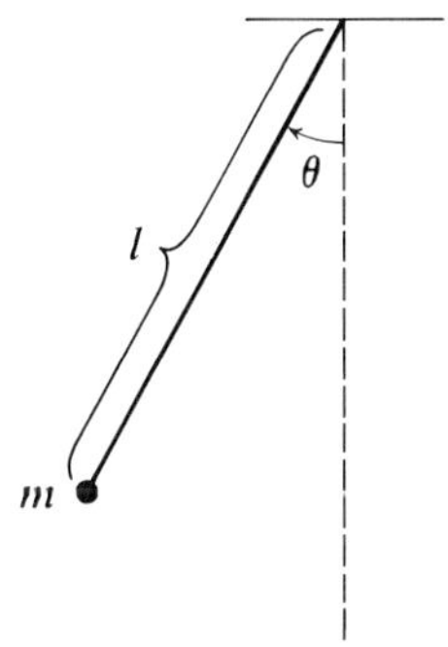

Figure 2.1

Because the length l is fixed (the bar does not stretch or contract), at any instant the motion of the bob must be in a direction perpendicular to the supporting bar to which it is attached. Thus, using Newton's law $F = ma$, we need only take into account the component of the force in the direction perpendicular to the bar, since the only nonzero component of acceleration is in this direction. Simple trigonometry (Figure 2.2) shows that the force due

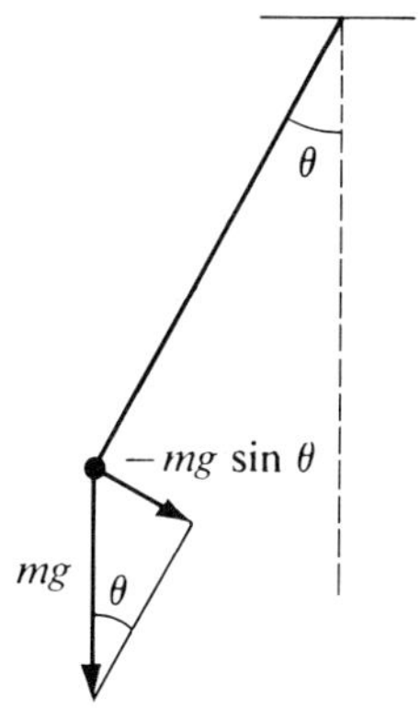

Figure 2.2

to gravity in this direction is $-mg \sin \theta(t)$, the negative sign being used as usual because the force is opposite the direction of displacement from the vertical. The actual distance of the bob from its lowest position, measured along the arc of its path, is $l\theta(t)$, since θ is measured in radians. Thus the acceleration is $l\ddot{\theta}$. If in addition to the gravitational force on the bob we also have a frictional force $-r\dot{\theta}$ and an external force $f(t)$, then the equation of motion of the bob is

$$ml\ddot{\theta} = -mg \sin \theta - r\dot{\theta} + f(t)$$

or

$$\ddot{\theta} + \eta\dot{\theta} + \frac{g}{l} \sin \theta = F(t),$$

where now $\eta = (r/ml)$ and $F(t) = f(t)/ml$. Because the transcendental function $\sin \theta$ occurs in the equation, this equation is *nonlinear*. Often in mathematical physics it is desirable to *linearize* such an equation in order to obtain a first approximation to the actual result. In this case, since

$$\sin \theta = \theta - \frac{\theta^3}{3!} + \frac{\theta^5}{5!} - \cdots,$$

we note that, if θ is very small, then $\sin \theta$ is very closely approximated by θ, so that *in the case when angular displacements are very small*, the pendulum equation becomes

$$\ddot{\theta} + \eta\dot{\theta} + \frac{g}{l}\theta = F(t).$$

Note that the form of this equation is exactly the same as the equation of the vibrating spring! Thus, solving either one of these equations will enable us to solve the other.

For initial conditions, as in the case of the spring, we will need to specify the initial value of θ and the initial value of $\dot{\theta}$, the angular velocity. Thus the initial value problem for the general pendulum equation is

$$\begin{gathered} \ddot{\theta} + \eta\dot{\theta} + \frac{g}{l} \sin \theta = F(t), \\ \theta(0) = \theta_0, \quad \dot{\theta}(0) = \omega_0. \end{gathered} \tag{2.2}$$

For the linearized equation it is

$$\ddot{\theta} + \eta\dot{\theta} + \frac{g}{l}\theta = F(t),$$
$$\theta(0) = \theta_0, \quad \dot{\theta}(0) = \omega_0. \tag{2.3}$$

In the general case, we made the approximations that the frictional term is linear and that it is effectively possible to constrain the pendulum's motion to two dimensions. Both of these are physically reasonable; in fact, *it is often possible to neglect friction entirely* since pendulums can be constructed so that friction at the point of suspension is virtually nonexistent.

In the linear case, we assumed that the angular displacement of the pendulum is small, which again can be realized by a pendulum in which l is very large and the linear displacement is a very small fraction of l. In future chapters we shall solve the linearized pendulum equation without much difficulty. However, not until Chapter 6 will we be able to attack and "solve" the nonlinear equation, and we shall do this numerically. In Chapter 6 we shall also consider a nonlinear equation that is somewhat related to the pendulum equation, except that the "frictional" term depends on the solution and can have either algebraic sign. This is the so-called Van der Pol equation, which arises out of the study of circuits containing vacuum tubes.

Now let us consider the problem of a vibrating spring with constant k_1 and a mass m_1 attached, to which is attached a second spring with constant k_2 and mass m_2 (Figure 2.3). Two variables $x_1(t)$ and $x_2(t)$ will be necessary to describe

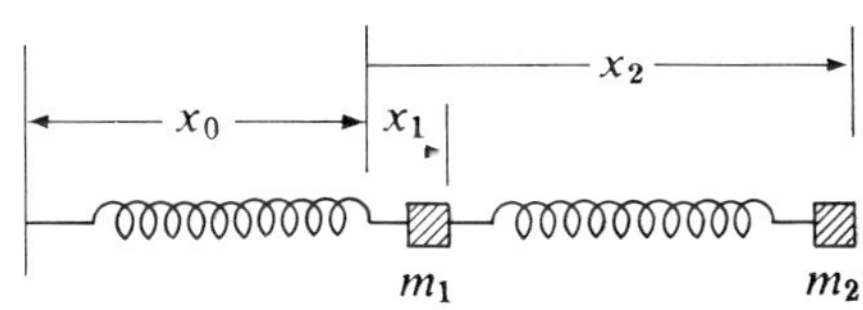

Figure 2.3

the positions of the two masses, and we shall measure both of these with respect to the equilibrium position of the first mass. We shall assume that the second spring is in equilibrium when $x_2 - x_1 = l$, a fixed constant. In order to make the derivation as simple as possible, we shall neglect friction entirely and assume that there are no external forces being imposed on the system.

These are two forces on the mass m_1, one due to the first spring and one due to the second; the first of these is opposite the displacement x_1, and the second is in the same direction if $x_2 - x_1 - l$ is positive. Thus for the first mass we have

$$m_1\ddot{x}_1 = -k_1x_1 + k_2(x_2 - x_1 - l).$$

The only force on the second mass is that due to the second spring, so

$$m_2\ddot{x}_2 = -k_2(x_2 - x_1 - l).$$

As initial conditions we shall need both the initial positions $x_1(0)$ and $x_2(0)$ and the initial velocities $\dot{x}_1(0)$ and $\dot{x}_2(0)$. Hence

$$\begin{aligned} &m_1\ddot{x}_1 + k_1x_1 - k_2(x_2 - x_1) = -k_2l, \\ &m_2\ddot{x}_2 + k_2(x_2 - x_1) = k_2l, \\ &x_1(0) = x_{01}, \quad x_2(0) = x_{02}, \quad \dot{x}_1(0) = v_{01}, \quad \dot{x}_2(0) = v_{02}, \end{aligned} \tag{2.4}$$

is the initial value problem describing the motion of the two masses. These equations are valid as long as the displacements are relatively small and as long as friction is neglected, which are somewhat optimistic assumptions.

The equations describing the flow of current through an electrical circuit are closely related to (2.1) and (2.4). The quantity of charge $Q = Q(t)$ is the unknown function analogous to the displacement in the previous problems. The current $I = I(t)$, which is the rate of flow of charge $[I(t) = \dot{Q}(t)]$, is analogous to the velocity. The electromotive force, or voltage, $E = E(t)$, is analogous to the external physical force in problems of mechanics. A resistor with resistance R is analogous to a frictional force in that it causes a voltage drop of $E = RI = R\dot{Q}$; that is, the "force" is proportional to the "velocity." (The expression $E = RI$ is known as Ohm's law.) An induction coil with inductance L produces an "acceleration force," $E = L\dot{I} = L\ddot{Q}$. A capacitor with capacitance C produces a "linear force" like the spring, $E = (1/C)Q$.

In describing the flow of charge through an electric circuit, we shall also need to use the so-called "Kirchhoff's laws." The first law says that whenever a number of wires are joined at one point, the total current into the junction equals the total current out. This first law is equivalent to the conservation of energy; the total amount of energy remains constant. The second law says that the sum of all voltage changes around a closed circuit, or around a closed portion of a larger circuit, is zero, which is equivalent to saying that all the forces are in equilibrium in a statics problem.

I have developed the analogy with mechanical systems here to make the study of circuits more intuitively obvious. It is not possible in a short mathematical text to develop the theory of circuits (or of mechanics, for that matter) in detail. I leave this to the physics and engineering courses. Therefore, we are taking for granted the physical laws and experimental observations of the sciences in order to develop our differential equations as rapidly as possible.

To start, let us consider the simple circuit illustrated in Figure 2.4. This circuit consists of an electrical generator or other source producing an electro-

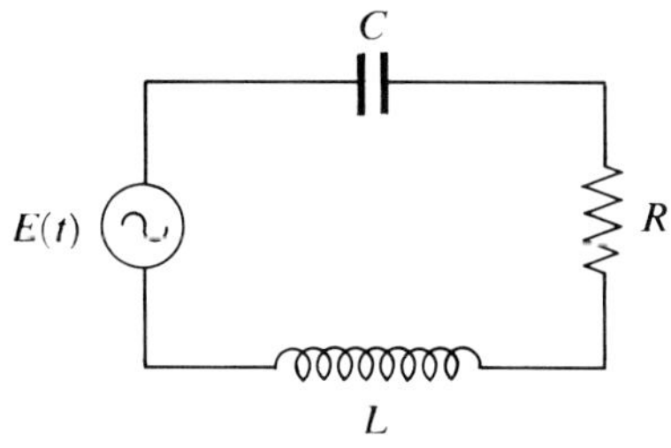

Figure 2.4

motive force $E(t)$ wired in series with a capacitance C, a resistance R, and an inductance L. To derive the differential equation for this circuit, we note that the generator produces an electromotive force $E(t)$ which is balanced by the voltage drops across the capacitor, resistor, and inductor of $-(1/C)Q$, $-R\dot{Q}$, and $L\ddot{Q}$, respectively. Thus by Kirchhoff's second law

$$E(t) - (1/C)Q - R\dot{Q} - L\ddot{Q} = 0$$

or

$$L\ddot{Q} + R\dot{Q} + (1/C)Q = E(t)$$

is the resulting differential equation for the charge on the capacitor $Q(t)$; to find the current through the circuit, we would first solve for Q, then differentiate it with respect to t in order to find I. Alternatively, one could differentiate the equation itself with respect to t, obtaining

$$L\ddot{I} + R\dot{I} + (1/C)I = E'(t).$$

As initial conditions, we need to know the initial charge on the capacitor $Q(0)$ and the initial current through the circuit $I(0)$. Thus

$$\begin{aligned} &L\ddot{Q} + R\dot{Q} + (1/C)Q = E(t), \\ &Q(0) = Q_0, \quad I(0) = \dot{Q}(0) = I_0, \end{aligned} \tag{2.5}$$

is the initial value problem for $Q(t)$. Note again that the form of this equation is the same as the form of the equation of the vibrating spring and the linearized pendulum equation.

Let us now derive the differential equations for the more complicated circuit in Figure 2.5. Here we have two circuits like the one of Figure 2.4 wired

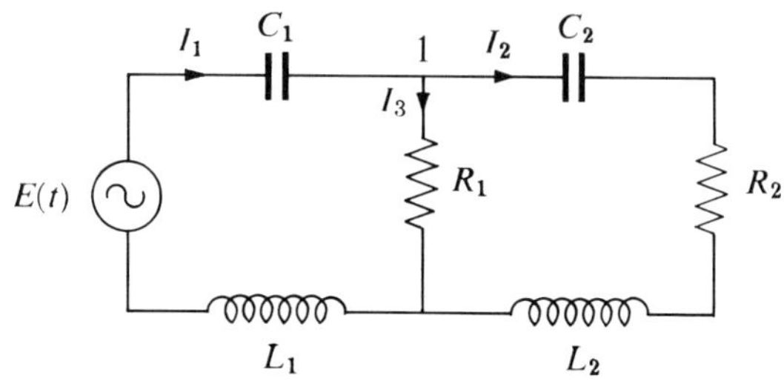

Figure 2.5

in parallel. There are now three unknowns, the currents $I_1(t)$, $I_2(t)$, and $I_3(t)$, related by the equation

$$I_1 = I_2 + I_3$$

derived by applying Kirchhoff's first law to junction 1.

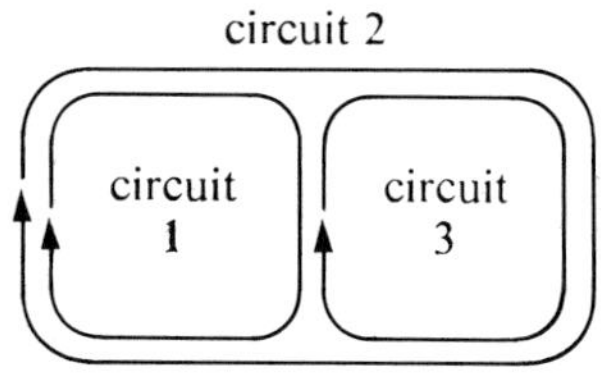

Figure 2.6

Since there are three unknowns, we shall need three equations, of which the above is one. However, we can obtain three additional equations by considering the three circuits of Figure 2.5, schematically shown in Figure 2.6. Thus, these four equations had better be consistent, or we are in trouble. Looking at circuit 1, we obtain

$$E(t) - (1/C_1)Q_1 - R_1 I_3 - L_1 \dot{I}_1 = 0,$$

by equating the total electromotive force to zero. For circuit 2 we obtain

$$E(t) - (1/C_1)Q_1 - (1/C_2)Q_2 - R_2 I_2 - L_2 \dot{I}_2 - L_1 \dot{I}_1 = 0,$$

and finally, for circuit 3,

$$-(1/C_1)Q_2 - R_2 I_2 - L_2 \dot{I}_2 + R_1 I_3 = 0,$$

where the last term differs in sign from the first three because the current flows through R_1 in the opposite direction from the direction of the circuit. Rewriting these three equations in terms of the charges, we have

$$L_1\ddot{Q}_1 + R_1\dot{Q}_3 + (1/C_1)Q_1 = E(t),$$
$$L_2\ddot{Q}_2 + L_1\ddot{Q}_1 + R_2\dot{Q}_2 + (1/C_1)Q_1 + (1/C_2)Q_2 = E(t),$$
$$L_2\ddot{Q}_2 + R_2\dot{Q}_2 - R_1\dot{Q}_3 + (1/C_2)Q_2 = 0.$$

Upon inspecting this system of equations, we note that if we subtract the third equation from the second, then we obtain the first. Thus, these three equations are dependent, and we need only two of them (say the first and third) together with the first Kirchhoff's law equation in order to describe the system. Since, as initial conditions, we shall need $Q_1(0)$, $Q_2(0)$, $Q_3(0)$, $I_1(0)$, and $I_2(0)$, we obtain the initial value problem

$$\begin{aligned} &\dot{Q}_1 - \dot{Q}_2 - \dot{Q}_3 = 0, \\ &L_1\ddot{Q}_1 + R_1\dot{Q}_3 + (1/C_1)Q_1 = E(t), \\ &L_2\ddot{Q}_2 + R_2\dot{Q}_2 - R_1\dot{Q}_3 + (1/C_2)Q_2 = 0, \\ &Q_1(0) = Q_{01}, \quad Q_2(0) = Q_{02}, \quad Q_3(0) = Q_{03}, \quad \dot{Q}(0) = I_{01}, \quad \dot{Q}(0) = I_{02}. \end{aligned} \tag{2.6}$$

Problems

1. Write the equation of the simple pendulum in the case in which there is no external force, when there is an additional frictional term depending on the cube of the velocity.

2. Write the equation of the simple pendulum in the case when friction is absent and when the external force is sinusoidal; i.e., $f(t) = k \sin \omega t$. Write the same equation in linearized form.

3. What is a solution of the equation of problem 1 with initial conditions $\theta(0) = 0$, $\dot{\theta}(0) = 0$? Justify your answer physically. Would this also be the solution of the equations of problem 2 with the same initial conditions? Why or why not?

4. Derive the initial value problem whose solutions describe the motion of the problem of two springs when x_2 is measured from the position of mass 1 (Figure 2.7). How could we arrive at this system from (2.4) without a second derivation?

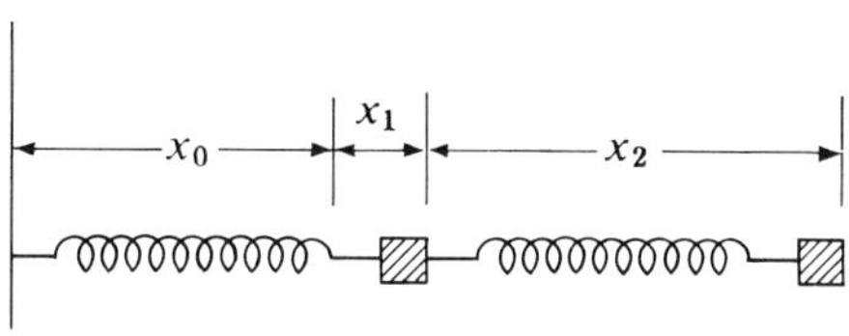

Figure 2.7

5. Derive the initial value problem for the circuit illustrated in Figure 2.8.

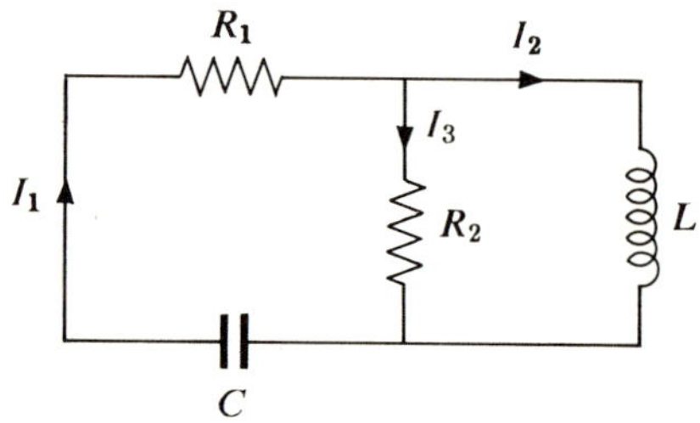

Figure 2.8

2.2 Preliminary Definitions and Theorems

Let us recall that the general linear equation of order n is*

$$a_n(x)D^n y + a_{n-1}(x)D^{n-1}y + \cdots + a_1(x)Dy + a_0(x)y = f(x). \tag{2.7}$$

The same equation, with the right side replaced by zero,

$$a_n(x)D^n y + a_{n-1}(x)D^{n-1}y + \cdots + a_1(x)Dy + a_0(x) = 0, \tag{2.8}$$

is called the *associated homogeneous equation* of (2.7). Much of our study of the general linear equation (2.7) will consist of seeking out the properties of the associated homogeneous equation (2.8), a somewhat simpler task by virtue of the zero on the right side.

In order to obtain some idea of what to expect, let us examine the simplest linear equation of order n,

$$D^n y = f(x), \tag{2.9}$$

and its associated homogeneous equation,

$$D^n y = 0. \tag{2.10}$$

We can obtain the solutions of these two equations by simple integration repeated n times. Thus, looking first at (2.10), we obtain, by repeated integrations,

$$D^{n-1}y = b_{n-1},$$

$$D^{n-2}y = b_{n-1}x + b_{n-2},$$

$$D^{n-3}y = \tfrac{1}{2}b_{n-1}x^2 + b_{n-2}x + b_{n-3},$$

* Throughout our study of linear equations in the next few chapters, particularly those with constant coefficients, we shall use the notation Dy for dy/dx or $y'(x)$, and $D^n y$ for $d^n y/dx^n$ or $y^{(n)}(x)$. This so-called "operator" notation is standard in the study of linear equations; the reasons for this will become apparent.

and, proceeding in this way, we eventually find that

$$y = \frac{b_{n-1}}{(n-1)!}x^{n-1} + \frac{b_{n-2}}{(n-2)!}x^{n-2} + \cdots + b_1 x + b_0,$$

or, renaming the constants, we have that

$$y = c_{n-1}x^{n-1} + c_{n-2}x^{n-2} + \cdots + c_1 x + c_0$$

is the general solution of (2.10). This solution involves n arbitrary constants $c_0, c_1, \ldots, c_{n-1}$ because each integration introduces another constant, and there were n integrations. We should suspect from this that the general solution of the general homogeneous equation (2.8) would also involve n arbitrary constants, because somehow solving it must also involve n integrations. In fact, this suspicion turns out to be true, as we shall see later.

If we go on to solve (2.9), we obtain, after n integrations, that

$$y = F(x) + c_{n-1}x^{n-1} + c_{n-2}x^{n-2} + \cdots + c_1 x + c_0,$$

where $F(x)$ is any function for which $D^n F = f$, i.e.,

$$F(x) = \underbrace{\int\int\cdots\int}_{n\text{ times}} f(x)dx\,dx\cdots dx.$$

Thus, we obtain the solution of the inhomogeneous equation by integrating $f(x)$ n times and adding an arbitrary constant at each integration. The effect of this process is the same as if we had added the general solution of the homogeneous equation to any function whose nth derivative is f. This property holds for all linear equations, and we shall prove it later in this section.

To begin our study of the general equations (2.7) and (2.8) we shall prove the following:

Theorem 2.1 (First Principle of Superposition)
If $y_1(x)$ and $y_2(x)$ are any two solutions of the linear homogeneous equation (2.8) and if c_1 and c_2 are any two arbitrary constants, then $c_1 y_1 + c_2 y_2$ is also a solution of (2.8).

Proof

The following string of equalities, based on familiar properties of the derivative, will suffice to prove this statement:

$$a_n D^n(c_1 y_1 + c_2 y_2) + a_{n-1}D^{n-1}(c_1 y_1 + c_2 y_2) + \cdots + a_1 D(c_1 y_1 + c_2 y_2) \\ + a_0(c_1 y_1 + c_2 y_2)$$

$$= a_n c_1 D^n y_1 + a_n c_2 D^n y_2 + \cdots + a_1 c_1 D y_1 + a_1 c_1 D y_1 + a_1 c_2 D y_2 + a_0 c_1 y_1 + a_0 c_2 y_2$$

$$= a_n c_1 D^n y_1 + \cdots + a_1 c_1 D^n y_1 + a_0 c_1 y_1 + a_n c_2 D^n y_2 + \cdots + a_1 c_2 D^n y_2 + a_0 c_2 y_2$$

$$= c_1(a_n D y_1 + \cdots + a_1 D y_1 + a_0 y_1) + c_2(a_n D y_2 + \cdots + a_1 D y_2 + a_0 y_2)$$

$$= c_1 \cdot (0) + c_2 \cdot (0) = 0.$$

Theorem 2.1 can be stated in a somewhat different way if we first pose the following:

Definition

If $y_1(x), y_2(x), \ldots, y_k(x)$ are any k functions and if $c_1, c_2, \ldots, c_k$ are any k constants, then the expression

$$c_1 y_1 + c_2 y_2 + \cdots + c_k y_k \tag{2.11}$$

is called a *linear combination* of $y_1, y_2, \ldots, y_k$. If the constants are all zero then (2.11) is called the *trivial linear combination.*

We can then put theorem 2.1 as follows: *Every linear combination of solutions of* (2.8) *is also a solution of* (2.8). Actually, we have only proved this statement for two solutions, not for any number; however, the proof can easily be completed by induction, an exercise which is left to the reader.

We now proceed to

Theorem 2.2 (Second Principle of Superposition)

If $y_h(x)$ is any solution of the homogeneous linear equation (2.8) and if $y_p(x)$ is any solution of the inhomogeneous linear equation (2.7), then $y_h + y_p$ is also a solution of the inhomogeneous equation.

Proof

This is done as before, by a string of equalities:

$$a_n D^n(y_h + y_p) + \cdots + a_1 D(y_h + y_p) + a_0(y_h + y_p)$$

$$= a_n D^n y_h + a_n D^n y_p + \cdots + a_1 D y_h + a_1 D y_p + a_0 y_h + a_0 y_p$$

$$= (a_n D^n y_h + \cdots + a_1 D y_h + a_0 y_h) + (a_n D^n y_p + \cdots + a_1 D y_p + a_0 y_p)$$

$$= (0) + [f(x)] = f(x).$$

We now have proved for (2.7) part of the general statement made of the special linear equation (2.9), but not the whole statement, because we do not yet know precisely what constitutes a general solution of the homogeneous equation. Our next step in pursuit of the proof is the concept of linear dependence and linear independence.

Definition
A set of two or more functions $y_1(x), \ldots, y_k(x)$ is said to be *linearly dependent* if there exist constants $c_1, \ldots, c_k$ *not all zero* such that

$$c_1 y_1(x) + \cdots + c_k y_k(x) \equiv 0.$$

In other words, $y_1, \ldots, y_k$ are linearly dependent if there is a nontrivial linear combination of them which vanishes identically. On the other hand, $y_1, \ldots, y_k$ are *linearly independent* if

$$c_1 y_1(x) + \cdots + c_k y_k(x) \equiv 0$$

implies that $c_1 = \cdots = c_k = 0$, i.e., the *only* linear combination of $y_1, \ldots, y_k$ which vanishes identically is the trivial one.

Since this is a very important concept, let us drive it home with a few examples.

(1) Zero and any set of one or more functions $y_1(x), \ldots, y_k(x)$ are linearly dependent. For, the linear combination

$$1 \cdot 0 + 0 \cdot y_1(x) + \cdots + 0 \cdot y_k(x) \equiv 0$$

vanishes identically, and the constant coefficients $(1, 0, \ldots, 0)$ are *not all zero*, as the definition requires!

(2) x, $3 + x$, and $1 - x$ are linearly dependent, since for example

$$-4(x) + 1(3 + x) - 3(1 - x) \equiv 0.$$

(3) The set of functions $1, x, x^2, \ldots, x^k$ are linearly independent for any $k \geqq 1$. For, if

$$c_0 + c_1 x + c_2 x^2 + \cdots + c_k x^k \equiv 0, \tag{2.12}$$

then we can prove that the constants $c_1, \ldots, c_k$ all vanish by either of two methods: the first is based on the algebraic theorem which says that an algebraic polynomial $c_0 + c_1 x + \cdots + c_k x^k$ with nonzero coefficients can have at most k roots; that is, there are at most k values of x for which the polynomial vanishes, and hence the identity (2.12) cannot hold unless $c_1 = c_2 = \cdots = c_k = 0$. The second method of proof is based on induction. First, 1 and x are linearly independent, since if

$$c_0 + c_1 x \equiv 0,$$

then by differentiating we have

$$c_1 = 0,$$

and hence $c_0 = 0$ also. Next, if we assume that $1, x, x^2, \ldots, x^{k-1}$ are linearly independent, for $k \geqq 2$, then the expression

$$c_0 + c_1 x + \cdots + c_k x^k \equiv 0 \tag{2.13}$$

yields, upon differentiation,

$$c_1 + 2c_2 x + \cdots + kc_k x^{k-1} \equiv 0,$$

which is a linear combination of $1, \ldots, x^{k-1}$, so that by the induction hypothesis,

$$c_1 = 2c_2 = \cdots = kc_k = 0$$

or

$$c_1 = c_2 = \cdots = c_k = 0.$$

Hence, substituting in (2.13), we obtain also that $c_0 = 0$. Thus, since *all* the coefficients must vanish, we conclude that $1, x, \ldots, x^k$ are linearly independent.

(4) The functions $e^{m_1 x}, e^{m_2 x}, \ldots, e^{m_k x}$ are linearly independent if $m_i \neq m_j$ when $i \neq j$, i.e., if no two of the m_i's are equal. We shall prove this for $k = 2$ and leave the remainder of the proof as an exercise. For $k = 2$,

$$c_1 e^{m_1 x} + c_2 e^{m_2 x} \equiv 0$$

implies by differentiation that

$$m_1 c_1 e^{m_1 x} + m_2 c_2 e^{m_2 x} \equiv 0,$$

and if we multiply the first equation by m_1 and subtract the second, we have

$$(m_1 - m_2) c_2 e^{m_2 x} \equiv 0,$$

which implies that $c_2 = 0$, since $m_1 - m_2 \neq 0$ and $e^{m_2 x} \neq 0$. Then we have $c_1 e^{m_1 x} \equiv 0$, which in turn implies that $c_1 = 0$, since $e^{m_1 x} \neq 0$.

(5) As our next example we shall show that $\sin k_1x$ and $\sin k_2x$ are linearly independent if $k_1^2 \neq k_2^2$ and neither k_1 nor k_2 is 0. If

$$c_1 \sin k_1x + c_2 \sin k_2x \equiv 0,$$

then after differentiating twice we have

$$-c_1k_1^2 \sin k_1x - c_2k_2^2 \sin k_2x \equiv 0;$$

multiplying the first equation by k_1^2 and adding, we obtain

$$c_2(k_1^2 - k_2^2)\sin k_2x \equiv 0,$$

and since $k_1^2 - k_2^2 \neq 0$ and $\sin k_2x$ is zero only for some values of x but not identically zero, we have $c_2 = 0$. Hence $c_1 \sin k_1x \equiv 0$, which in turn implies that $c_1 = 0$ and hence that the two functions are linearly independent.

(6) Another way of determining whether n functions $f_1(x), f_2(x), \ldots, f_n(x)$ which possess at least $n - 1$ derivatives are linearly independent is by means of the *Wronskian determinant*, which we shall define in the following. First, let us suppose that the functions $f_1, f_2, \ldots, f_n$ are linearly dependent for all x in the interval $a \leqq x \leqq b$, for example. Then there exist n constants $c_1, c_2, \ldots, c_n$ *not all zero* such that

$$c_1f_1(x) + c_2f_2(x) + \cdots + c_nf_n(x) \equiv 0$$

for all x in the interval $[a, b]$.* Then the following system of equations is valid for all x in the interval, since the successive equations are obtained from the first by repeated differentiation:

$$\begin{aligned}
c_1f_1(x) + c_2f_2(x) + \cdots + c_nf_n(x) &\equiv 0,\\
c_1f_1'(x) + c_2f_2'(x) + \cdots + c_nf_n'(x) &\equiv 0,\\
c_1f_1''(x) + c_2f_2''(x) + \cdots + c_nf_n''(x) &\equiv 0,\\
&\vdots\\
c_1f_1^{(n-1)}(x) + c_2f_2^{(n-1)}(x) + \cdots + c_nf_n^{(n-1)}(x) &\equiv 0.
\end{aligned}$$

We can, for each value of x, consider this to be a linear homogeneous system of n equations in the n unknowns $c_1, c_2, \ldots, c_n$; and by hypothesis we know that it has a nontrivial solution for each x in $[a, b]$: that is, it has a solution

* The reader is reminded that the notation $[a, b]$ means "the set of all x such that $a \leqslant x \leqslant b$." $a < b$ is assumed. $[a, b]$ is called the closed interval from a to b.

set $c_1, c_2, \ldots, c_n$ such that at least one c_i is nonzero. From the elementary theory of linear equations, we know that a homogeneous system of linear equations has a nontrivial solution only if its determinant vanishes; and since the determinant of this system is a function of x, it must vanish identically:

$$\begin{vmatrix} f_1 & f_2 & \cdots & f_n \\ f_1' & f_2' & \cdots & f_n' \\ \vdots & \cdot & & \\ f_1^{(n-1)} & f_2^{(n-1)} & \cdots & f_n^{(n-1)} \end{vmatrix} \equiv 0.$$

This determinant, usually denoted by $W[f_1, f_2, \ldots, f_n]$, is called the *Wronskian* of the set of functions $f_1, f_2, \ldots, f_n$. Conversely, if the Wronskian is nonzero even at a single point in $[a, b]$, then the only possible values for the c_i's are all zero. Thus we have proved the following:

Theorem 2.3

Let $f_1, f_2, \ldots, f_n$ be a set of functions defined on an interval $[a, b]$, each possessing $n - 1$ derivatives there. Then

(a) If $f_1, f_2, \ldots, f_n$ are linearly dependent, we have $W[f_1, f_2, \ldots, f_n] \equiv 0$, that is, the Wronskian vanishes identically.

(b) If $W[f_1, f_2, \ldots, f_n] \not\equiv 0$, that is, if the Wronskian is nonzero at even one point, the $f_1, f_2, \ldots, f_n$ are linearly independent.

As an example of the use of the Wronskian to prove linear independence, let us again consider the set of functions $1, x, x^2, \ldots, x^n$ from example (1). Then, since

$$W[1, x, x^2, \ldots, x^n] = \begin{vmatrix} 1 & x & x^2 & x^3 \cdots x^n \\ 0 & 1 & 2x & 3x^2 \cdots nx^{n-1} \\ 0 & 0 & 2! & 3 \cdot 2x \cdots n(n-1)x^{n-2} \\ 0 & 0 & 0 & 3! \cdots n(n-1)(n-2)x^{n-3} \\ & & & \vdots \\ 0 & 0 & 0 & 0 \cdots n! \end{vmatrix} = 0!1!2! \cdots n! \neq 0,$$

we have immediately that these functions must be linearly independent for *all* values of x.

Many more examples of linearly dependent and linearly independent sets of functions occur in the problems at the end of this section.

The importance of the concept of linearly independent sets of functions is that it enables us to recognize the general solution of the linear homogeneous equation (2.8). Any linear combination of solutions of the homogeneous equation is itself a solution; but if the set of solutions with which we start is a linearly dependent set, then at least one solution is a linear combination of the others, so that we ultimately have fewer arbitrary constants than the number of solutions in the set. For example, using the functions $y_1 = x$, $y_2 = 3 + x$, $y_3 = 1 - x$ from example 2, we have $y_3 = \frac{1}{3}y_2 - \frac{4}{3}y_1$, so that the expression $c_1y_1 + c_2y_2 + c_3y_3$ equals

$$\begin{aligned} &c_1y_1 + c_2y_2 + c_3(\tfrac{1}{3}y_2 - \tfrac{4}{3}y_1) \\ &= (c_1 - \tfrac{4}{3}c_3)y_1 + (c_2 + \tfrac{1}{3}c_3)y_2 = k_1y_1 + k_2y_2; \end{aligned}$$

and we see that there are only two arbitrary constants present in $c_1y_1 + c_2y_2 + c_3y_3$ because of the linear dependence. Consequently, if we have an nth-order equation and n solutions, then in order to obtain the general solution by superposition, these n functions must be linearly independent. Otherwise we could not have n arbitrary constants in the general solution, contrary to our observation that n integrations produce n arbitrary constants. Thus:

Theorem 2.4
If each of the functions $y_1(x), y_2(x), \ldots, y_n(x)$ is a solution of the homogeneous linear equation (2.8) of order n and if $y_1, y_2, \ldots, y_n$ is a linearly independent set of functions, then

$$y_h(x) = c_1y_1(x) + c_2y_2(x) + \cdots + c_ny_n(x),$$

with the c_i's constants, is the general solution of (2.8).

From this theorem, which we shall prove in the next section, we have immediately that

Theorem 2.5
If $y_p(x)$ is any particular solution of the inhomogeneous linear equation (2.7) of order n and if $y_1(x), \ldots, y_n(x)$ is a linearly independent set of solutions of the associated homogeneous equation (2.8), then

$$y(x) = y_p(x) + c_1y_1(x) + \cdots + c_ny_n(x) = y_p(x) + y_h(x),$$

with the c_i's arbitrary constants, is the general solution of the inhomogeneous linear equation (2.7).

To interpret the principles of superposition physically, let us consider a special case of the equation of the vibrating spring,

$$\ddot{x} + x = 0.$$

Here we are assuming that there is no friction or externally applied force, and that the ratio k/m is unity. In this case we would expect that once the spring had been started in motion, it would continue indefinitely, because of the lack of friction. Experimental observation leads us also to the conclusion that the motion of the mass is oscillatory, or periodic, in nature.

It is not difficult to verify that $x_1(t) = \sin t$ is a solution of this equation; it satisfies the initial conditions $x_1(0) = 0$, $\dot{x}_1(0) = 1$. $x_2(t) = \cos t$ is also a solution, this time with $x_2(0) = 1$ and $\dot{x}_2(0) = 0$. In fact, one could find these solutions using the method of "independent variable missing" described in Section 1.7. x_1 describes what happens when we impart unit velocity to the spring when it is initially at rest; x_2 describes its motion when it is released at rest after an initial displacement of 1 unit. From what we have seen previously, then, the function

$$x_h(t) = c_1x_1(t) + c_2x_2(t) = c_1 \sin t + c_2 \cos t$$

is the general solution of the equation. Physically this means that every possible motion of the spring is a superposition of these two basic motions; it has what we might call a component of each type. We notice that every possible motion is periodic with period 2π.

Provided that c_1 and c_2 are not zero, we can write $x_h(t)$ in the form

$$x_h(t) = \sqrt{c_1^2 + c_2^2}\left(\frac{c_1}{\sqrt{c_1^2 + c_2^2}} \sin t + \frac{c_2}{\sqrt{c_1^2 + c_2^2}} \cos t\right).$$

The sum of the squares of the coefficients of the quantity in parentheses is 1; therefore it is possible to define an angle δ by

$$\sin \delta = \frac{c_1}{\sqrt{c_1^2 + c_2^2}}, \qquad \cos \delta = \frac{c_2}{\sqrt{c_1^2 + c_2^2}},$$

and thus

$$x_h(t) = \sqrt{c_1^2 + c_2^2}\,(\sin \delta \sin t + \cos \delta \cos t).$$

Writing $\sqrt{c_1^2 + c_2^2} = A$, we thus see that

$$x_h(t) = A \cos(t - \delta)$$

is the general solution. A is the amplitude of the motion, representing the maximum distance the mass moves from equilibrium, and the angle δ is called the phase angle because it represents how much the motion is out of phase with $A \cos t$.

In case there is an external driving force $f(t)$, our differential equation becomes

$$\ddot{x} + x = f(t).$$

If $x_p(t)$ is any solution of this equation, then by superposition

$$x(t) = A\cos(t - \delta) + x_p(t)$$

is the general solution. Thus there is a component of the motion caused directly by the driving force, but superimposed on this motion is the periodic motion caused by the peculiar characteristics of the spring itself. $x_p(t)$ alone represents the motion of the spring if it had started from rest at the equilibrium position. The superimposed (or, as we say in mathematics, superposed) periodic motion is the effect of imparting an initial displacement and/or velocity to the mass.

Finally, to say that two solutions of the homogenous equation are linearly independent simply means that the motions represented by the two solutions are out of phase, and that the difference between their phase angles is not an integer multiple of π.

Problems

In the following problems, determine whether or not the set of functions given is linearly independent; if it is dependent, give a nontrivial linear combination which is identically zero.

1. $3,\ 1 + x,\ 2 - x + x^2,\ x^2 + 1$.
2. $\sin 2x,\ \cos 2x$.
3. $x,\ \sin x,\ \cos x$.
4. $\cos^2 x,\ \sin^2 x,\ \cos 2x$.
5. $x,\ \cos 2x,\ e^{-x},\ e^{2x}$.
6. $e^{2x},\ \sinh 2x,\ \cosh 2x,\ \cos x$.
7. $e^{-3x},\ x^2e^{-3x},\ e^x,\ e^x\cos 3x$.
8. $1,\ \cos 2x,\ \sin^2 x,\ x,\ x^2,\ x^3$.
9. $\cos(\ln x),\ \cos(2\ln x),\ x^3$.
10. $\sin x,\ \sin 2x,\ \sin 3x$.
11. Prove the statement: if $y_1(x), \ldots, y_m(x)$ are all solutions of the homogeneous equation (2.8), then so is any linear combination of them, for any value of m.
12. Complete the proof of example 4: that $e^{m_1x}, e^{m_2x}, \ldots, e^{m_kx}$ are linearly independent provided that $m_i \neq m_j$ when $i \neq j$.

13. Prove that the set of functions $\sin k_1x, \sin k_2x, \ldots, \sin k_mx$ is linearly independent provided that $k_i^2 \neq k_j^2$ for $i \neq j$ and that $k_i \neq 0$ $(i = 1, 2, \ldots, m)$. Show that $\sin k_1x$ and $\sin k_2x$ are linearly dependent if either $k_i = 0$ or $k_1 = \pm k_2$.

14. Prove that the set of functions $\cos k_1x, \cos k_2x, \ldots, \cos k_mx$ is linearly independent provided that $k_i^2 \neq k_j^2$ for $i \neq j$. Why do we not need the proviso that $k_i \neq 0$ in this case?

15. Prove that the set of functions $\sin k_1x, \cos k_1x, \sin k_2x, \cos k_2x, \ldots, \sin k_mx, \cos k_mx$ is linearly independent provided that $k_i^2 \neq k_j^2$ for $i \neq j$ and that $k_i \neq 0$ $(i = 1, 2, \ldots, m)$.

16. Prove that the set of functions $e^{mx}, xe^{mx}, \ldots, x^ke^{mx}$ is linearly independent if $k \geqq 1$.

17. Prove that the set of functions

$$e^{m_1x}, xe^{m_1x}, \ldots, x^{k_1}e^{m_1x},$$
$$e^{m_2x}, xe^{m_2x}, \ldots, x^{k_2}e^{m_2x},$$
$$e^{m_rx}, xe^{m_rx}, \ldots, x^{k_r}e^{m_rx}$$

is linearly independent provided that $m_i \neq m_j$ for $i \neq j$.

18. Do the same thing where e^{m_ix} is replaced by $\sin m_ix$ $(i = 1, 2, \ldots, r)$, and where $m_i^2 \neq m_j^2$ for $i \neq j$ and $m_i \neq 0$ $(i = 1, 2, \ldots, r)$.

19. Prove that if $y_1(x), y_2(x), \ldots, y_k(x)$ is a linearly dependent set of functions and if $u_1(x), u_2(x), \ldots, u_n(x)$ is any other set of functions (dependent or independent), then the set $y_1, y_2, \ldots, y_k, u_1, u_2, \ldots, u_n$ is linearly dependent.

20. If $y_1, \ldots, y_k$ and $u_1, \ldots, u_n$ are two linearly independent sets, is the set $y_1, \ldots, y_k, u_1, \ldots, u_n$ necessarily linearly independent? Prove your statement.

2.3 Some Remarks about the Existence and Uniqueness of Solutions of Linear Equations; more on Wronskians

So far our remarks about the form of solutions of linear equations must seem rather artificial, in that in the absence of knowledge about whether or not such solutions exist, any discussion of their properties is about as futile as shouting in a vacuum. In this section we shall discuss the known existence and uniqueness theorems relating to linear equations and systems. Let us first consider a linear first-order system; and then by using the mathematicians' technique of reducing everything to the previous case, we shall show how all problems of interest can be reformulated as first-order systems.

A set of equations of the form

$$\begin{aligned} y_1' &= f_{11}(x)y_1 + f_{12}(x)y_2 + \ldots + f_{1n}(x)y_n + g_1(x) \\ y_2' &= f_{21}(x)y_1 + f_{22}(x)y_2 + \ldots + f_{2n}(x)y_n + g_2(x) \\ &\vdots \\ y_n' &= f_{n1}(x)y_1 + f_{n2}(x)y_2 + \ldots + f_{nn}(x)y_n + g_n(x), \end{aligned} \tag{2.14}$$

where each g_i and each f_{ij} $(1 \leqq i \leqq n, \quad 1 \leqq j \leqq n)$ is defined and continuous for all x in an interval $[a, b]$, is called a *linear first-order system* of ordinary differential equations for the n unknown functions $y_1, y_2, \ldots, y_n$. An initial value problem for this system involves in addition the n initial conditions

$$y_1(a) = c_1, y_2(a) = c_2, \ldots, y_n(a) = c_n. \tag{2.15}$$

The condition that the f_{ij}'s be continuous is important because it guarantees that the right sides of (2.14) will be Lipschitz continuous in the y_i's; and we noted in Chapter 1 that the existence and uniqueness proof depends on this hypothesis. For example, if we write

$$f_1(x, y_1, y_2, \ldots, y_n) = f_{11}(x)y_1 + f_{12}(x)y_2 + \ldots + f_{1n}(x)y_n + g_1(x),$$

then calculate the difference between the values of f_1 at two different points $(x, y_1, y_2, \ldots, y_n)$ and $(x, \bar{y}_1, y_2, \ldots, y_n)$, where only the second coordinates differ, we have

$$f_1(x, y_1, y_2, \ldots, y_n) - f_1(x, \bar{y}_1, y_2, \ldots, y_n) = f_{11}(x)(y_1 - \bar{y}_1)$$

and hence

$$|f_1(x, y_1, \ldots, y_n) - f_1(x, \bar{y}_1, \ldots, y_n)| \leqq |f_{11}(x)|\,|y_1 - \bar{y}_1|.$$

Since f_{11} is continuous on $[a, b]$, it is bounded there, and therefore f_1 is Lipschitz continuous in y_1. Similarly it [and the other right sides of (2.14)] is Lipschitz continuous in all the other y_i's. Using this fact, we can prove the following theorem in much the same way as one proves the existence theorem in Chapter 1:

Theorem 2.6
There exists one and only one solution set $(y_1(x), y_2(x), \ldots, y_n(x))$, defined on $[a, b]$, satisfying the linear system (2.14) and the initial conditions (2.15), provided that all of the functions f_{ij} and g_i are continuous on $[a, b]$.

Let us now see how we can reduce the initial value problem for the single equation of order n

$$\begin{aligned} a_n(x)y^{(n)} + a_{n-1}(x)y^{(n-1)} + \ldots + a_0(x)y &= f(x), \\ y(a) = c_1, y'(a) = c_2, \ldots, y^{(n-1)}(x) &= c_n, \end{aligned} \tag{2.16}$$

to a linear system. To do this, we simply define $y_1(x) = y(x)$. Then we further define functions $y_2, y_3, \ldots, y_n$ by means of the equations

$$\begin{aligned} y_1' &= y_2, \\ y_2' &= y_3, \\ y_3' &= y_4, \\ &\vdots \\ y_n' &= -\frac{a_{n-1}}{a_n}y_{n-1} - \frac{a_{n-2}}{a_n}y_{n-2} - \cdots - \frac{a_0}{a_n}y_1 + f, \end{aligned}$$

which is possible wherever $a_n(x) \neq 0$. We see that $y_k(x) \equiv y^{(k-1)}(x)$ for $k = 1, 2, \ldots, n$ and that this first-order system plus the initial conditions

$$y_1(a) = c_1, y_2(a) = c_2, \ldots, y_n(a) = c_n$$

is precisely the same as the original initial value problem (2.16). By applying our previous theorem to this new system, we conclude that the following is true:

Theorem 2.7

Let $[a, b]$ be an interval in which the a_i $(1 \leqq i \leqq n)$ and f are continuous and in which $a_n(x) \neq 0$ for any x. Then the initial value problem (2.16) possesses one and only one solution $y(x)$ defined on $[a, b]$.

Note how crucial it is that $a_n(x)$ not vanish in the interval $[a, b]$. If it did, then the right side of the last equation would not be defined; because of this, any point x_0 for which $a_n(x_0) = 0$ is called a *singular point* of equation (2.16). Sometimes solutions exist near a singular point and sometimes not; we shall discuss this in more detail later.

To see how a more complicated problem might be reduced to a first-order system, let us consider as an example the system (2.6) derived earlier in this chapter:

$$\begin{aligned} &\dot{Q}_1 - \dot{Q}_2 - \dot{Q}_3 = 0, \\ &L_1\ddot{Q}_1 + R_1\dot{Q}_3 + K_1Q_1 = E \quad (K_1 = 1/C_1), \\ &L_2\ddot{Q}_2 + R_2\dot{Q}_2 - R_1\dot{Q}_3 + K_2Q_2 = 0 \quad (K_2 = 1/C_2), \\ &Q_1(0) = Q_{01}, \quad Q_2(0) = Q_{02}, \quad Q_3(0) = Q_{03}, \quad \dot{Q}_1(0) = I_{01}, \quad \dot{Q}_2(0) = I_{02}. \end{aligned}$$

In order to convert this second-order system to a first-order system, we define $y_1 \equiv Q_1, y_2 \equiv Q_2, y_3 \equiv Q_3$, then write

$$\begin{aligned}
\dot{y}_1 &= y_4,\\
\dot{y}_2 &= y_5,\\
\dot{y}_3 &= y_4 - y_5,\\
\dot{y}_4 &= -(R_1/L_1)(y_4 - y_5) - (K_1/L_1)y_1 + E/L_1,\\
\dot{y}_5 &= -(R_2/L_2)y_5 + (R_1/L_2)(y_4 - y_5) - K_2y_2,\\
y_1(0) &= Q_{01}, \quad y_2(0) = Q_{02}, \quad y_3(0) = Q_{03}, \quad y_4(0) = I_{01}, \quad y_5(0) = I_{02}.
\end{aligned}$$

It is easy to see that this first-order system is equivalent to the original problem, (2.6). Provided that the inductances do not vanish and that all the coefficients are constant, or, if variable, continuous, then once again we can apply the existence and uniqueness theorems of this chapter and conclude that (2.6) possesses a unique solution set Q_1, Q_2, Q_3. In a precisely similar way it is always possible to convert any higher-order linear system into a first-order linear system.

Now that we have stated some existence and uniqueness theorems for solutions of linear equations, it is possible to make some further observations about the Wronskian of a set of solutions of a given linear equation. Although we state and prove the following set of theorems for second-order linear equations, they are true theorems for linear equations of any order:

Theorem 2.8

Let y_1 and y_2 be two solutions of the equation $a_2(x)y'' + a_1(x)y' + a_0(x)y = 0$ on some interval $[a, b]$, and suppose $a_2(x) \neq 0$ for x in $[a, b]$. Then either $W[y_1, y_2] \equiv 0$ on $[a, b]$ or else $W[y_1, y_2]$ never vanishes on $[a, b]$.

Proof

First note that

$$W[y_1, y_2] = y_1y_2' - y_2y_1'$$

and hence

$$W'[y_1, y_2] = y_1y_2'' - y_2y_1'' + y_1'y_2' - y_1'y_2' = y_1y_2'' - y_2y_1''.$$

Furthermore, because

$$y_i'' = -(a_1/a_2)y_i' - (a_0/a_2)y_i \quad (i = 1, 2),$$

and since each y_i is a solution of the differential equation and $a_2 \neq 0$, we have

$$\begin{aligned}
W'[y_1, y_2] &= y_1[-(a_1/a_2)y_2' - (a_0/a_2)y_2] - y_2[-(a_1/a_2)y_1' - (a_0/a_2)y_1]\\
&= -(a_1/a_2)y_1y_2' + (a_1/a_2)y_2y_1' = -(a_1/a_2)(y_1y_2' - y_2y_1')\\
&= -(a_1/a_2)W[y_1, y_2].
\end{aligned}$$

Thus W is a solution of the ordinary differential equation

$$W' + (a_1/a_2)W = 0,$$

which is a linear equation of the first order.

If we define

$$A(x) = \int_a^x \frac{a_1(t)}{a_2(t)}\,dt,$$

then the integrating factor for this equation is

$$v(x) = e^{A(x)}$$

and we have

$$(e^{A(x)}W)' = 0$$

so that the general solution is

$$W = ce^{-A(x)}.$$

Clearly this expression for W is never zero unless $c = 0$, and then it is identically zero. This completes the proof.

It is now possible to prove the following criterion for linear dependence of solutions:

Theorem 2.9

If y_1 and y_2 are two solutions of the equation $a_2(x)y'' + a_1(x)y' + a_0(x)y = 0$ on some interval $[a, b]$, if a_2 is never zero on $[a, b]$, and if there exists a point x_0 in $[a, b]$ where $W[y_1(x_0), y_2(x_0)] = 0$, then $y_1(x)$ and $y_2(x)$ are linearly dependent on $[a, b]$. Thus, if y_1 and y_2 are linearly independent, we must have $W[y_1, y_2] \neq 0$ for all x in $[a, b]$.

Proof

At the point x_0, we have

$$W[y_1(x_0), y_2(x_0)] = y_1(x_0)y_2'(x_0) - y_1'(x_0)y_2(x_0) = 0,$$

and hence the pair of linear equations

$$\begin{aligned} c_1y_1(x_0) + c_2y_2(x_0) &= 0, \\ c_1y_1'(x_0) + c_2y_2'(x_0) &= 0 \end{aligned}$$

has a zero determinant. Therefore it has an infinite number of nontrivial solution pairs (c_1, c_2). We now choose one such pair and define $y_3(x) = c_1y_1(x) + c_2y_2(x)$. Then, because c_1 and c_2 are solutions of the above system of equations, we see that $y_3(x_0) = 0$ and $y_3'(x_0) = 0$. By the first principle of superposition, y_3 is a solution of our differential equation since it is a linear combination of two solutions. But clearly $y \equiv 0$ is also a solution of the differential equation and the initial conditions $y(x_0) = 0, y'(x_0) = 0$. By theorem 2.7, there can be at most one solution to this problem; hence

$$y_3(x) \equiv 0 \equiv c_1y_1(x) + c_2y_2(x),$$

and therefore y_1 and y_2 are linearly dependent.

Our existence and uniqueness theorem, theorem 2.7, enables us to prove theorem 2.4 of the previous section, which we shall restate as follows:

Consider the homogeneous linear equation

$$a_n(x)y^{(n)} + a_{n-1}(x)y^{(n-1)} + \cdots + a_1(x)y' + a_0(x)y = 0,$$

and suppose that x_0 is a point where $a_n(x_0) \neq 0$. Then this differential equation possesses n solutions $y_1, y_2, \ldots, y_n$ which are linearly independent in some interval around x_0; and any other solution y defined near x_0 is a linear combination of $y_1, y_2, \ldots, y_n$. In other words, there are no more than n linearly independent solutions.

As proof, we define y_1 to be the unique solution of the equation which satisfies the initial condition

$$y(x_0) = 1, \quad y'(x_0) = 0, \quad y''(x_0) = 0, \ldots, y^{(n-1)}(x_0) = 0.$$

y_2 is the solution of the equation satisfying

$$y(x_0) = 0, \quad y'(x_0) = 1, \quad y''(x_0) = 0, \ldots, y^{(n-1)}(x_0) = 0.$$

And so on; y_k will be the solution whose $(k-1)$st derivative at x_0 has the value 1, while all the others vanish.

Because of the way we have constructed the y_k's, their Wronskian at the point x_0 does not vanish; indeed, it has the value 1. Therefore, by theorem 2.8, there is a whole interval around x_0 where this Wronskian does not vanish, and hence the y_k's are linearly independent in this interval by theorem 2.9. Now, suppose $y(x)$ is any other solution of the equation defined at and near x_0, and further suppose that

$$y(x_0) = c_1, \quad y'(x_0) = c_2, \ldots, y^{(n-1)}(x_0) = c_n.$$

Then another solution of the equation which has these same initial values is

$$c_1y_1 + c_2y_2 + \cdots + c_ny_n.$$

By the uniqueness theorem, however, there can be only one solution of the equation assuming given initial values, and therefore

$$y(x) \equiv c_1y_1(x) + c_2y_2(x) + \cdots + c_ny_n(x).$$

Hence every other solution of the equation is a linear combination of the n linearly independent solutions constructed as above.

Another interesting sidelight of this discussion is that given *one* solution y_1 of the linear differential equation

$$a_2y'' + a_1y' + a_0y = 0,$$

it is always possible to find a second linearly independent solution by using the equation

$$W[y_1, y_2] = e^{-A(x)},$$

where as before, $A(x)$ is an indefinite integral of a_1/a_2, and where we have chosen $c = 1$ for a particular solution. This last equation is simply

$$y_1y_2' - y_2y_1' = e^{-A(x)},$$

a *first-order* linear equation for y_2. Thus, by use of the Wronskian and one known solution of our given equation, we have succeeded in finding an equation of lower order for the other; this process is called the *reduction of order*.

For example, consider the equation

$$x^2y'' - 2xy' + 2y = 0,$$

and suppose we are given the particular solution $y_1 = x$. Then the second solution can be found by computing

$$A(x) = \int \frac{a_1(x)}{a_2(x)}dx = -\int \frac{2x}{x^2}dx = -2\ln x = \ln x^{-2}$$

and then setting up the equation

$$y_1y_2' - y_2y_1' = \exp[-A(x)]$$

which in this case is

$$xy_2' - y_2 = \exp(-\ln x^{-2}) = x^2.$$

Upon putting this equation in the standard form,

$$y_2' - (1/x)y_2 = x,$$

we see that the integrating factor is

$$v(x) = \exp\left[\int (-1/x)\, dx\right] = \exp(-\ln x) = x^{-1}.$$

Therefore

$$[(1/x)y_2]' = 1$$

and hence, integrating, we obtain

$$(1/x)y_2 = x.$$

Thus $y_2 = x^2$ is a second linearly independent solution.
As a second example, $y_1 = x - 1$ is one solution of

$$xy'' + (1 - x)y' + y = 0.$$

Here $A(x) = \int[(1 - x)/x]\, dx = \ln x - x$, so we have

$$(x - 1)y_2' - y_2 = xe^{-x},$$

a linear equation of the first order which can be solved by the usual methods to give a second linearly independent solution.

Problems

Use the method of reduction of order to find a second, linearly independent solution of each of the following equations with y_1 given.

1. $(x - 1)^2 y'' + (x - 1)y' - y = 0, \quad y_1 = x - 1.$
2. $x(2x + 1)y'' + 2(x + 1)y' - 2y = 0, \quad y_1 = x + 1.$
3. $y'' + y = 0, \quad y_1 = \sin x.$
4. $(x^2 + 2x + 4)y'' - 2(x + 1)y' + 2y = 0, \quad y_1 = 2x + 1.$
5. $y'' - 4y' + 4y = 0, \quad y_1 = e^{2x}.$
6. $x^2 y'' + xy' + y = 0, \quad y_1 = \cos(\ln x).$

7. $x^2y'' + 5xy' + 4y = 0, \quad y_1 = x^{-2}$.
8. $2(x+3)^2y'' - 3(x+3)y' - 3y = 0, \quad y_1 = (x+3)^3$.
9. $y'' + y' = 0, \quad y_1 = 1$.
10. $y'' - 2y' + y = 0, \quad y_1 = e^x \sin x$.

Reduce each of the following equations or systems to a first-order system.

11. $x^2y'' + xy' + (x^2 - v^2)y = 0$.
12. $y''' - 3y'' - y' + 3y = 4xe^x - \sin x$.
13. $\dot{x} - \ddot{y} = 4t, 5\dot{x} + 4\dot{y} = 0$.
14. $\dot{x} + \dot{y} = 4e^{2t} - 6t, \dot{y} - 2y - \ddot{x} = 6$.
15. $\ddot{y} - 3\dot{x} + \ddot{z} = 3t, \dot{z} - \ddot{x} + y = 0, 2\ddot{x} - \ddot{y} + \dot{z} = 2te^t$.

2.4 Linear Differential Operators

Throughout this chapter we have been using the notation Df to mean "the derivative of f with respect to x." The symbol D by itself is meaningless out of context; it is meaningful only when we prefix it to some differentiable function (known or unknown), whereupon it becomes an entity which produces from $f(x)$ the new function

$$Df(x) = \lim_{h \to 0} \frac{f(x+h) - f(x)}{h}.$$

The notation $Df(x)$ does not mean that we multiply $f(x)$ by D, but rather that D is something which, when "applied to" f, produces a new function, namely the derivative function of f. Thus the notation $f(x)D$ has no real meaning. However, the expression $f(x)Dg(x)$ certainly has a meaning: $f(x)$ multiplied by the derivative of g at x. In mathematics, we are always free to take an expression such as $f(x)D$, which is devoid of meaning, and assign a meaning to it, and we shall proceed to do so, quite generally, as follows:

Definition

An abstract expression of the form

$$L = L(D) = a_n(x)D^n + a_{n-1}(x)D^{n-1} + \cdots + a_1(x)D + a_0(x),$$

where $a_n(x) \not\equiv 0$, is called a *linear differential operator of order n*. L is defined to be that entity which, when applied to a function f, possessing n derivatives, produces the new function

$$L[f] = a_nD^nf + a_{n-1}D^{n-1}f + \cdots + a_1Df + a_0f.$$

A linear differential operator $L(D)$ is thus a formal algebraic polynomial in D with functions of x as coefficients. Algebraic polynomials can be added (or subtracted) and multiplied, thus producing new polynomials. However, we shall see that linear differential operators cannot usually be multiplied according to the normal algebraic rules—precisely because an operator is an abstract entity defined only in terms of what it does to a function. Thus we shall have to define addition and multiplication of operators also in terms of what the result does to a function, as follows:

Definition
If $L_1(D)$ and $L_2(D)$ are two linear differential operators then the *operator sum* $L_1(D) + L_2(D)$ is defined to be that linear operator which, when applied to a function $f(x)$, yields $L_1 f(x) + L_2 f(x)$. The *operator product* $L_1(D)L_2(D)$ is defined to be that operator which, when applied to a function $f(x)$, yields $L_1[L_2 f(x)]$.*

We cannot emphasize too strongly that, although we have used the usual algebraic notation for the algebraic sum and product, the operator sum and product is quite a different thing. We shall see in the problems that the operator sum of two linear operators $L_1(D)$ and $L_2(D)$ can be computed using the usual algebraic rule (add the coefficients of like powers of D). However, to calculate the operator product of two operators, we must use the calculus; and the result that we obtain is not what we would get by using algebra alone. To illustrate this, suppose $L_1 = xD + 1$ and $L_2 = D - x$; then

$$\begin{aligned}(L_1L_2)f &= L_1(L_2 f) = L_1(Df - xf) = (xD + 1)(Df - xf) \\ &= xD(Df - xf) + 1(Df - xf) = xD^2 f + x(-f - xDf) + Df - xf \\ &= xD^2 f + (1 - x^2)Df - 2xf = [xD^2 + (1 - x^2)D - 2x]f.\end{aligned}$$

On the other hand,

$$\begin{aligned}(L_2L_1)f &= L_2(L_1 f) = L_2(xDf + f) = (D - x)(xDf + f) \\ &= D(xDf + f) - x(xDf + f) = xD^2 f + Df + Df - x^2Df - xf \\ &= xD^2 f + (2 - x^2)Df - xf = [xD^2 + (2 - x^2)D - x]f.\end{aligned}$$

Thus we see that

$$L_1L_2 = xD^2 + (1 - x^2)D - 2x \neq xD^2 + (2 - x^2)D - x = L_2L_1,$$

so that we do not get the same result by the operator multiplication L_1L_2

* Naturally we always assume that $f(x)$ possesses as many derivatives as necessary to make the expressions meaningful.

as we do by the operator multiplication L_2L_1. In mathematical terms, we express this by saying that operator multiplication is *not commutative*. Since we know that algebraic multiplication *is* a commutative operation, we certainly have something quite different here.

An important special case of linear differential operators is the case in which all the coefficients $a_0, a_1, \ldots, a_n$ are constants. In this case the multiplication of two linear operators *is* commutative, and the result is exactly the same as we would get by formal algebraic multiplication. Let us demonstrate this in the simple case where $L_1 = aD + b$, $L_2 = cD + d$, where a, b, c, d are constants. Then algebraically,

$$L_1L_2 = (aD + b)(cD + d) = acD^2 + (ad + bc)D + bd;$$

operationally,

$$\begin{aligned}(L_1L_2)f &= (aD + b)(cDf + df) = aD(cDf + df) + b(cDf + df)\\ &= acD^2f + adDf + bcDf + bdf = [acD^2 + (ad + bc)D + bd]f,\end{aligned}$$

and

$$\begin{aligned}(L_2L_1)f &= (cD + d)(aDf + bf) = cD(aDf + bf) + d(aDf + bf)\\ &= acD^2f + bcDf + adDf + bdf = [acD^2 + (ad + bc)D + bd]f.\end{aligned}$$

The last two are equal and are the same as the algebraic result. The reason that algebraic multiplication and operator multiplication coincide in the case of operators with constant coefficients is that constants "pass through" D, that is, $D(cf) = cDf$ if c is a constant. In a sense, one might say that D and its powers commute with constants, though this is not really correct since cD is a differential operator, whereas Dc is a function (the zero function, in fact). D and its powers do commute with constants, though, when applied to a function: $D(cf) = (cD)f$, which is what we mean by the above statement.

On the other hand, operator multiplication is not commutative when the coefficients are not constant because D does not "commute" with functions (in the above sense):

$$a(x)Df(x) \neq D[a(x)f(x)] = f(x)Da(x) + a(x)Df(x).$$

The product rule for differentiation thus destroys any hope of commutativity of operator multiplication in the general case because of the extra term $f(x)Da(x)$.

Problems

In each of the following, find $L_1 + L_2$, L_1L_2, L_2L_1.

1. $L_1 = xD + 2$, $L_2 = 3D + x^2$.
2. $L_1 = x^2D^2 + 2xD$, $L_2 = 3D - x$.
3. $L_1 = (\cos x)D - \sin x$, $L_2 = e^xD - 1$.
4. $L_1 = 3D - 2$; $L_2 = 4D^2 - 5D + 1$.
5. $L_1 = \dfrac{1}{1 + x^2}D - 1$; $L_2 = \dfrac{1}{x}$.
6. $L_1 = (\sinh x)D^2 + (\cosh x)D - e^x$; $L_2 = 2D - 1$.
7. $L_1 = (x^2 + 2x)D + (x^2 + 1)$; $L_2 = (x - 1)D - x^3$.
8. $L_1 = 2D^2 + 3D - 1$; $L_2 = D^2 - 2D + 3$.
9. $L_1 = D^3 - 1$; $L_2 = D + x$.
10. $L_1 = D + 3x$; $L_2 = D + 2x$.
11. If

$$L_1 = a_n(x)D^n + \cdots + a_1(x)D + a_0(x)$$

and

$$L_2 = b_m(x)D^m + \cdots + b_1(x)D + b_0(x),$$

with $n \geqq m$, prove that $L_1 + L_2 = L_2 + L_1$ and that the result obtained is what we would expect if we added L_1 and L_2 algebraically.

12. If $L_1 = aD^n$ and $L_2 = bD^m$, with a and b constants, show that $L_1L_2 = L_2L_1 = abD^nD^m = abD^{n+m}$.

13. Using the result of problem 12, prove that if we multiply the two operators with *constant* coefficients $L_1 = a_nD^n + \cdots + a_1D + a_0$, $L_2 = b_mD^m + \cdots + b_1D + b_0$, then $L_1L_2 = L_2L_1$, and the result is the same as the algebraic result obtained from the usual rules of multiplication.

2.5 More on Linear Operators with Constant Coefficients

We have seen in the previous section that two (or more) linear operators with constant coefficients can be multiplied together just as if they were algebraic polynomials in the variable D. It follows from this that we can also reverse the process and *factor* linear operators with constant coefficients as if they were polynomials in the variable D; and the order in which the factors are written is irrelevant. Thus, for example,

$$D^2f + (a + b)Df + abf = [D^2 + (a + b)D + ab]f$$
$$= (D + a)(D + b)f = (D + b)(D + a)f,$$

and similarly for other such "polynomials" in D. Remember that this string of equalities is not a statement about polynomials, but rather a statement about operators. The statement says that the functions produced by applying the operators $(D + a)(D + b)$, $(D + b)(D + a)$, and $[D^2 + (a + b)D + ab]$ to the same function f are all identical.

To discover some further properties of these entities, let us write

$$L(D) = a_nD^n + a_{n-1}D^{n-1} + \cdots + a_1D + a_0$$

for a general linear operator with constant coefficients, and see what the effect of this formal polynomial in D is when it is applied to the familiar function e^{mx}. We first note that $De^{mx} = me^{mx}$ and $D^2e^{mx} = D(De^{mx}) = D(me^{mx}) = m^2e^{mx}$. We might suspect that in general $D^re^{mx} = m^re^{mx}$ for any integer r, and indeed, this is easy to prove by induction: If we assume that $D^ke^{mx} = m^ke^{mx}$, then to prove the same thing for $r = k + 1$, we see that $D^{k+1}e^{mx} = D(D^ke^{mx}) = D(m^ke^{mx}) = m^{k+1}e^{mx}$, thus completing the induction proof. Therefore

$$\begin{aligned} L(D)e^{mx} &= a_nD^ne^{mx} + a_{n-1}D^{n-1}e^{mx} + \cdots + a_1De^{mx} + a_0e^{mx} \\ &= a_nm^ne^{mx} + a_{n-1}m^{n-1}e^{mx} + \cdots + a_1me^{mx} + a_0e^{mx} \\ &= (a_nm^n + a_{n-1}m^{n-1} + \cdots + a_1m + a_0)e^{mx} = L(m)e^{mx}. \end{aligned}$$

By eliminating all the intermediate steps, we have $L(D)e^{mx} = L(m)e^{mx}$. In other words, *the effect of applying a linear operator $L(D)$ with constant coefficients to the exponential e^{mx} is the same as if we evaluate the formal polynomial $L(D)$ for $D = m$ and multiply the result times e^{mx}.*

In particular, if $t = m$ happens to be a root of the equation $L(t) = 0$, then $L(D)e^{mx} = L(m)e^{mx} = 0$. Hence $y = e^{mx}$ is a solution of the homogeneous equation $L(D)y = 0$ if $t = m$ is a root of the so-called "auxiliary equation" $L(t) = 0$. Thus in this one aspect, at least, the process of solving a homogeneous equation with constant coefficients is reduced to the algebraic problem of finding the roots of a polynomial. We shall have more to say about this in the next chapter.

Let us next seek to determine the effect of certain linear operators with constant coefficients on the functions x^ke^{mx}. First, we see that

$$D(x^ke^{mx}) = mx^ke^{mx} + kx^{k-1}e^{mx},$$

so that

$$D(x^ke^{mx}) - mx^ke^{mx} = (D - m)x^ke^{mx} = kx^{k-1}e^{mx}.$$

Thus the effect of the operator $(D - m)$ on $x^k e^{mx}$ is to differentiate the factor x^k, leaving the e^{mx} unchanged. If we apply D^2 to $x^k e^{mx}$, we obtain that

$$\begin{aligned} D^2(x^k e^{mx}) &= D(Dx^k e^{mx}) = D(mx^k e^{mx} + kx^{k-1}e^{mx}) \\ &= m^2 x^k e^{mx} + 2mkx^{k-1}e^{mx} + k(k-1)x^{k-2}e^{mx}. \end{aligned}$$

Applying the formula $(D - m)x^k e^{mx} = kx^{k-1}e^{mx}$, we have

$$D^2(x^k e^{mx}) = m^2 x^k e^{mx} + 2m(D - m)x^k e^{mx} + k(k-1)x^{k-2}e^{mx}$$

so that

$$[D^2 + 2m(D - m) - m^2]x^k e^{mx} = k(k-1)x^{k-2}e^{mx}$$

or, finally,

$$(D - m)^2 x^k e^{mx} = k(k-1)x^{k-2}e^{mx}.$$

Thus a second application of the linear operator $(D - m)$ differentiates x^{k-1} and leaves the e^{mx} factor unchanged, as before. We therefore begin to suspect that

$$(D - m)^l x^k e^{mx} = k(k-1)\cdots(k-l+1)x^{k-1}e^{mx}$$

is true, and indeed, this formula can be verified by induction, an exercise which we leave to the reader. In particular, if $l = k$ we have

$$(D - m)^l x^k e^{mx} = k!e^{mx}$$

and, if $l = k + 1$,

$$(D - m)^{k+1} x^k e^{mx} = 0.$$

From this point on, successive applications of the operator $(D - m)$ will continue to produce 0. This result we can state as follows:

$$(D - m)^l x^k e^{mx} = 0 \qquad \text{for } l > k.$$

Consequently the above formula implies that if the operator $L(D)$ has a factor $(D - m)^l$, then the functions $e^{mx}, xe^{mx}, \ldots, x^{l-1}e^{mx}$ are all solutions of the homogeneous equation $L(D)y = 0$.

We have assumed all along, of course, that $L(D)$ is a polynomial in D with real coefficients. Naturally this does not necessarily mean that the polynomial $L(t)$ has all real roots, corresponding to factors of the form $(t - m)$ or $(t - m)^l$ with m real. For example, we know that $L(D) = D^2 + 1$ cannot be written as a product of linear factors with real coefficients. However, the theory of equations tells us that if a polynomial with real coefficients has a complex root $a + ib$, then the complex conjugate $a - ib$ is also a root; in other words, the polynomial $L(D)$ will have a quadratic factor with real coefficients of the form $(D - a)^2 + b^2$ for every complex conjugate pair of roots $a \pm ib$. We shall take this result for granted; we can then formulate the following principle:

Every linear operator $L(D)$ with real, constant coefficients and $a_n = 1$ can be factored completely into real linear factors $(D - m)$ and real quadratic factors $(D - a)^2 + b^2$.

We have so far seen that the linear factors and their powers produce zero when applied to properly chosen exponentials or exponentials multiplied by certain powers of x. Mathematically we say that the linear factors *annihilate* these functions. Now, how about the quadratic factors of $L(D)$? For which functions y do we have

$$[(D - a)^2 + b^2]y = 0?$$

Let us first consider the simpler problem of the case when $a = 0$, so that we are considering second-order operators of the form $D^2 + b^2$.*

In examining such operators we must bear in mind why exponentials worked for linear operators; the operator D, applied to an exponential e^{mx}, reproduces the exponential except for a multiplicative constant, me^{mx}. Can we think of a function, then, which the operator D^2 reproduces, except for a multiplicative constant? Well, surely the exponential works, since $D^2e^{mx} = m^2e^{mx}$; the problem is that $(D^2 + b^2)e^{mx} = (m^2 + b^2)e^{mx}$, which never vanishes, because the sum of two squares $m^2 + b^2$ is invariably positive except in the trivial case when $b = 0$, $m = 0$. So we shall have to think of another function which is reproduced by D^2, but with a *change in sign*, since we want $D^2y = -b^2y$. The trigonometric functions are well-known functions with this property: We recall that $D^2 \sin mx = -m^2 \sin mx$ and $D^2 \cos mx = -m^2 \cos mx$, so that undoubtedly the sine and cosine are the functions we are looking for. In fact,

$$(D^2 + b^2)\sin mx = (-m^2 + b^2)\sin mx$$

and

$$(D^2 + b^2)\cos mx = (-m^2 + b^2)\cos mx;$$

* An alternative derivation of the functions annihilated by operators of this sort is given in the next section using complex variables.

thus the operator $D^2 + b^2$ annihilates $\sin mx$ and $\cos mx$ if $b^2 - m^2 = 0$, or if $m = \pm b$. Since the sign of m is irrelevant, we now know that if

$$L(D)y = 0$$

and if $L(D)$ has a factor $(D^2 + b^2)$, then $y = \sin bx$ and $y = \cos bx$ are two solutions of the equation.

In order to consider the case $(D - a)^2 + b^2$, we are first going to derive an important formula for the higher derivatives of a product. We know from the elementary calculus that

$$D(fg) = f\,Dg + gDf.$$

Applying D a second time, we obtain

$$D^2(fg) = D(f\,Dg + gDf) = f\,D^2g + 2(Df)(Dg) + (D^2f)g.$$

Applying the D once more, we see that

$$\begin{aligned} D^3(fg) &= D[f\,D^2g + 2(Df)(Dg) + D^2f \cdot g] \\ &= f\,D^3g + 3(Df)(D^2g) + 3(D^2f)(Dg) + (D^3f)g. \end{aligned}$$

By now a pattern is emerging; as the order of the derivative of g decreases by one from one term to the next, the order of the derivative of f increases by one. The sum of the orders of the two is the same in each term, namely the order of the derivative on the left side; and the numerical coefficients (1, 1), (1, 2, 1), and (1, 3, 3, 1) look like the coefficients of a binomial expansion, usually denoted by the notation

$$\binom{n}{k} = \frac{n!}{k!(n-k)!}.$$

Thus we might suspect that

$$D^n(fg) = \sum_{k=0}^{n} \binom{n}{k} (D^kf)(D^{n-k}g),$$

where we adopt the notation $D^0f \equiv f$. Let us see whether we can prove this formula by induction, first assuming that it is true for $n = l$:

$$D^l(fg) = \sum_{k=0}^{l} \binom{l}{k} (D^kf)(D^{l-k}g),$$

then attempting to prove it for $n = l + 1$. We have

$$
\begin{aligned}
D^{l+1}(fg) &= D \sum_{k=0}^{l} \binom{l}{k} (D^k f)(D^{l-k} g) = \sum_{k=0}^{l} \binom{l}{k} D[(D^k f)(D^{l-k} g)] \\
&= \sum_{k=0}^{l} \binom{l}{k} (D^{k+1} f)(D^{l-k} g) + \sum_{k=0}^{l} \binom{l}{k} (D^k f)(D^{l-k+1} g) \qquad (2.17) \\
&= \sum_{k=1}^{l+1} \binom{l}{k-1} (D^k f)(D^{l-k+1} g) + \sum_{k=0}^{l} \binom{l}{k} (D^k f)(D^{l-k+1} g),
\end{aligned}
$$

where we have changed the index of summation from k to $k - 1$ in the first sum with the result that the derivatives of f and g are of the same order in both sums. Now

$$
\begin{aligned}
\binom{l}{k-1} + \binom{l}{k} &= \frac{l!}{(k-1)!(l-k+1)!} + \frac{l!}{k!(l-k)!} \\
&= \frac{l!k}{k!(l-k+1)!} + \frac{l!(l-k+1)}{k!(l-k+1)!} = \frac{l!k + l!(l-k+1)}{k!(l-k+1)!} \\
&= \frac{l!(k+l-k+1)}{k!(l-k+1)!} = \frac{l!(l+1)}{k!(l-k+1)!} = \frac{(l+1)!}{k!(l-k+1)!} \\
&= \binom{l+1}{k}.
\end{aligned}
$$

(This is the well-known "Pascal's triangle" property of the binomial coefficients.) Hence we can combine the two summations in (2.17) as a single sum from 0 to $l + 1$, and we have

$$
D^{l+1}(fg) = \sum_{k=0}^{l+1} \binom{l+1}{k} (D^k f)(D^{l-k+1} g),
$$

which completes the induction proof.

Why do we need this product rule for derivatives of higher order than the first? Primarily because we wish to see what happens when one of the factors $f(x)$ and $g(x)$ is an exponential, a function which we already know plays a special and important role in the theory of linear equations with constant coefficients. So let us evaluate the expression $D^l(e^{mx} f)$:

$$
\begin{aligned}
D^l(e^{mx} f) &= \sum_{k=0}^{l} \binom{l}{k} (D^k e^{mx})(D^{l-k} f) = \sum_{k=0}^{l} \binom{l}{k} m^k e^{mx} D^{l-k} f \\
&= e^{mx} \sum_{k=0}^{l} \binom{l}{k} m^k D^{l-k} f = e^{mx} \sum_{k=0}^{l} \left[\binom{l}{k} m^k D^{l-k} \right] f.
\end{aligned}
$$

The last expression consists of e^{mx} times a certain linear operator with constant coefficients applied to $f(x)$. This operator, in the large parentheses, can immediately be factored—for, by the binomial theorem, we have formally

$$\sum_{k=0}^{l} \binom{l}{k} m^k D^{l-k} = (D + m)^l.$$

Hence

$$D^l(e^{mx}f) = e^{mx}(D + m)^l f.$$

In this manner an exponential can be "shifted" to the left through a derivative of any order, and the effect of the shift is to transform the operator D into the operator $D + m$ on the remaining function. This phenomenon is called the *exponential shift*. For a general linear operator

$$L(D) = a_n D^n + a_{n-1} D^{n-1} + \cdots + a_1 D + a_0$$

we have

$$\begin{aligned} L(D)[e^{mx}f(x)] &= (a_n D^n + \cdots + a_1 D + a_0)[e^{mx}f(x)] \\ &= a_n D^n[e^{mx}f(x)] + \cdots + a_1 D[e^{mx}f(x)] + a_0[e^{mx}f(x)] \\ &= e^{mx}[a_n(D + m)^n f(x) + \cdots + a_1(D + m)f(x) + a_0 f(x)], \end{aligned}$$

so that

$$L(D)[e^{mx}f(x)] = e^{mx}L(D + m)f(x),$$

and, conversely,

$$e^{mx}L(D)f(x) = L(D - m)[e^{mx}f(x)].$$

The usefulness of the exponential shift becomes readily apparent when we apply it to the known expressions

$$\begin{aligned} D^l x^k &= k(k - 1)\cdots(k - l + 1)x^{k-l} \qquad &&\text{for } l \leqq k, \\ D^l x^k &= 0 &&\text{for } l > k. \end{aligned}$$

Multiplying by e^{mx}, we have

$$\begin{aligned} e^{mx}D^l x^k &= k(k - 1)\cdots(k - l + 1)x^{k-l}e^{mx} \qquad &&\text{for } l \leqq k, \\ e^{mx}D^l x^k &= 0 &&\text{for } l > k. \end{aligned}$$

Applying the exponential shift to each of these, we obtain

$$(D - m)^l(x^k e^{mx}) = k(k - 1)\cdots(k - l + 1)x^{k-l}e^{mx} \qquad \text{for } l \leqq k,$$

$$(D - m)^l(x^k e^{mx}) = 0 \qquad \text{for } l > k,$$

two formulas which we earlier left to the student to derive directly by induction.

The immediate application of this which we wish to perform is in connection with the second-order operator $D^2 + b^2$. If we multiply the two known formulas

$$(D^2 + b^2)\sin bx = 0, \quad (D^2 + b^2)\cos bx = 0$$

on the left by e^{ax},

$$e^{ax}(D^2 + b^2)\sin bx = 0, \quad e^{ax}(D^2 + b^2)\cos bx = 0,$$

and perform the exponential shift, we have

$$[(D - a)^2 + b^2](e^{ax}\sin bx) = 0, \quad [(D - a)^2 + b^2](e^{ax}\cos bx) = 0.$$

Thus by the application of this device, we have discovered without further effort that if $L(D)$ has a factor $(D - a)^2 + b^2$, then the equation

$$L(D)y = 0$$

has $e^{ax}\cos bx$ and $e^{ax}\sin bx$ as two of its solutions. It can also be shown that corresponding to higher powers of the factor $(D - a)^2 + b^2$ we will have solutions of the form $x^k e^{ax}\cos bx$ and $x^k e^{ax}\sin bx$; the proof of this will be left to the student in the problems.

Problems

1. Suppose $L(D^2) = b_n D^{2n} + b_{n-1}D^{2n-2} + \cdots + b_1 D^2 + b_0$ is a linear operator with constant coefficients containing only even powers of D. Prove that $L(D^2)\sin mx = L(-m^2)\sin mx$ and $L(D^2)\cos mx = L(-m^2)\cos mx$.

2. Suppose $L(D) = c_n D^{2n+1} + c_{n-1}D^{2n-1} + \cdots + c_1 D$ is a linear operator with constant coefficients containing only odd powers of D. Show that we can write $L(D)$ in the form $DL^*(D^2)$, where L^* contains only even powers of D, and show that $L(D)\sin mx = mL^*(-m^2)\cos mx$, and that $L(D)\cos mx = -mL^*(-m^2)\sin mx$.

3. Now let $L(D) = a_n D^n + a_{n-1} D^{n-1} + \cdots + a_1 D + a_0$. Show that $L(D)$ can be written as $L_1(D^2) + DL_2(D^2)$, where L_1 and L_2 contain only even powers of D, and show that

$$L(D)\sin mx = L_1(-m^2)\sin mx + mL_2(-m^2)\cos mx$$

and

$$L(D)\cos mx = L_1(-m^2)\cos mx - mL_2(-m^2)\sin mx.$$

Hence conclude that $y = \cos mx$ and $y = \sin mx$ are solutions of the equation $L(D)y = 0$ if and only if $L_1(-m^2) = mL_2(-m^2) = 0$.

4. Verify by direct computation that $y = x\cos mx$ and $y = x\sin mx$ are solutions of $(D^2 + m^2)^2 y = 0$; then use the exponential shift to show that $y = e^{ax} x\cos mx$ and $y = e^{ax} x \sin mx$ are solutions of $((D - a)^2 + m^2)^2 y = 0$.

5. Show that

$$(D^2 + m^2)(\cos mx)[f(x)] = \cos mx\, D^2 f - 2m\sin mx\, Df.$$

$$(D^2 + m^2)(\sin mx)[f(x)] = \sin mx\, D^2 f + 2m\cos mx\, Df.$$

Use this result to compute $(D^2 + m^2)^k(\cos mx)[f(x)]$ and

$$(D^2 + m^2)^k(\sin mx)[f(x)] \text{ for } k = 2, 3, 4.$$

6. Show by induction that in the expressions $(D^2 + m^2)^k(\sin mx)[f(x)]$ and $(D^2 + m^2)^k(\cos mx)[f(x)]$, the *lowest*-order derivative of f which appears is $D^k f$. Using this, prove that

$$(D^2 + m^2)^k x^l \cos mx = 0,$$

$$(D^2 + m^2)^k x^l \sin mx = 0,$$

provided that $k > l$.

7. Using the result of problem 6 and the exponential shift, display $2k$ solutions of the equation $[(D - a)^2 + m^2]^k y = 0$.

8. Use induction to verify the formula $(D - m)^l x^k e^{mx} = k(k - 1)\cdots(k - l + 1)x^{k-l}e^{mx}$.

2.6 Complex Solutions

In this section we shall consider an alternative approach to finding functions which are annihilated by operators of the form $(D - a)^2 + b^2$. To do this we shall work with complex numbers and functions of complex variables. We note that formally the operator $(D - a)^2 + b^2$ can be factored into the two linear factors

$$(D - a + ib)(D - a - ib)$$

where i, the imaginary unit, is used to designate one solution of the equation $i^2 = -1$. Again, speaking purely formally,* since we know that a factor of the form $(D - m)$ annihilates the function e^{-mx}, then $D - (a + ib)$ should annihilate $e^{(a+ib)x}$ and $D - (a - ib)$ should annihilate $e^{(a-ib)x}$. If we can by some means define what we mean by these complex exponentials, we shall have found an alternative method of formulating solutions corresponding to quadratic factors of linear operators with constant coefficients.

First, because we would certainly want complex exponentials to have all the usual properties of real exponentials, we must demand that

$$e^{(a+ib)x} = e^{ax+ibx} = e^{ax}e^{ibx}.$$

The expression e^{ax} is, of course, well defined. It remains to define the meaning of e^{ibx}.

To do this, we recall that it is possible to represent e^z for any real value of z in terms of its Maclaurin series:

$$e^z = 1 + z + \frac{z^2}{2!} + \frac{z^3}{3!} + \cdots + \frac{z^n}{n!} + \cdots = \sum_{k=0}^{\infty} \frac{z^k}{k!},$$

where the series converges absolutely for all finite values of z. It certainly seems reasonable, then, that this formula hold for imaginary values of z as well, say $z = i\theta$. Then we would have

$$e^{i\theta} = 1 + (i\theta) + \frac{(i\theta)^2}{2!} + \frac{(i\theta)^3}{3!} + \cdots = \sum_{k=0}^{\infty} \frac{(i\theta)^k}{k!}.$$

Now it is easy to see that since $i^2 = -1$, we have $i^3 = -i$, $i^4 = 1$, and so on, so that in general

$$(i)^{2n} = (-1)^n \text{ and } (i)^{2n+1} = (-1)^n i$$

* Note: When a mathematician says "formally," he often means what an English professor would call "informally"!

for all values of n. Therefore, if we split the series for $e^{i\theta}$ up into the two separate series for even and odd values of k, we have

$$e^{i\theta} = \sum_{n=0}^{\infty} \frac{(i\theta)^{2n}}{(2n)!} + \sum_{n=0}^{\infty} \frac{(i\theta)^{2n+1}}{(2n+1)!} = \sum_{n=0}^{\infty} \frac{(-1)^n \theta^{2n}}{(2n)!} + i \sum_{n=0}^{\infty} \frac{(-1)^n \theta^{2n+1}}{(2n+1)!}.$$

This rearrangement of terms is justified because of the absolute convergence of each series. We recognize that the first of the two series on the right is the Maclaurin series for $\cos\theta$, while the second is the Maclaurin series for $\sin\theta$. Thus we have shown formally that

$$e^{i\theta} = \cos\theta + i\sin\theta.$$

To strengthen our conviction that this formula (*de Moivre's formula*, as it is called) is reasonable, let us compute $e^{i\theta_1}e^{i\theta_2}$:

$$\begin{aligned} e^{i\theta_1}e^{i\theta_2} &= (\cos\theta_1 + i\sin\theta_1)(\cos\theta_2 + i\sin\theta_2) \\ &= (\cos\theta_1\cos\theta_2 - \sin\theta_1\sin\theta_2) + i(\sin\theta_1\cos\theta_2 + \sin\theta_2\cos\theta_1) \\ &= \cos(\theta_1 + \theta_2) + i\sin(\theta_1 + \theta_2), \end{aligned}$$

and from this we see that $e^{i\theta_1}e^{i\theta_2} = e^{i(\theta_1+\theta_2)}$, so that complex exponentials preserve the same properties as real ones.

Therefore, the two expressions $e^{(a+ib)x}$ and $e^{(a-ib)x}$ which are annihilated by the operator $(D - a)^2 + b^2$ are well defined; indeed,

$$\begin{aligned} e^{(a+ib)x} &= e^{ax}(\cos bx + i\sin bx), \\ e^{(a-ib)x} &= e^{ax}(\cos bx - i\sin bx). \end{aligned}$$

Furthermore, these two functions are linearly independent because

$$\begin{aligned} W[e^{(a+ib)x}, e^{(a-ib)x}] &= \begin{vmatrix} e^{(a+ib)x} & e^{(a-ib)x} \\ (a+ib)e^{(a+ib)x} & (a-ib)e^{(a-ib)x} \end{vmatrix} \\ &= (a - ib)e^{(a+ib)x}e^{(a-ib)x} - (a + ib)e^{(a+ib)x}e^{(a-ib)x} \\ &= -2ibe^{2ax}, \end{aligned}$$

which is never zero.

In fact, because of the fact that

$$\tfrac{1}{2}(e^{(a+ib)x} + e^{(a-ib)x}) = e^{ax}\cos bx$$

and

$$\tfrac{1}{2}i(e^{(a+ib)x} - e^{(a-ib)x}) = e^{ax} \sin bx,$$

we see that we can obtain the two solutions which we discovered in the previous section as particular linear combinations of these two complex exponential solutions.

There are several reasons for the importance of complex exponential solutions of linear equations with complex coefficients:

(1) We shall see that with the aid of complex exponential representation, every solution of a homogeneous linear equation with constant coefficients can be written as a linear combination of exponential functions, perhaps multiplied by polynomials.

(2) Exponential functions, even complex ones, are often easier to deal with than sines and cosines because they are reproduced by differentiation.

(3) Complex exponentials are quite useful in the study of electric circuits because they contain more information than real exponentials or sines or cosines. For instance, if one finds the complex solution I of an equation of the form

$$L(D)I = e^{i\omega t} = \cos \omega t + i \sin \omega t,$$

then its real part I_R is the solution of the equation

$$L(D)I_R = \cos \omega t,$$

and its imaginary part I_I is the solution of

$$L(D)I_I = \sin \omega t.$$

Thus two complete solutions are contained in one expression.

Problems

1. Show that $\cos x = \frac{1}{2}(e^{ix} + e^{-ix})$.
2. Show that $\sin x = \frac{1}{2}i(e^{ix} - e^{-ix})$.
3. Show that $\cos ix = \cosh x$.
4. Show that $\sin ix = i \sinh x$.
5. Derive the addition formulas for sine and cosine using the complex exponential definition.
6. Show that the exponential shift holds for complex exponentials as well as real ones.

7. Show that the four solutions e^{ix}, e^{-ix}, $\sin x$, and $\cos x$, of the equation $(D^2 + 1)y = 0$ are linearly dependent, provided that we allow our constants to be complex.

3

Solving Linear Equations with Constant Coefficients (Operator and Matrix Methods)

Thanks to the discussion of linear operators with constant coefficients, we are now ready to sit down and solve homogeneous equations, a process which we shall shortly see to be almost a purely algebraic one. From homogeneous equations to inhomogeneous equations is then a matter of applying some of the other techniques of Chapter 2 to the question of finding particular solutions. With what we learn here, we shall be able to solve the equation of the vibrating spring and the linearized pendulum equation, as well as many electrical circuit problems.

3.1 Homogeneous Equations with Constant Coefficients

In order to solve the homogeneous equation

$$L(D)y = a_nD^ny + a_{n-1}D^{n-1}y + \cdots + a_1Dy + a_0y = 0, \qquad (3.1)$$

the crucial step, and the only difficult part, is to factor the linear operator $L(D)$ into the linear factors $(D - m)$ and quadratic factors $(D - a)^2 + b^2$ with real coefficients. Let us first divide (3.1) through by the leading coefficient a_n, which is not zero if the equation is truly of order n, thus obtaining

$$(1/a_n)L(D)y = D^ny + b_{n-1}D^{n-1}y + \cdots + b_1Dy + b_0y = 0,$$

where

$$b_k = a_k/a_n, \qquad n = 0, 1, \ldots, n - 1. \qquad (3.2)$$

Every solution of (3.2) is a solution of (3.1) and conversely; and the new linear operator

$$P(D) = D^n + b_{n-1}D^{n-1} + \cdots + b_1 D + b_0$$

is somewhat easier to factor than the original, if we wish to use synthetic division in the process.

The process of finding the n roots of the polynomial equation

$$P(t) = t^n + b_{n-1}t^{n-1} + \cdots + b_1 t + b_0 = 0 \tag{3.3}$$

is equivalent to factoring $P(t)$, because for every real root m there will be a factor of the form $t - m$, and for every complex conjugate pair of roots $a \pm ib$ there will be a factor of the form $(t - a)^2 + b^2$. It is far from a trivial task, however, to find the roots of a general algebraic polynomial. We spend much time in algebra studying the carefully contrived, easily factored polynomials that occur in textbooks. While this time is not completely wasted, we should realize that such polynomials are rare indeed. In actual scientific practice, polynomials will have noninteger coefficients that are known only to a certain number of decimal places; in most such cases the only practicable method of finding the roots is to use numerical methods on an electronic computer. Naturally any roots obtained in this way are only approximations, and, often enough, poor approximations. Furthermore, it is not possible to develop mathematical formulas for the roots of polynomials of degree five and higher,* so that numerical approximation is the only way of obtaining roots for most polynomials.

With this disclaimer aside, we shall proceed to do just as we did in courses in elementary algebra: study polynomials which are easily factored into linear and quadratic factors, primarily because the facilities for factoring polynomials numerically are not readily available to undergraduate students. To this end, let us recall some algebraic rules which facilitate the process of finding the roots of polynomials with real coefficients, in which, as in (3.2), the leading coefficient is 1:

(1) If $P(t_1)$ and $P(t_2)$ are of opposite signs, then $P(t) = 0$ has a root between t_1 and t_2. This result is based on the intermediate value theorem of the calculus. Thus, for example, if $P(t) = t^4 + 4t^2 - 4$, then $P(0) = -4 < 0$ and $P(1) = 1 > 0$, so $P(t) = 0$ has a root between 0 and 1.

(2) Every polynomial $P(t)$ of odd degree has at least one real root. This is because $P(t) < 0$ for very large negative values of t,† while $P(t) > 0$ for very

* For a proof of this fact, see, for instance, I. N. Herstein, *Topics in Algebra* (Blaisdell, New York, 1964), pp. 208–214.

† Recall that the leading coefficient is 1.

large positive values of t, so that there must be a value of t in between for which $P(t) = 0$.

(3) If m is a real root of

$$P(t) = t^n + b_{n-1}t^{n-1} + \cdots + b_1 t + b_0 = 0,$$

if $a + ib$ is a complex root of $P(t) = 0$, and if m, a, b, and b_0 are all integers, then m and $a^2 + b^2$ are divisors of b_0. This is because $P(t)$ has a factor of $(t - m)$ for every real root m, and a factor of $[(t - a)^2 + b^2] = (t^2 - 2at + a^2 + b^2)$ for every complex root $a + ib$; and the product of the constant terms of all such factors, which obviously must contain m and $a^2 + b^2$ as factors, is equal to b_0.

Using these three rules, we shall be able to factor all polynomials that occur in this textbook. Let us begin by attempting to solve the homogeneous equation

$$(D^3 - 4D^2 + D + 6)y = 0.$$

The polynomial $P(t) = t^3 - 4t^2 + t + 6$ is of odd degree, so it must have at least one real zero, and if this zero is an integer it has to be ± 1, ± 2, ± 3, or ± 6 (one of the factors of 6, the constant term). We begin by trying $+1$ by synthetic division. We write down 1) and then, in descending order, the coefficients of $P(t)$:

$$1)\quad 1 \quad -4 \quad 1 \quad 6$$

Then we carry down the leading coefficient, multiply it by the prospective root 1, and add to the second coefficient:

$$\begin{array}{rrrrr} 1) & 1 & -4 & 1 & 6 \\ & & 1 & & \\ \hline & 1 & -3 & & \end{array}$$

We multiply this result by the prospective root 1, put it below the third coefficient, add, and continue the process to the end:

$$\begin{array}{rrrrr} 1) & 1 & -4 & 1 & 6 \\ & & 1 & -3 & -2 \\ \hline & 1 & -3 & -2 & \multicolumn{1}{|r}{4} \end{array}$$

The first three numbers are the coefficients of the quotient of $t^3 - 4t^2 + t + 6$

when divided by $t - 1$, the factor corresponding to a root of 1. The fourth number is the remainder, the value of $P(1)$. Since this is not zero, $t - 1$ can not be a factor of $P(t)$. (Note that the process of synthetic division does the same thing as long division, saving time because we write only the coefficients and not the entire polynomials). We now try $t = -1$:

$$\begin{array}{r|rrr|r} -1) & 1 & -4 & 1 & 6 \\ & & -1 & 5 & -6 \\ \hline & 1 & -5 & 6 & 0 \end{array}$$

Since the remainder is zero, we have

$$t^3 - 4t^2 + t + 6 = (t^2 - 5t + 6)(t + 1).$$

The quadratic factor can be seen to be $(t - 2)(t - 3)$ by inspection, so that

$$P(t) = t^3 - 4t^2 + t + 6 = (t - 2)(t - 3)(t + 1),$$

and hence the zeros of $P(t)$ are $t = -1, 2, 3$. Thus $P(D)y = (D - 2)(D - 3)(D + 1)y = 0$ has, corresponding to these three roots, solutions e^{-t}, e^{2t}, e^{3t}. Therefore the general solution of $P(D)y = 0$ is $y = c_1e^{-t} + c_2e^{2t} + c_3e^{3t}$, since these three exponential functions are linearly independent.

From this example we can see that finding the solutions of a linear homogeneous equation with constant coefficients involves finding the factors of $P(D)$, and then writing down a linearly independent solution corresponding to each one, and finally superposing them to obtain the general solution. If $P(D)$ decomposes into linear factors $(D - m_1)\cdots(D - m_n)$, each of which has a different root, then the exponentials $e^{m_1x}, \ldots, e^{m_nx}$ (each of which we know is a solution because of our groundwork in Chapter 2) form a linearly independent set of functions (which we know again by virtue of our work in Chapter 2). Hence the general solution is $c_1e^{m_1x} + \cdots + c_ne^{m_nx}$, a general linear combination of this set of particular solutions.

Let us examine another example:

$$P(D)y = (D^4 - 10D^2 + 24)y = 0.$$

This can be factored by inspection:

$$\begin{aligned}(D^4 - 10D^2 + 24)y = (D^2 - 6)(D^2 - 4)y = (D + \sqrt{6})(D - \sqrt{6})(D + 2) \\ (D - 2)y = 0.\end{aligned}$$

The linearly independent set of solutions is thus $e^{\sqrt{6}x}, e^{-\sqrt{6}x}, e^{2x}, e^{-2x}$, so the general solution is $y = c_1e^{\sqrt{6}x} + c_2e^{-\sqrt{6}x} + c_3e^{2x} + c_4e^{-2x}$.

We now see how easy it is to solve an equation $P(D)y = 0$ when all the roots of $P(t) = 0$ are real and unequal. All the work is algebraic; once $P(D)$ is factored, we simply write down a linearly independent set of solutions by inspection. Now, what happens when roots are repeated? Let us consider

$$P(D)y = (D^4 - 6D^3 + 13D^2 - 12D^2 + 4)y = 0.$$

The possible integral zeros of $P(t)$ are ± 1, ± 2, ± 4. Trying 1, we have

$$\begin{array}{rrrrr|r} 1) & 1 & -6 & 13 & -12 & 4 \\ & & 1 & -5 & 8 & -4 \\ \hline & 1 & -5 & 8 & -4 & 0 \end{array}$$

so

$$(D^4 - 6D^3 + 13D^2 - 12D + 4) = (D^3 - 5D^2 + 8D - 4)(D - 1).$$

± 1, ± 2, and ± 4 are still possible zeros of the cubic factor. Trying 1 again,

$$\begin{array}{rrrr|r} 1) & 1 & -5 & 8 & -4 \\ & & 1 & -4 & 4 \\ \hline & 1 & -4 & 4 & 0 \end{array}$$

so

$$(D^3 - 5D^2 + 8D - 4)(D - 1) = (D^2 - 4D + 4)(D - 1)^2 = (D - 2)^2(D - 1)^2.$$

The zeros are 1 and 2, each repeated twice. But the set of functions e^x, e^x, e^{2x}, e^{2x} is not linearly independent! What do we do? Well, we know from Chapter 2 that $(D - m)^k(x^l e^{mx}) = 0$ whenever $k > l$, and from this we see that xe^x and xe^{2x} are also solutions of the equation under consideration. The set of solutions e^x, xe^x, e^{2x}, xe^{2x} is a linearly independent set, and therefore

$$y = c_1 e^x + c_2 xe^x + c_3 e^{2x} + c_4 xe^{2x} = (c_1 + c_2 x)e^x + (c_3 + c_4 x)e^{2x}$$

is the general solution.

We can generalize the result above by saying that corresponding to a factor $(D - m)^k$ of $P(D)$ there are k linearly independent solutions of the equation $P(D)y = 0$, namely $e^{mx}, xe^{mx}, x^2e^{mx}, \ldots, x^{k-1}e^{mx}$. For example, suppose that $P(D)y = 0$ factors into

$$(D - 3)^4(D + 1)^3(D - 2)(D - 1)y = 0.$$

Then a linearly independent set of solutions is $e^{3x}, xe^{3x}, x^2e^{3x}, x^3e^{3x}, e^{-x}, xe^{-x}, x^2e^{-x}, e^{2x}, e^x$.

Naturally $P(D)$ may also have quadratic factors, in which case we shall need to use sine and cosine functions (or complex exponential functions) as well. For example, let us examine

$$P(D)y = (D^5 + 3D^4 + 7D^3 + 5D^2)y = 0.$$

First of all, $D^5 + 3D^4 + 7D^3 + 5D^2 = D^2(D^3 + 3D^2 + 7D + 5)$. The cubic factor must have a real zero, and if it is an integer, then it must be ± 1 or ± 5. Because the sign of each coefficient is positive, we see that a positive zero is impossible; so let us try -1:

$$\begin{array}{r|rrr|r} -1) & 1 & 3 & 7 & 5 \\ & & -1 & -2 & -5 \\ \hline & 1 & 2 & 5 & 0 \end{array}$$

Hence $P(D) = D^2(D + 1)(D^2 + 2D + 5) = D^2(D + 1)[(D + 1)^2 + 2^2]$. Corresponding to the real roots $0, 0, -1$ are the solutions $1, x, e^{-x}$; and corresponding to the quadratic factor are the solutions $e^{-x}\cos 2x$ and $e^{-x}\sin 2x$. Hence the general solution is

$$y = c_1 + c_2x + c_3e^{-x} + c_4e^{-x}\cos 2x + c_5e^{-x}\sin 2x,$$

since these five functions are linearly independent. (Alternatively, we could write the last two solutions in complex exponential form as $\exp[-(1 + 2i)x]$ and $\exp[-(1 - 2i)x]$.)

If a linear operator has a quadratic factor raised to a power higher than one, then, as with exponentials, we can use the sine and cosine functions times powers of x, as we showed in the problems at the end of Chapter 2. For example, the equation

$$(D - 1)(D + 2)(D^2 + 1)^2[(D - 1)^2 + 1^2]^3y = 0$$

has solutions $e^x, e^{-2x}, \sin x, \cos x, x\sin x, x\cos x, e^x\sin x, e^x\cos x, xe^x\sin x, xe^x\cos x, x^2e^x\sin x$, and $x^2e^x\cos x$. Therefore the general solution is

$$\begin{aligned} y &= c_1e^x + c_2e^{-2x} + (c_3 + c_4x)\sin x + (c_5 + c_6x)\cos x \\ &+ (c_7 + c_8x + c_9x^2)e^x\sin x + (c_{10} + c_{11} + c_{12}x^2)e^x\cos x. \end{aligned}$$

Alternatively, the trigonometric portions of the solutions could be represented by means of complex exponentials.

These examples show that out theoretical work in Chapter 2 was very worthwhile; there is almost no work at all involved in solving homogeneous equations with constant coefficients; once the operator $P(D)$ has been factored, we can simply read off the solutions of the equation corresponding to each factor, then superpose these solutions to get the general solution.

Problems

In the following set of problems, read off the linearly independent set of solutions.

1. $D^5 y = 0.$
2. $(D + 2)^4 y = 0.$
3. $(D + 1)^3 (D - 1)^2 D^3 y = 0.$
4. $(D^2 + 4)(D - 3)^3 y = 0.$
5. $D(D - 2)[(D + 2)^2 + 6]^2 y = 0.$
6. $(D^2 + 16)^5 [(D - 1)^2 + 25]^3 y = 0.$
7. $D(D - 1)^2 (D - 2)^3 (D^2 + 4)^4 y = 0.$
8. $(D + 17)^2 (D - 13)^3 [(D - 5)^2 + 36]^5 y = 0.$
9. $(D - 1.56)^3 (D + 3.74)^3 [D^2 + (9.43)^2]^3 y = 0.$
10. $(D^2 + 6)(D^2 + 7)^4 y = 0.$
11. $(D + \sqrt{2})^2 (D - \sqrt{3})^3 (D^2 + 3)^4 y = 0.$
12. $[(D - \pi)^2 + (\pi/2)^2]^3 (D - 3)^3 = 0.$
13. $(D + 3.472)^2 [(D - 1.453)^2 + (2.05)^2] y = 0.$
14. $(D + a)^3 (D + b)^4 [(D - c) + d^2]^2 y = 0, a \neq b.$
15. $(D^2 + 6D + 10) y = 0.$
16. $(D^2 + 2D + 1)(D - 1) y = 0.$
17. $(D^2 + 5D + 6) y = 0.$
18. $(D^2 - 2D + 1)(D^2 + 4D + 4) y = 0.$
19. $(D^2 + 4D + 5)^3 (D^3 + 3D^2 + 3D + 1) y = 0.$
20. $(D^2 - 1)^2 (D^2 + 2D + 2)^4 y = 0.$

In the following set of problems, find the general solution in real or complex form (if appropriate) of the given equation:

1. $(D^3 - 6D^2 + 11D - 6) y = 0.$
2. $(D^3 + 3D^2 - 10D - 24) y = 0.$
3. $(D^3 + 4D) y = 0.$
4. $(D^3 + 6D^2 + D - 34) y = 0.$
5. $(D^3 + 9D^2 + 27D + 27) y = 0.$
6. $(D^4 - 16) y = 0.$
7. $(D^4 + 12D^3 + 39D^2 + 8D - 60) y = 0.$

8. $(D^4 + 3D^3 - D^2 - 9D - 18)y = 0.$
9. $(D^4 + 4D^2 + 16)y = 0.$
10. $(D^5 + D^4 - D^3 + D^2 + 4D + 2)y = 0.$

3.2 Applications of Homogeneous Equations

If we look back on Chapter 2, we recognize that the equation of the vibrating spring (2.1) is a linear equation with constant coefficients. If there is no external force applied to the spring, then this equation is homogeneous, and the initial value problem becomes

$$\begin{aligned} &mD^2x + rDx + kx = 0, \\ &x(0) = x_0, \quad Dx(0) = v_0. \end{aligned} \tag{3.4}$$

(Here D represents differentiation with respect to time.) The polynomial $mD^2 + rD + k$ has zeros at

$$-\frac{r}{2m} \pm \frac{\sqrt{r^2 - 4mk}}{2m},$$

and the solution of the equation will be quite different according as these roots are both complex or both real, and if real, whether or not they are equal. Let us examine these cases.

Case 1 $\quad r^2 - 4mk < 0$

Then both roots are complex, and the general solution of the differential equation is

$$x(t) = \exp\left(-\frac{rt}{2m}\right)\left(c_2 \cos \frac{\sqrt{4mk - r^2}}{2m} t + c_2 \sin \frac{\sqrt{4mk - r^2}}{2m} t\right)$$

or, for simplicity, if we write $r/2m = \alpha$ and $\sqrt{4mk - r^2}/2m = \omega$, then

$$x(t) = e^{-\alpha t}(c_1 \cos \omega t + c_2 \sin \omega t).$$

We can rewrite this in an even simpler form by dividing and multiplying by $\sqrt{c_1^2 + c_2^2}$,

$$x(t) = \sqrt{c_1^2 + c_2^2}\, e^{-\alpha t}\left(\frac{c_1}{\sqrt{c_1^2 + c_2^2}} \cos \omega t + \frac{c_2}{\sqrt{c_1^2 + c_2^2}} \sin \omega t\right);$$

we then note that the sum of the squares of the coefficients of $\cos \omega t$ and $\sin \omega t$ is 1, so we can define an angle δ by $\cos \delta = c_1/\sqrt{c_1^2 + c_2^2}$, and $\sin \delta = c_2/\sqrt{c_1^2 + c_2^2}$; and if we put $A = \sqrt{c_1^2 + c_2^2}$, we have

$$x(t) = Ae^{-\alpha t}(\cos\delta\cos\omega t + \sin\delta\sin\omega t).$$

Recognizing the quantity in parentheses as being $\cos(\omega t - \delta)$, we finally have

$$x(t) = Ae^{-\alpha t}\cos(\omega t - \delta). \tag{3.5}$$

The two new arbitrary constants A and δ are called the *amplitude* and the *phase angle* of the motion, respectively. The known constant ω is called the *circular frequency*. Thus, if $r^2 < 4km$, that is, if the friction is relatively small in comparison with the mass m and spring constant k, then the motion of the spring is an exponentially damped, out-of-phase cosine wave (Figure 3.1). If there is no friction present ($r = 0$), then $x(t) = A\cos(\omega t - \delta)$ and the motion is described by an out-of-phase cosine wave alone (Figure 3.2). The mass simply oscillates back and forth about its equilibrium position forever.

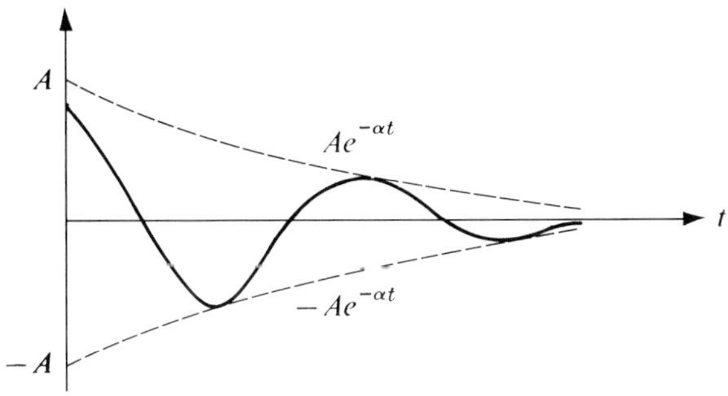

Figure 3.1

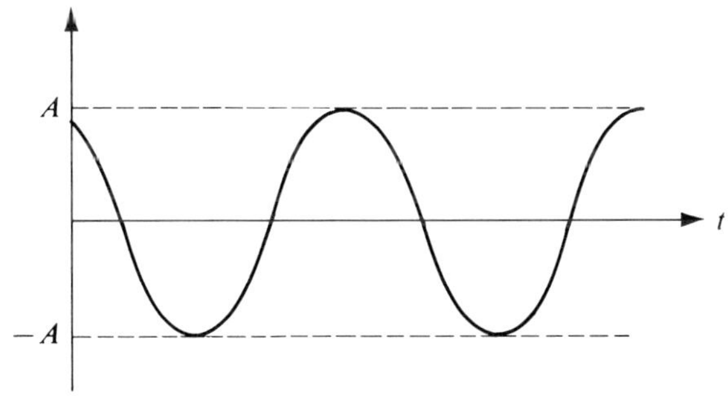

Figure 3.2

The arbitrary constants c_1 and c_2 (or A and δ) can be evaluated by using the initial conditions.

Case 2 $r^2 = 4km$
In this case, the friction, in a sense, exactly balances the mass and spring constant, and both zeros of the polynomial are real and equal. The solution is

$$x(t) = (c_1 + c_2 t)e^{-\alpha t},$$

where α is as before. Let us evaluate the arbitrary constants c_1 and c_2 in this case. We have

$$x(0) = c_1 = x_0;$$

and $Dx = [c_2 - \alpha(c_1 + c_2 t)]e^{-\alpha t}$, so $Dx(0) = v_0 = c_2 - \alpha c_1 = c_2 - \alpha x_0$; hence $c_2 = v_0 + \alpha x_0$, and we have

$$x(t) = [x_0 + (v_0 + \alpha x_0)t]e^{-\alpha t}.$$

The exact behavior of this function depends on the initial values x_0 and v_0. However, a linear function multiplied by a decaying exponential still goes to zero as $t \to \infty$. Figure 3.3 gives an idea of what $x(t)$ looks like. In this case no

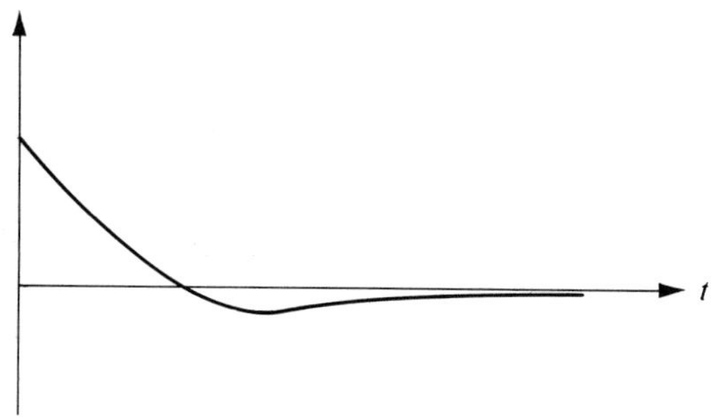

Figure 3.3

oscillation takes place; the mass may pass once through the equilibrium position, but it then settles back to equilibrium. The friction, in other words, is just strong enough to damp the oscillation; for this reason we say that when $r^2 = 4km$, then the spring has been *critically damped.*

Case 3 $r^2 > 4km$

We now have two real, unequal zeros

$$-\alpha_1 = -\frac{r}{2m} - \frac{\sqrt{r^2 - 4km}}{2m}, \qquad -\alpha_2 = -\frac{r}{2m} - \frac{\sqrt{r^2 - 4km}}{2m},$$

both of which are negative. The general solution,

$$x(t) = c_1 \exp(-\alpha_1 t) + c_2 \exp(-\alpha_2 t),$$

is the superposition of two decaying exponentials. Once again there is no oscillation; the spring returns to its equilibrium position. Cases 2 and 3 might arise, for example, if the spring and mass are contained in a viscous fluid, which creates considerable frictional force. The damped oscillation of case 1 is what normally occurs when the spring is vibrating in air and frictional forces are very small.

The linearized pendulum equation with no external forces

$$D^2\theta + \eta D\theta + \frac{g}{l}\,\theta = 0$$

is in form identical to the equation of the vibrating spring. Having solved the spring equation, we can now make some general observations about the solutions of the pendulum equation. In the case where friction is lacking, i.e., $\eta = 0$, we have

$$\theta(t) = A\cos(\sqrt{g/l}\,t - \delta),$$

so that an undamped cosine wave describes the pendulum's motion. Its circular frequency is $\sqrt{g/l}$, so that the period of the pendulum (the time it takes to perform one complete swing back and forth) is $2\pi\sqrt{l/g}$, whence the interesting fact that the period of a frictionless pendulum with small oscillations is proportional to $\sqrt{l}$ only, and does not depend at all on the mass of the bob! This is a fact often verified experimentally in the laboratory in an elementary physics course.

In case friction is present, the motion of the pendulum is a damped oscillation, critically damped, or a superposition of two decaying exponentials, according as η^2 is less than, equal to, or greater than $4g/l$. We shall see in Chapter 6 that the nonlinear pendulum works in the same way

The equation for a simple electrical circuit with no impressed electromotive force

$$LD^2Q + RDQ + \kappa Q = 0 \qquad (\kappa = C^{-1})$$

is also the same as the equation of the vibrating spring and the linearized pendulum equation. In the exercises we shall ask the student to give an analysis of the various possible solutions of this case. Meanwhile, let us consider this equation from a different point of view by examining its complex exponential solutions and seeing the useful information that can be gleaned from them.

First, let us suppose that the current through a circuit is described by an exponential function

$$I = I_0 e^{st},$$

where I_0 is the current at time $t = 0$. (For the moment we shall make no assumption about the parameter s.) Ohm's law tells us that the voltage drop across a resistance R is

$$E = IR = I_0 R e^{st}.$$

The voltage drop across an inductance L is

$$E = L\frac{dI}{dt} = sI_0 L e^{st}.$$

Finally, the voltage drop across a capacitance C is

$$E = \frac{1}{C}\int I\,dt = \frac{I}{sC}I_0 e^{st}.$$

Because I is an exponential function, its form is unaltered by either differentiation or integration. Thus with an exponential current the voltage drop across a resistor, inductor, and capacitor, respectively, is

$$E = RI, \qquad E = sLI, \qquad \text{and} \quad E = \frac{1}{sC}I.$$

If a total resistance of R, inductance of L, and capacitance of C are wired in series, then the total voltage drop is

$$E = \left(R + sL + \frac{1}{sC}\right)I = ZI.$$

The quantity Z is called the *impedance* of the circuit. It represents the total amount that the current is impeded as it passes through the circuit, and as we shall shortly see, it gives us information about how the current and voltage are out of phase.

Impedance is a relatively straightforward concept when s is a real variable; for example, when s is negative, we have a current which dies down like a decaying exponential, and Z is simply a real constant analogous to resistance. It becomes somewhat more difficult and interesting when s is a complex variable. Then the current I is a complex-valued function, the impedance Z becomes a complex impedance, and the voltage drop E also becomes complex-valued. These complex quantities contain considerable information about the circuit which is not readily available otherwise in the real notation.

For example, consider a circuit which has a current function $I = \cos \omega t$ and a resistance of R, inductance of L, and capacitance of C. We ask (1) What is the voltage drop through this circuit? and (2) What is the impedance of the circuit? To answer this, we note that cos ωt is the real part of $e^{i\omega t} = \cos \omega t + i \sin \omega t$; hence we shall first consider a current of the form $e^{i\omega t}$ and then later extract the cosine portion. Then, since

$$E = \left(R + sL + \frac{1}{sC}\right)I,$$

we have

$$E = \left(R + i\omega L + \frac{1}{i\omega C}\right)e^{i\omega t}$$

$$= \left[R + i\left(\omega L - \frac{1}{\omega c}\right)\right]e^{i\omega t}.$$

Now, considering only the cosine part of $e^{i\omega t}$, we have

$$E = \left[R + i\left(\omega L - \frac{1}{\omega C}\right)\right]\cos \omega t.$$

Interestingly enough, the voltage drop and impedance as given by this equation are both complex-valued quantities. How are we to interpret these? The answer is that the inductance and capacitance portions of the impedance throw the voltage out of phase with the current fluctuations. To see this we rewrite

$$E = \left[R + i\left(\omega L - \frac{1}{\omega C}\right)\right]e^{i\omega t}$$

as

$$E = \left[R + i\left(\omega L - \frac{1}{\omega C}\right)\right](\cos \omega t + i \sin \omega t)$$

$$= R \cos \omega t - \left(\omega L - \frac{1}{\omega C}\right)\sin \omega t + i\left[R \sin \omega t + \left(\omega L - \frac{1}{\omega C}\right)\cos \omega t\right].$$

Taking the real part of both sides, we have

$$E_R = R \cos \omega t - \left(\omega L - \frac{1}{\omega C}\right)\sin \omega t$$

$$= A \cos(\omega t + \theta),$$

where

$$A = \left[R^2 + \left(\omega L - \frac{1}{\omega C}\right)^2\right]^{\frac{1}{2}}$$

and

$$\theta = \arctan \frac{\omega L - 1/\omega C}{R}.$$

E_R is precisely the voltage drop due to the real current $\cos \omega t$, and we note that it is out of phase with the current by an amount θ; this angle θ is called the *angle of impedance.* Geometrically, it is the angle made by a radius vector from the origin to the point with coordinates $(R, \omega L - 1/\omega C)$; the amplitude A is the length of this impedance vector. [This identification of the complex number $a + ib$ with the point in the Cartesian plane whose coordinates are (a, b) is standard in mathematics; we then speak of "the complex plane".]

Thus the advantage of the complex notation in circuit analysis is that it contains all the above information in a compact and readily accessible form; in addition, the imaginary part of the complex equation contains information about the voltage drop caused by a real current of $\sin \omega t$. For this reason it is customary to use complex exponentials in circuit analysis. (Electrical engineers generally use j for the complex unit because they use a lowercase i to represent the current.)

Problems

Find the solutions of each of the following initial value problems.

1. $(D^2 + 2D + 1)x = 0;\ x(0) = 1, Dx(0) = -1.$
2. $(D^2 + 1)x = 0;\ x(0) = 1, Dx(0) = -1.$
3. $(D^2 - 4)x = 0;\ x(0) = 1, Dx(0) = 6.$
4. $(D^2 + 4D + 4)x = 0;\ x(0) = 1, Dx(0) = -1.$
5. $(D^2 - 4D + 8)x = 0;\ x(0) = 1, Dx(0) = 2.$
6. $(D^2 - 6D + 9)x = 0;\ x(0) = 0, Dx(0) = 2.$
7. $(D^2 + 6D + 13)x = 0;\ x(0) = 1, Dx(0) = 1.$
8. $(D^3 - 1)x = 0;\ x(0) = 5, Dx(0) = 5, D^2x(0) = 5.$
9. $(D^3 + 3D^2 + 3D + 1)x = 0;\ x(0) = 0, Dx(0) = 1, D^2x(0) = 0.$
10. $(D^4 + 8D^2 + 16)x = 0;\ x(0) = 0;\ Dx(0) = 8, D^3x(0) = 12.$
11. Using the complex notation wherever possible, solve the initial value problem for a simple circuit, equation (2.5), and determine the current $I(t)$ as a function of time in each of the possible cases, assuming that the impressed

electromotive force $E(t)$ is zero. Describe physically what is happening in each case.

3.3 Inhomogeneous Equations: Particular Solutions by Inspection

Our next task is to devise methods for solving the inhomogeneous linear equation with constant coefficients

$$P(D)y = (a_n D^n + \cdots + a_1 D + a_0)y = f(x). \tag{3.6}$$

As we saw in the previous chapter, the general solution of (3.6) is of the form $y = y_h + y_p$, where y_h is the *general* solution of the homogeneous equation $P(D)y = 0$, and y_p is any *particular* solution of the inhomogeneous equation. We have already cracked the homogeneous problem wide open by reducing it to a simpler problem in basic algebra. Thus we have only to consider how to find a particular solution of (3.6). We shall do this first for rather simple forms of the inhomogeneous term $f(x)$, and then extend our results to more complicated forms.

To find a function $y_p(x)$ such that $P(D)y_p = f(x)$ is to answer the question, "What is a function that will give us $f(x)$ if we apply the operator $P(D)$ to it?" In many cases, this question can be answered almost by inspection. Let us consider a couple of simple examples which demonstrate this: first,

$$(D^2 + 2)y = x - 1.$$

To what kind of function y_p must we apply the operator $D^2 + 2$ to get the polynomial $x - 1$? Clearly the function y_p could also be a polynomial; its degree could not be two or higher, because otherwise we would get a polynomial of degree two or higher on the right. So if we assume a solution of the form $y_p = ax + b$, we note that the term D^2 annihilates y_p; hence only the constant term affects the result. Since we have

$$2(ax + b) = x - 1,$$

we must have $2a = 1$, $2b = -1$, and therefore

$$y_p = \tfrac{1}{2}(x - 1).$$

It is easy to generalize from this to say that if $f(x)$ is a polynomial, and if the lowest derivative in $P(D)$ is sufficient to annihilate a polynomial of this degree, then $y_p = (1/a_0)f(x)$, assuming $a_0 \neq 0$. Thus, for example, a particular solution of

$$(D^4 + D^3 - 3)y = 2x^2 - x + 1$$

is

$$y_p = -\tfrac{1}{3}(2x^2 - x + 1).$$

A second example is

$$(D^2 + 2)y = \sin x.$$

To what function y_p can $D^2 + 2$ be applied to yield a sine function? The answer is, of course, a sine, since only derivatives of even order appear in the operator. If we assume a solution of the form $y_p = A \sin x$, then

$$(D^2 + 2)A \sin x = (-1 + 2)A \sin x = A \sin x = \sin x,$$

so $A = 1$ and $y_p = \sin x$. Again we can generalize from this to say that if $P(D^2)y = a \sin mx$, then

$$y_p = \frac{1}{P(-m^2)} a \sin mx,$$

assuming that $P(-m^2) \neq 0$. Hence, for example, a particular solution of

$$(D^4 - 3D^2 + 1)y = 3 \sin 2x$$

is

$$y_p = \frac{1}{(-4)^2 - 3(-4) + 1} 3 \sin 2x = \tfrac{3}{29} \sin 2x.$$

An analogous rule holds if the right side is a cosine function.

One should be able to guess immediately that if

$$P(D)y = ae^{kx},$$

then a particular solution is

$$y_p = \frac{a}{P(k)} e^{kx}$$

provided that $P(k) \neq 0$. Thus

$$(D^3 - D^2 + 3D - 1)y = 5e^{2x}$$

has the particular solution

$$y_p = \frac{5}{2^3 - 2^2 + 3 \cdot 2 - 1} e^{2x} = \frac{5}{8 - 4 + 6 - 1} e^{2x} = \frac{5}{9} e^{2x}.$$

We have thus discerned a number of simple special cases in which it is possible to read off a particular solution with very little work. With a modicum of common sense we can even extend this result to somewhat more complicated cases; for instance if

$$(D^2 + 1)y = 2 - e^x,$$

we should be able to see that $y_p = 2 - \frac{1}{2}e^x$. How do we do this? Basically, we see that 2 is a particular solution of $(D^2 + 1)y = 2$ and $-\frac{1}{2}e^x$ is a particular solution of $(D^2 + 1)y = -e^x$, then we add these two results. We can formalize this as follows:

Theorem 3.1

A particular solution of $P(D)y = f(x) + g(x)$ is $y_p = y_1 + y_2$, where $P(D)y_1 = f(x)$ and $P(D)y_2 = g(x)$.

Proof

We simply add the latter two equations, and obtain

$$P(D)y_1 + P(D)y_2 = P(D)(y_1 + y_2) = f(x) + g(x),$$

which completes the proof.

Naturally theorem 3.1 holds for a sum of any number of functions on the right side. Thus a particular solution of

$$(D^2 - 1)y = 1 + 2x - 5e^{2x} + \sin x$$

is

$$y_p = -(1 + 2x) - \tfrac{5}{3}e^{2x} + \tfrac{1}{2}\sin x,$$

as can be found by looking at each of the terms on the right side individually.

Problems

In each of the following problems, find a particular solution by inspection.

1. $(D^4 - D^2 + 4)y = 3x - 2.$
2. $(D^4 + D^2 - 2)y = \sin 2x.$
3. $(D^4 - 3)y = x^3 - 3x^2 + 7x + 1.$
4. $(D^3 - 2D^2 + D + 1)y = 2e^{3x} - 1.$
5. $(2D^6 - D^4 + 1)y = 3x^3 + x - \cos x.$
6. $(D^4 + 1)y = \sin x - \frac{1}{2}\cos x.$
7. $(D^4 - D^3 + 2)y = x^2 + 2x - e^x.$
8. $(3D^2 - 2)y = 3\sin 4x.$
9. $(D^2 - D + 1)y = 3 - 2e^{2x}.$
10. $(D^4 + 2D^2 + 1)y = 5\sin 2x.$
11. $(D^6 - 2D^2 + 3)y = 5x - 1 + e^{2x}.$
12. $(4D^4 + 2D^3 - 7D^2 + D - 3)y = 2e^x.$
13. $(4D^4 + 2D^3 - 7D^2 + D - 3)y = 2e^{2x}.$
14. $(4D^8 + 2D^6 - 7D^4 + D^2 - 3)y = 2\cos x.$
15. $(2D^6 + D^4 - 2)y = 5x^2 - 2x + 5\sin x.$
16. $(D^2 + D)y = 3.$
17. $(D^3 - 2D)y = 2x - 1.$
18. $(D^4 + 3D^2 + 2)y = 3e^{-x} - \sin 2x.$
19. $(D^4 - 2D^2)y = 3 + 7\sin 3x.$
20. $(D^5 - 7D^4 + 1)y = 17e^{2x} - \frac{5}{2}\sin x.$
21. If $DP(D^2)y = a \sin mx$, show that a particular solution is

$$y_p = \frac{a}{mP(-m^2)}\cos mx, \qquad \text{provided } P(-m^2) \neq 0.$$

22. If $DP(D^2)y = a\cos mx$, show that a particular solution is

$$y_p = -\frac{a}{mP(-m^2)}\sin mx, \qquad \text{provided that } P(-m^2) \neq 0.$$

Use the results of problems 21 and 22 to find a particular solution for each of the following problems:

23. $(D^3 + D)y = \sin 2x.$
24. $(D^5 + 2D^3 - D)y = 3\cos x.$
25. $(D^5 - D^3 + D)y = 5\sin x + \cos 2x.$
26. $(D^5 + D)y = 4\sin 2x - \cos x.$
27. $(3D^5 - 2D^3 + 3D)y = \cos x - \frac{1}{2}\sin x.$

3.4 Inhomogeneous Equations: the Annihilator Method

We have seen how, in some cases, we can simply read off the particular solutions of an inhomogeneous equation by inspection. However, this method does not always work, as we see by the following example:

$$(D^2 + 1)y = \sin x.$$

If we attempt to put $y = A \sin x$, we have

$$(D^2 + 1)y = A(-1 + 1) \sin x = 0.$$

Thus there is no particular solution of the inhomogeneous equation of the form $A \sin x$. Two questions arise: Why should this be so? and: What do we have to do to get a particular solution?

In answer to the first question, $A \sin x$ is a solution of the homogeneous equation $(D^2 + 1)y = 0$, so obviously it cannot also be a particular solution of the inhomogeneous equation $(D^2 + 1)y = \sin x$. This is true in general: if $P(D)y = f(x)$ and if $P(D)f(x) = 0$, that is, if the inhomogeneous term is itself a solution of the homogeneous equation, then a particular solution will be of a different form from what one might guess by inspection of $f(x)$. Consequently, in trying to determine the form of y_p and thus answer our second question, we shall have to devise some more general technique than just inspection.

Our new technique is based on the following observation: if y_h is the general solution of the homogeneous equation and y_p is a particular solution of the inhomogeneous equation, then

$$P(D)(y_h + y_p) = f(x). \tag{3.7}$$

Assuming that $f(x)$ is one of the combinations of exponentials, sines and cosines, and powers of x which we know to be a solution of a homogeneous equation, we can reverse the procedure of Section 3.1 and determine for $f(x)$ an *operator* $Q(D)$ such that $Q(D)f(x) = 0$: In other words, given a solution, find an equation for it. [$Q(D)$ is called an *annihilator* of $f(x)$; hence the name of this method.] Then applying $Q(D)$ to both sides of (3.7), we have

$$Q(D)P(D)(y_h + y_p) = 0. \tag{3.8}$$

The new, higher-order equation (3.8) is a *homogeneous equation which has* $y_h + y_p$ *as a particular solution.* Using the methods of Section 3.1, we can determine the general solution of (3.8), say $y_h + y_g$; y_h, of course, will be a part of this solution because of the presence of $P(D)$; the rest of the solution y_g comes from the new factor $Q(D)$. It is clear by comparing $y_h + y_p$ and $y_h + y_g$

that y_p must be obtainable from y_g by a particular choice of the arbitrary constants in y_g. To determine the value of these constants, we simply calculate

$$P(D)y_g = f(x)$$

and equate the coefficients of both sides. For this reason, this technique is also sometimes called the *method of undetermined coefficients.*

Let us return to our example

$$(D^2 + 1)y = \sin x \tag{3.9}$$

and see how we could apply the preceding method in this case. As we noted before, $\sin x$ is annihilated by $Q(D) = D^2 + 1$; applying this operator to both sides of (3.9), we obtain

$$(D^2 + 1)^2 y = 0,$$

for which the general solution is

$$y = c_1 \sin x + c_2 \cos x + c_3 x \sin x + c_4 x \cos x.$$

Since $y_h = c_1 \sin x + c_2 \cos x$ for equation (3.9), we have $y_g = c_3 x \sin x + c_4 x \cos x$. y_p must be obtainable from y_g by a proper choice of the constants c_3 and c_4. We must insert this function into (3.9) to determine the proper values. First we compute that

$$Dy_g = (c_3 - c_4 x) \sin x + (c_4 + c_3 x) \cos x$$

and

$$D^2 y_g = -(2c_4 + c_3 x) \sin x + (2c_3 - c_4 x) \cos x;$$

hence, from (3.9),

$$(D^2 + 1)y_g = -2c_4 \sin x + 2c_3 \cos x = \sin x;$$

and by comparing the coefficients of the (linearly independent) sine and cosine terms on both sides of this equation, we obtain $c_4 = -\frac{1}{2}$ and $c_3 = 0$. Therefore

$$y_p = -\tfrac{1}{2} x \cos x$$

is a particular solution of (3.9).

An alternative way of approaching this problem is available by using the complex exponential notation. We note that the imaginary part of the equation for a new (complex) variable $\tilde{y}$,

$$(D^2 + 1)\tilde{y} = e^{ix} = \cos x + i \sin x \tag{3.10}$$

is our original equation (3.9). Thus, if we can find a complex-valued particular solution of (3.10) and then take its imaginary part, we shall have a particular solution of the original equation. Note also that the *real* part of a particular solution of (3.10) will be a particular solution of

$$(D^2 + 1)y = \cos x.$$

Thus, as usual, the complex solution contains considerably more information than the real one. Furthermore, exponentials are always easier to treat than other functions.

To solve (3.10), then, we note that $(D - i)$ annihilates the right side:

$$(D - i)(D^2 + 1)\tilde{y} = (D - i)e^{ix} = 0.$$

The solution of this equation is

$$\tilde{y} = c_1 e^{ix} + c_2 e^{-ix} + c_3 x e^{ix} = \tilde{y}_h + \tilde{y}_g.$$

Since $\tilde{y}_g = c_3 x e^{ix}$, we shall look for a particular solution of this form. Now, since

$$D(\tilde{y}_g) = D(c_3 x e^{ix}) = c_3 e^{ix} + ic_3 x e^{ix},$$

$$D^2(\tilde{y}_g) = D(c_3 e^{ix} + ic_3 x e^{ix}) = 2ic_3 e^{ix} - c_3 x e^{ix},$$

we must have

$$(D^2 + 1)\tilde{y}_g = (2ic_3 e^{ix} - c_3 x e^{ix}) + c_3 x e^{ix} = e^{ix}.$$

Hence $2ic_3 = 1$ and therefore

$$c_3 = \frac{1}{2i} = -\frac{i}{2}.$$

Thus

$$\tilde{y}_p = -\frac{i}{2} x e^{ix} = -\frac{i}{2}(x \cos x - ix \sin x) = -\frac{1}{2} x \sin x - \frac{i}{2} x \cos x.$$

The imaginary part of this solution is

$$y_p = -\tfrac{1}{2}x \cos x,$$

as required.

Another example is the following:

$$(D^3 - 3D^2 - 4D)y = x^2 - 3x + 1. \tag{3.11}$$

The operator D^3 will annihilate the right side, so we have

$$D^3(D^3 - 3D^2 - 4D)y = D^3 \cdot D(D - 4)(D + 1)y = 0.$$

From this, we see that

$$y_h = c_1 e^{4x} + c_2 e^{-x} + c_3,$$

and

$$y_h + y_g = c_1 e^{4x} + c_2 e^{-x} + c_3 + c_4 x + c_5 x^2 + c_6 x^3,$$

so that

$$y_g = c_4 x + c_5 x^2 + c_6 x^3.$$

Now

$$Dy_g = c_4 + 2c_5 x + 3c_6 x^2,$$

$$D^2 y_g = 2c_5 + 6c_6 x, \qquad D^3 y_g = 6c_6,$$

so that when we put y_g into (3.11) we get

$$-12c_6 x^2 + (-8c_5 - 18c_6)x + (-4c_4 - 6c_5 + 6c_6) = x^2 - 3x + 1.$$

Equating the coefficients of like powers of x on both sides, we have

$$-12c_6 = 1, \ -8c_5 - 18c_6 = -3, \ -4c_4 - 6c_5 + 6c_6 = 1$$

and hence

$$c_6 = -\tfrac{1}{12}, \ c_5 = \tfrac{9}{16}, \ c_4 = -\tfrac{39}{32}.$$

Therefore

$$y_p = -\tfrac{1}{12}x^3 + \tfrac{9}{16}x^2 - \tfrac{39}{32}x$$

is a particular solution of (3.11).

Sometimes the exponential shift can be used to good advantage in finding particular solutions of inhomogeneous equations. To illustrate this, let us consider

$$(D^2 + 3D + 2)y = 2e^x \cos x. \tag{3.12}$$

Multiplying both sides by e^{-x}, we have

$$e^{-x}(D^2 + 3D + 2)y = 2\cos x,$$

and, shifting e^{-x} to the right through the operator $(D^2 + 3D + 2)$, we obtain

$$[(D + 1)^2 + 3(D + 1) + 2]e^{-x}y = (D^2 + 5D + 6)u = 2\cos x, \tag{3.13}$$

where the new function $u = e^{-x}y$. Once we have found a particular solution u_p for (3.13), we will have y_p because $e^x u_p = y_p$. And finding u_p will be somewhat easier than the process of finding y_p from scratch would be. An annihilator of the right side of (3.13) is $(D^2 + 1)$, so

$$(D^2 + 1)(D^2 + 4D + 6)u = (D^2 + 1)(D + 2)(D + 3)u = 0.$$

Since

$$u_h = c_1 e^{-2x} + c_2 e^{-3x}$$

and

$$u_h + u_g = c_1 e^{-2x} + c_2 e^{-3x} + c_3 \cos x + c_4 \sin x,$$

we have

$$u_g = c_3 \cos x + c_4 \sin x.$$

To find u_p we need to evaluate c_3 and c_4 using (3.13); since

$$\begin{aligned} Du_g &= -c_3 \sin x + c_4 \cos x, \\ D^2 u_g &= -c_3 \cos x - c_4 \sin x, \end{aligned}$$

we have from (3.13)

$$(5c_4 + 5c_3)\cos x + (5c_4 - 5c_3)\sin x = 2\cos x,$$

and, equating coefficients of like terms, we have

$$5c_4 + 5c_3 = 2, \quad 5c_4 - 5c_3 = 0,$$

so that $c_3 = c_4 = \frac{1}{5}$, and hence

$$u_p = e^{-x}y_p = \tfrac{1}{5}(\cos x + \sin x);$$

therefore a particular solution of (3.13) is

$$y_p = \tfrac{1}{5}e^x(\cos x + \sin x).$$

The exponential shift effects considerable simplification; it should be used whenever there is an exponential factor in the inhomogeneous term. The exponential shift may also, of course, be used with complex exponentials. The student should also bear in mind one of the simplifications from the previous section, that an equation whose inhomogeneous term is the sum of two or more different expressions can be solved by breaking it up into two or more simpler equations. For example, to find a particular solution of

$$(D^2 + 3D - 2)y = x^2 + 2 + 3e^x \cos 2x,$$

we simply add the particular solutions of the two equations

$$(D^2 + 3D - 2)y = x^2 + 2,$$

$$(D^2 + 3D - 2)y = 3e^x \cos 2x.$$

Note that the exponential shift can be used to facilitate the solution of the second of these equations.

Problems

Find a particular solution of each of the following equations.

1. $(D^2 + D - 2)y = 3e^x$.
2. $(D^2 - 3D + 5)y = x^3 - 3x + 2$.
3. $(D^2 + 2D + 2)y = 2\cos 2x$.
4. $(D^2 - 4D + 8)y = 3e^{2x}\cos x$.
5. $(D^2 + 2D + 6)y = 4xe^{-x}$.

6. $(D^2 - 2D + 6)y = 4xe^{-x}$.
7. $(D^4 + 2D^2 + 1)y = 3\cos x$.
8. $(D^3 - 2)y = x^5 - 1$.
9. $(D^3 - 2D^2 + 3D + 4)y = 4\cos x - 1$.
10. $(D^4 + 3D^3 - D^2 + 5D - 1)y = 3x^3e^x$.
11. $(3D^3 - 2D^2 + 5D + 3)y = 2x + 1 - 3e^{2x}\cos 3x$.
12. $(4D^2 - 5D + 3)y = 2\sinh x$.*
13. $(D^2 - 2D + 1)y = (3 + x)e^x$.
14. $(D^2 - 5D + 6)y = e^{2x} + \cos 4x$.
15. $(D^2 - 5D + 6)y = e^{2x} + \cos 4x$.

3.5 Particular Solutions by Variation of Parameters

The method of the previous section always works when the right side is a combination of functions which are themselves solutions of linear equations with constant coefficients: powers of x, exponentials, sines, and cosines. This method works precisely because such functions have only a finite number of linearly independent derivatives. (Think about this!) It would certainly not work on an equation of the form

$$(D^2 + 1)y = \cos(\cos x)$$

since $\cos(\cos x)$ has an infinite number of linearly independent derivatives. (Check this!)

To find a particular solution of an equation of this type, we use a method called *variation of parameters*. This works as follows: First, the associated homogeneous equation $(D^2 + 1)y = 0$ has as its general solution $y_h = c_1 \cos x + c_2 \sin x$. Here c_1 and c_2 are arbitrary constants, or *parameters*, to be determined by the problem. The insight of the method of variation of parameters is to allow c_1 and c_2 to *vary*, as functions of x, in order to see if there is a particular solution of the form

$$y_p(x) = c_1(x)\cos x + c_2(x)\sin x.$$

First, we find the first two derivatives

$$\begin{aligned} Dy_p &= -c_1(x)\sin x + c_2(x)\cos x + c_1'(x)\cos x + c_2'(x)\sin x, \\ D^2y_p &= -c_1(x)\cos x - c_2(x)\sin x - c_1'(x)\sin x + c_2'(x)\cos x \\ &\quad + [c_1'(x)\cos x + c_2'(x)\sin x]'. \end{aligned}$$

* Recall: $\sinh x = \frac{1}{2}(e^x - e^{-x})$, $\cosh x = \frac{1}{2}(e^x + e^{-x})$.

Hence, adding D^2y_p to y_p, we obtain

$$[c_1'(x)\cos x + c_2'(x)\sin x]' - c_1'(x)\sin x + c_2'(x)\cos x = \cos(\cos x).$$

This can certainly be satisfied if we can find functions $c_1(x)$ and $c_2(x)$ such that

$$c_1' \cos x + c_2' \sin x = 0,$$
$$-c_1' \sin x + c_2' \cos x = \cos(\cos x).$$

This system of two linear equations in two unknowns can be solved by the usual methods for c_1' and c_2', yielding

$$c_1' = -\sin x \cos(\cos x),$$
$$c_2' = \cos x \cos(\cos x).$$

Therefore we have

$$c_1 = -\int \sin x \cos(\cos x)dx = \sin(\cos x),$$

and

$$c_2 = \int \cos x \cos(\cos x)dx,$$

which cannot be integrated in closed form. Hence we have that

$$y_p = c_1 \cos x + c_2 \sin x = \cos x \sin(\cos x) + \sin x \int \cos x \cos(\cos x)dx.$$

Naturally variation of parameters also works on problems of the sort considered in the previous section. For example, let us examine

$$(D^2 - 3D + 2)y = 2e^x.$$

Then since $y_h = c_1e^x + c_2e^{2x}$, we try a particular solution of the form

$$y_h = c_1(x)e^x + c_2(x)e^{2x}.$$

Therefore

$$Dy_h = (c_1e^x + 2c_2e^{2x}) + (c_1'e^x + c_2'e^{2x})$$
$$D^2y_h = (c_1e^x + 4c_2e^{2x}) + (c_1'e^x + 2c_2'e^{2x}) + (c_1'e^x + c_2'e^{2x})'.$$

Putting this into the differential equation, we obtain

$$\begin{aligned}(c_1e^x + 4c_2e^{2x}) - 3(c_1e^x + 2c_2e^{2x}) + 2(c_1e^x + c_2e^{2x})\\ + (c_1'e^x + 2c_2'e^{2x}) - 3(c_1'e^x + c_2'e^{2x}) + (c_1'e^x + c_2'e^{2x})' = 2e^x,\end{aligned}$$

or, finally,

$$(-2e^xc_1' - e^{2x}c_2') + (e^xc_1' + e^{2x}c_2')' = 2e^x,$$

which yields the pair of equations

$$\begin{aligned}e^xc_1' + e^{2x}c_2' &= 0,\\ -2e^xc_1' - e^{2x}c_2' &= 2e^x.\end{aligned}$$

Solving this pair of simultaneous equations, we have

$$c_1' = -2, \qquad c_2' = 2e^{-x},$$

and therefore $c_1 = -2x$ and $c_2 = -2e^{-x}$. Hence

$$y_p = c_1e^x + c_2e^{2x} = -2xe^x - 2e^x = -2(x + 1)e^x.$$

Since $-2e^x$ is a solution of the associated homogeneous equation, $-2xe^x$ is another possible particular solution.

Although it is just as simple to work every problem in variation of parameters from scratch as we have done above, it is also possible to derive a formula for c_1' and c_2'. This is done as follows (for simplicity we restrict ourselves to second-order equations).

Given the differential equation

$$a_2D^2y + a_1Dy + a_0y = f(x),$$

and that $y_h = c_1y_1 + c_2y_2$, we form

$$y_p = c_1(x)y_1(x) + c_2(x)y_2(x),$$

then calculate

$$\begin{aligned}Dy_p &= (c_1y_1' + c_2y_2') + (c_1'y_1 + c_2'y_2)\\ D^2y_p &= (c_1y_1'' + c_2y_2'') + (c_1'y_1' + c_2'y_2') + (c_1'y_1 + c_2'y_2)'.\end{aligned}$$

Then, putting this into the differential equation, we obtain

$$\begin{aligned}
&a_2(c_1y_1'' + c_2y_2'') + a_2(c_1'y_1' + c_2'y_2') + a_2(c_1'y_1 + c_2'y_2)' \\
&+ a_1(c_1y_1' + c_2y_2') + a_1(c_1'y_1 + c_2'y_2) + a_0(c_1y_1 + c_2y_2) \\
&= c_1(a_2y_1'' + a_1y_1' + a_0y_1) + c_2(a_2y_2'' + a_1y_2' + a_0y_2) \\
&+ a_2(c_1'y_1 + c_2'y_2) + a_2(c_1'y_1 + c_2'y_2)' + a_1(c_1'y_1 + c_2'y_2) = f(x).
\end{aligned}$$

In the last expression, the first two terms vanish because y_1 and y_2 are each solutions of the homogeneous equation. Hence we are left with

$$a_2(c_1'y_1' + c_2'y_2') + a_2(c_1'y_1 + c_2'y_2)' + a_1(c_1'y_1 + c_2'y_2) = f(x)$$

which will hold if we can find c_1' and c_2' such that

$$\begin{aligned}
c_1'y_1 + c_2'y_2 &= 0, \\
c_1'y_1' + c_2'y_2' &= f(x)/a_2, \qquad a_2 \neq 0.
\end{aligned}$$

The determinant of this system is

$$\begin{vmatrix} y_1 & y_2 \\ y_1' & y_2' \end{vmatrix} = W[y_1, y_2]$$

which, by theorem 2.9, cannot be indentically zero if y_1 and y_2 are linearly independent. Hence we can solve for c_1' and c_2' and obtain

$$c_1' = \frac{\begin{vmatrix} 0 & y_2 \\ f(x)/a_2 & y_2' \end{vmatrix}}{W[y_1, y_2]}, \qquad c_2' = \frac{\begin{vmatrix} y_1 & 0 \\ y_1' & f(x)/a_2 \end{vmatrix}}{W[y_1, y_2]}.$$

Any indefinite integrals of these expressions will give values for c_1 and c_2, and thus a particular solution is always at hand in the form $c_1y_1 + c_2y_2$.

Problems

1–15. Solve problems 1–15 of the previous section by means of the technique of variation of parameters.

Find particular solutions of the following, using variation of parameters:

16. $D^2y - 3Dy + 2y = \cos e^{-x}$.
17. $D^2y + y = \sec x$.

18. $D^2y + 2Dy + y = 4e^{-x} \ln x.$
19. $D^2y + 2Dy + y = e^{-x} \ln x.$
20. $D^2y + y = \cos^2 x.$

3.6 Applications of Methods of Solution of Inhomogeneous Equations

We derived a number of inhomogeneous linear equations with constant coefficients in previous chapters, which are now amenable to solution by the methods of the preceding sections. Before we consider some of the equations which we have not previously solved, let us reexamine the traffic problem

$$\left(D + \frac{1}{k}\right)x_2 = \frac{1}{k}\left(\frac{1}{2}a_1t^2 - 11\right).$$

$$x_2(0) = -11.$$

We first note immediately that

$$(x_2)_h = c\exp(-t/k)$$

is the general solution of the associated homogeneous equation; and since D^3 will annihilate the right side, we have

$$D^3\left(D + \frac{1}{k}\right)x_2 = 0,$$

so that x_2 must be of the form

$$ce^{-(1/k)t} + b_1x^2 + b_2x + b_3,$$

and hence we have that $(x_2)_p$ must be of the form $b_1x^2 + b_2x + b_3$. Applying the operator $D + 1/k$ to this, we obtain

$$2b_1x + b_2 + \frac{1}{k}(b_1x^2 + b_2x + b_3) = \frac{1}{k}\left(\frac{1}{2}a_2t^2 - 11\right),$$

whence

$$\frac{b_1}{k} = \frac{a_1}{2k}, \quad 2b_1 + \frac{b_2}{k} = 0, \quad b_2 + \frac{b_3}{k} = -\frac{11}{k}.$$

Thus

$$b_1 = \tfrac{1}{2}a_1, \quad b_2 = -ka_1, \quad b_3 = k^2a_1 - 11.$$

Hence the general solution of our traffic equation is

$$x_2 = ce^{-(1/k)t} + a_1t^2 - ka_1t + (k^2a_1 - 11),$$

and the initial condition $x_2(0) = -11$ gives

$$-11 = c + k^2a_1 - 11,$$

so $c = -k^2a_1$ and we have

$$x_2(t) = -k^2a_1e^{-(1/k)t} + a_1t^2 - ka_1t + (k^2a_1 - 11),$$

exactly as before. Because this method of solution does not require laborious integration by parts, it is simpler than the previous method, and is therefore preferable.

An example from Chapter 2 of an inhomogeneous linear equation with constant coefficients is the linearized pendulum–vibrating spring–simple circuit equation. We shall consider this equation in the form

$$D^2x + \omega^2x = f(t),$$

where the fractional term (resistance, in the case of the electrical circuit) is lacking. We shall consider, in particular, the case in which $f(t) = A \sin \omega' t$, so that the external force (or electromotive force) is itself a sinusoidal function of time. For example, in the electrical circuit, an electromotive force which is of sinusoidal form represents an alternating current. A general alternating current is the superposition of a number of solutions of equations of this type:

$$(D^2 + \omega^2)x = A \sin \omega' t.$$

We know that $x_h = c_1 \cos \omega t + c_2 \sin \omega t$, and, since $(D^2 + \omega'^2)$ is the operator which annihilates the right side, we have

$$(D^2 + \omega'^2)(D^2 + \omega^2)x = 0.$$

There are two possible cases, depending on whether or not ω' and ω are equal.

Case 1 $\omega \neq \omega'$
We have

$$x_h + x_g = c_1 \cos \omega t + c_2 \sin \omega t + c_3 \cos \omega' t + c_4 \sin \omega' t,$$

so that x_p is of the form $c_3 \cos \omega' t + c_4 \sin \omega' t$. Inserting this into the original equation, we obtain

$$(-\omega'^2 + \omega^2)(c_3 \cos \omega' t + c_4 \sin \omega' t) = A \sin \omega' t,$$

which implies that $c_3 = 0$ and $c_4 = A/(\omega^2 - \omega'^2)$. Thus the general solution of the equation is

$$x(t) = c_1 \sin \omega t + c_2 \cos \omega t + \frac{A}{(\omega^2 - \omega'^2)} \sin \omega' t.$$

The constants c_1 and c_2 can as usual be evaluated by making use of the initial conditions.

Case 2 $\omega = \omega'$

In this case

$$x_h + x_g = c_1 \cos \omega t + c_2 \sin \omega t + t(c_3 \sin \omega t + c_4 \cos \omega t),$$

so that x_p is of the form $t(c_3 \sin \omega t + c_4 \cos \omega t)$. Since

$$D^2[t(c_3 \sin \omega t + c_4 \cos \omega t)] = 2(c_3\omega \cos \omega t - c_4\omega \sin \omega t)$$
$$- t(c_3\omega^2 \sin \omega t + c_4\omega^2 \cos \omega t),$$

we have

$$(2c_3\omega + c_4\omega^2 t)\cos \omega t + (-2c_4\omega + c_3\omega^2 t)\sin \omega t - t(c_3\omega^2 \sin \omega t$$
$$+ c_4\omega^2 \cos \omega t) = A \sin \omega t,$$

whence $c_3 = 0$, $c_4 = -A/2\omega$. Thus the general solution in this case is

$$x(t) = c_1 \sin \omega t + c_2 \cos \omega t - \frac{At}{2\omega} \cos \omega t.$$

Once again, the arbitrary constants c_1 and c_2 can be evaluated from the initial conditions. The interesting feature here, however, is the term $(-A/2\omega)t \cos \omega t$. Because of the factor of t, the amplitude of this term becomes larger and larger as t increases. This phenomenon is known as resonance: because the driving force has the same frequency as the solution, this force has a cumulative effect, adding this effect to the usual vibration. Eventually the spring would oscillate so wildly that it would be bent or broken; the pendulum would go through an arc so wide that the linearized equation would no longer be valid; and the current in the circuit would reach such a peak that something would melt. Thus *resonance* represents an unstable situation.

The analysis of this problem and the more general one with the friction (or resistance) term present is somewhat simpler and more satisfactory if one uses the method of complex exponentials, which is left to the reader in the problems at the end of this section.

At this point we now have the ammunition necessary to complete the discussion of Kepler's laws which we began in Section 1.3. The reader will recall that the vector equation of motion (in polar coordinates) of a particle of mass m at the point P moving under the influence of a central force emanating from the origin O was

$$-f(r)\boldsymbol{U} = m(D^2r - rD\theta^2)\boldsymbol{U} + m(rD^2\theta + 2DrD\theta)\boldsymbol{V}.$$

By integrating the second differential equation once, we discovered Kepler's second law, that the radius vector between O and P sweeps out equal areas in equal lengths of time.

$$r^2D\theta = c.$$

The first equation is of the form

$$m(D^2r - rD\theta^2) = -f(r).$$

Newton's law of gravitation states that $f(r)$ is proportional to the mass M at O and the mass m at P, and inversely proportional to the square of the distance r between. Hence

$$m(D^2r - rD\theta^2) = -\frac{kMm}{r^2}.$$

Cancelling m from both sides and noticing that $D\theta = c/r^2$, we finally have

$$D^2r - \frac{c^2}{r^3} = -\frac{kM}{r^2}.$$

This last equation is nonlinear with the independent variable t missing, so it could conceivably be solved by the methods of Section 1.7. The resulting solutions $[r(t), \theta(t)]$ will be parametric equations of the trajectory of the particle P.

However, in order to verify Kepler's laws, it would be somewhat easier if we had the equation of the trajectory in the nonparametric form $r = r(\theta)$.

To make this change of independent variables in the above equation, we note that

$$Dr = \frac{dr}{dt} = \frac{dr}{d\theta}\frac{d\theta}{dt} = \frac{c}{r^2}\frac{dr}{d\theta}$$

and

$$D^2r = \frac{d}{dt}\left(\frac{c}{r^2}\frac{dr}{d\theta}\right) = \frac{d}{d\theta}\left(\frac{c}{r^2}\frac{dr}{d\theta}\right)\frac{c}{r^2} = -\frac{2c^2}{r^5}\frac{dr}{d\theta} + \frac{c^2d^2r}{r^4d\theta^2}.$$

Thus the differential equation for r in terms of θ becomes

$$\frac{c^2}{r^4}D^2r - \frac{2c^2}{r^5}Dr - \frac{c^2}{r^3} = -\frac{kM}{r^2}.$$

Here the operator D now means differentiation with respect to θ. Putting all r dependence on the left side, we obtain

$$\frac{1}{r^2}D^2r - \frac{2}{r^3}(Dr)^2 - \frac{1}{r} = -\frac{kM}{c^2}.$$

Because of all the occurrences of r in the denominator, we are tempted to change dependent variables from r to $u = 1/r$. If we do so, then

$$Dr = D(1/u) = (-1/u^2)Du$$

and

$$Dr = -D[(1/u^2)Du] = (2/u^3)(Du)^2 - (1/u^2)D^2u.$$

Putting this into the equation, we obtain

$$u^2\left(-\frac{1}{u^2}D^2u + \frac{2}{u^3}Du^2\right) - 2u^3\left(\frac{1}{u^4}Du^2\right) - u = -\frac{kM}{c^2}.$$

The troublesome nonlinear term cancels and we finally have

$$D^2u + u = kM/c^2,$$

which is a linear equation with constant coefficients for u. As we know, its solution is

$$u = A\cos(\theta - \delta) + \frac{kM}{c^2}\cdot$$

The phase angle δ here represents merely a rotation of the coordinate system; hence without loss of generality, we can choose our coordinate system in such a way that $\delta = 0$. We then have that the equation of the trajectory of P is

$$r\left(A\cos\theta + \frac{kM}{c^2}\right) = 1.$$

Clearly if $A = 0$, this is the equation of a circle. Otherwise, it is the equation of a conic section with one focus at the origin, a fact which we leave to the reader in the problems. Hence we have derived *Kepler's first law*: The planets travel around the sun in elliptical orbits with the sun at one focus. In fact, we have even generalized this law from ellipses to other conic sections; as a matter of fact, comets and intruders from outside the solar system have been observed to traverse parabolic or hyperbolic orbits.

Kepler's third law states that the square of the period of revolution of a planet is proportional to the cube of the length of the semimajor axis of its orbit. To show this, we observe first that the area of the elliptical orbit is πab, where a is the length of the semimajor axis and b is the length of the semiminor axis. On the other hand, since during the period of time T which it takes the particle P to complete one revolution, the radius vector sweeps out the area of the ellipse, this area is also equal to $\frac{1}{2}cT$. (Recall from Section 1.3 that we showed that the area swept out is $\frac{1}{2}c$ times the elapsed time.) Hence we have

$$\tfrac{1}{2}cT = \pi ab.$$

In the problems we shall prove that $b = c\sqrt{a/kM}$, and hence we have

$$T = \frac{2\pi c^2 a^{3/2}}{\sqrt{kM}}$$

or, finally,

$$T^2 = \frac{4\pi^2 c^4}{kM}a^3,$$

which is precisely the statement of Kepler's third law.

Problems

1. In the case in which friction is present in the pendulum–spring–circuit equation, this equation can be written in the form

$$[(D + \alpha)^2 + \omega^2]x = f(t).$$

Investigate the possible solutions of this equation when $f(t) = Ae^{-\alpha' t} \sin \omega' t$ including the case in which ω and ω' both vanish. If $\alpha = \alpha'$ and $\omega = \omega'$, is there is a resonance phenomenon similar to the one which occurs when friction is absent? What affect does the presence of the frictional term $(2\alpha Dx)$ have on the resonance?

2. Consider the equation $(D^2 + \omega^2)x = f(t)$. For simplicity, suppose $\omega^2 = 1$, $x(0) = 0$, $x'(0) = 1$. Let $f(t)$ be a step function as follows:

$$f(t) = 1 \text{ for } n\pi \leqq t \leqq (n + 1)\pi \text{ when } n \text{ is even;}$$

$$f(t) = 0 \text{ for } n\pi < t < (n + 1)\pi \text{ when } n \text{ is odd.}$$

How do you solve this initial value problem? What happens to the solutions as $t \to \infty$? Would it have made any difference if in the definition of $f(t)$ we had used some number $\alpha < \pi$ instead of π? Explain the situation physically.

3. Prove that $r(A \cos \theta + kM/c^2) = 1$ is the equation of an ellipse, parabola, or hyperbola, according as $A^2 < k^2M^2/c^4$, $A^2 = k^2M^2/c^4$, or $A^2 > k^2M^2/c^4$, respectively. In the elliptical case, find the lengths of the semiaxes a and b, and prove that $b - c\sqrt{a/kM}$.

4. Consider problem 1 using the complex exponential method of solution. Be sure to include the case $\alpha = 0$.

3.7 Euler Differential Equations

A linear ordinary differential equation of the form

$$a_n x^n D^n y + a_{n-1} x^{n-1} D^{n-1} y + \cdots + a_1 x D y + a_0 y = f(x)$$

is called an *Euler equation.* Each coefficient is a constant times a power of x, and the power of x is equal to the order of the derivative which it multiplies. In particular, a second-order homogeneous Euler equation is of the form

$$a_2 x^2 D^2 y + a_1 x D y + a_0 y = 0.$$

By examining this equation for a moment we can surmise that its solutions might be powers of x. In fact, if y were a power of x, then the first derivative would reduce the exponent of x by one, and the second derivative by two, but the powers of x in the coefficients would raise the exponent back up again. Thus, if we assume a solution of the form $y = x^\alpha$, then we obtain

$$a_2x^2\alpha(\alpha - 1)x^{\alpha-2} + a_1x\alpha x^{\alpha-1} + a_0x^\alpha = 0,$$

or

$$x^\alpha[a_2\alpha^2 + (a_1 - a_2)\alpha + a_0] = 0.$$

Hence x will be a solution if α is a root of the algebraic equation

$$a_2\alpha^2 + (a_1 - a_2)\alpha + a_0 = 0.$$

Two possible problems arise from this approach: First, this equation might have equal roots, in which case we obtain only one solution by this method; and of course a second-order equation must have two independent solutions. Second, what would happen if α were complex? It is in answering this second question that we obtain a method for solving Euler equations.

We do this as follows: Certainly for real values of α it is true that

$$x^\alpha = e^{\ln x^\alpha} = e^{\alpha \ln x}$$

Let us postulate that this same relationship holds for complex α. Then, if $\alpha = a + ib$, we have

$$x^\alpha = e^{(a+ib)\ln x} = e^{a \ln x}e^{ib \ln x} = x^a(\cos(b \ln x) + i \sin(b \ln x)).$$

Once again the complex exponential notation comes to the rescue! But it is in going back to the original formula

$$x^\alpha = e^{\ln x^\alpha} = e^{\alpha \ln x} \text{ or } x = e^{(1/\alpha)\ln x^\alpha}$$

that we note a possible substitution that we can make in the Euler equation to simplify it. This substitution is

$$x = e^t \text{ or } t = \ln x.$$

This seems to be a reasonable substitution to make, since any power of x, including x itself, can be represented as an exponential.

If we make this change of independent variable, then

$$\frac{dy}{dx} = \frac{dy}{dt}\frac{dt}{dx} = \frac{\frac{1}{dx}}{dt}\frac{dy}{dt} = e^{-t}\frac{dy}{dt},$$

and

$$\frac{d^2y}{dx^2} = e^{-t}\frac{d}{dt}\left(e^{-t}\frac{dy}{dt}\right) = e^{-2t}\left(\frac{d^2y}{dt^2} - \frac{dy}{dt}\right),$$

and so on for higher derivatives. In particular, the second-order equation

$$a_2x^2D^2y + a_1xDy + a_0y = 0$$

becomes simply

$$a_2\frac{d^2y}{dt^2} + (a_1 - a_2)\frac{dy}{dt} + a_0y = 0$$

in terms of the new variable t, which is a linear equation with constant coefficients and hence can easily be solved for y as a function of t by the methods of this chapter. Then, the return substitution $t = \ln x$ gives the solution y as a function of x.

This substitution works for an Euler equation of any order. For instance, because we have

$$\frac{d^3y}{dt^3} = e^{-t}\frac{d}{dt}\left[e^{-2t}\left(\frac{d^2y}{dt^2} - \frac{dy}{dt}\right)\right] = e^{-3t}\left(\frac{d^3y}{dt^3} - 3\frac{d^2y}{dt^2} + 2\frac{dy}{dt}\right),$$

the third-order equation

$$a_3x^3D^3y + a_2x^2D^2y + a_1xDy + a_0y = 0$$

becomes

$$a_3\frac{d^3y}{dt^3} + (a_2 - 3a_3)\frac{d^2y}{dt^2} + (a_1 - a_2 + 2a_3)\frac{dy}{dt} + a_0y = 0$$

in terms of t, and so on.

For example, consider the equation

$$3x^3D^3y + 14x^2D^2y + 12xDy - 2y = 0.$$

After substituting $x = e^t$, the equation becomes

$$3\frac{d^3y}{dt^3} + 5\frac{d^2y}{dt^2} + 4\frac{dy}{dt} - 2y = 0,$$

or, in operator notation, using D to represent differentiation with respect to t,

$$(3D^3 + 5D^2 + 4D - 2)y = 0,$$

which factors into

$$(3D - 1)(D^2 + 2D + 2)y = 0$$

or, finally,

$$(3D - 1)[(D + 1)^2 + 1]y = 0.$$

This equation has as its general solution

$$y = c_1e^{1/3t} + e^{-t}(c_2 \cos t + c_3 \sin t),$$

and after putting in $t = \ln x$, we obtain

$$y = c_1x^{1/3} + x^{-1}[c_2 \cos(\ln x) + c_3 \sin(\ln x)].$$

We could equally well, of course, have left the second and third terms of the solution in complex exponential form, but it all comes to the same in the end.

Another example is the following:

$$x^2D^2y + 7xDy + 9y = 0.$$

Making the substitution $x = e^t$, this becomes

$$\frac{d^2y}{dt^2} + 6\frac{dy}{dt} + 9y = (D + 3)^2y = 0,$$

the general solution of which is

$$y = e^{-3t}(c_1 + c_2t),$$

which yields

$$y = x^{-3}(c_1 + c_2 \ln x).$$

It is this perfect equivalence with equations with constant coefficients that makes it an easy matter to solve Euler equations. These Euler equations are a special case of the more general linear equations which we shall be solving by series methods later in the text; the form of the solutions of Euler equations will give us considerable insight into the solution of these more general equations.

Incidentally, one can solve inhomogeneous Euler equations by means of this same substitution. For example, let us consider

$$x^2D^2y + 2xDy - 6y = 3x - x^3.$$

The substitution $x = e^t$ yields

$$\frac{d^2y}{dt^2} + \frac{dy}{dt} - 6y = (D + 3)(D - 2)y = 3e^t - e^{3t}.$$

By inspection $y_h = c_1e^{-3t} + c_2e^{2t}$, and only a little more work shows that

$$y_p = -\tfrac{3}{4}e^t - \tfrac{1}{6}e^{3t}.$$

Hence we have

$$y = y_h + y_p = c_1e^{-3t} + c_2e^{2t} - \tfrac{3}{4}e^t - \tfrac{1}{6}e^{3t},$$

and in terms of the original variable x we have

$$y = c_1x^{-3} + c_2x^2 - \tfrac{3}{4}x - \tfrac{1}{6}x^3.$$

A *partial differential equation* is an equation for an unknown function of two or more independent variables in which some of the partial derivatives of the unknown function appear. Such equations occur frequently in mathematical physics; the following problem leads to an application of Euler equations as well. We consider a metal disk of radius 1, centered at the origin, insulated on the top and bottom, and with a given temperature distribution on its circumference. Heat will flow through the disk until the temperature stabilizes at a constant value at each point, a value which no longer changes with time. Then the following partial differential equation and boundary condition holds for the temperature $u = u(r, \theta)$ (in polar coordinates)*:

$$\frac{\partial^2 u}{\partial r^2} + \frac{1}{r}\frac{\partial u}{\partial r} + \frac{1}{r^2}\frac{\partial^2 u}{\partial \theta^2} = 0, \qquad u(1, \theta) = f(\theta).$$

* See, for example, R. V. Churchill, *Fourier Series and Boundary Value Problems* (McGraw-Hill Book Company, New York, 1941).

Assuming that the function $f(\theta)$ in the boundary condition is periodic with period 2π and continuous, physical considerations lead us to the obvious requirement that the solution $u(r, \theta)$ be continuous and periodic with period 2π in θ throughout the disk.

In attacking this and similar problems in more advanced courses, the first step is to seek particular solutions of the differential equation which have the form $R(r)\Theta(\theta)$, that is, a function of r times a function of θ. This process is called the *separation of variables* (and not to be confused with the separation of variables in first-order ordinary differential equations). We ask only that R be continuous, that $\Theta(\theta) = \Theta(\theta + 2\pi)$ for all Θ, and that $R\Theta$ satisfy the partial differential equation.

In this case we will obtain

$$R''\Theta + \frac{1}{r}R'\Theta + \frac{1}{r^2}R\Theta'' = 0,$$

which, upon multiplication by r^2 and division by $R\Theta$, becomes

$$\frac{r^2R'' + rR'}{R} + \frac{\Theta''}{\Theta} = 0$$

or

$$\frac{r^2R'' + rR'}{R} = -\frac{\Theta''}{\Theta}.$$

On the left side we have a function of r alone; on the right, a function of θ alone. Since r and θ may vary independently of one another, the only way this equation can be true is for both sides to equal the same constant, say λ. Then we have

$$\frac{r^2R'' + rR'}{R} = \lambda, \qquad -\frac{\Theta''}{\Theta} = \lambda,$$

which leads to the pair of ordinary differential equations

$$r^2R'' + rR' - \lambda R = 0,$$
$$\Theta'' + \lambda\Theta = 0.$$

So far the constant λ is unknown. Our question must be: Are there values of λ such that these two equations have nontrivial (i.e. nonzero) solutions? And if so, what are these values and the solutions which correspond to them? If such values of λ exist, they are called the *eigenvalues* of the problem, a combination German-English word which means "proper" or "characteristic" values. The corresponding solutions are usually called *eigenfunctions.*

To find the eigenvalues in this problem, let us consider the second equation; it is

$$\Theta'' + \lambda\Theta = 0.$$

If λ is negative, say $\lambda = -\mu$, then the general solution of this equation is

$$\Theta = c_1 \exp(\sqrt{\mu}\theta) + c_2 \exp(-\sqrt{\mu}\theta);$$

this solution is continuous, but not periodic, and hence we must reject the possibility that $\lambda > 0$. If $\lambda \leqq 0$, however, the solution is

$$\Theta = c_1 \cos \sqrt{\lambda}\theta + c_2 \sin \sqrt{\lambda}\theta.$$

This solution is indeed periodic; in fact, it is periodic with period 2π if and only if $\sqrt{\lambda}$ is a nonnegative integer n; then we have the eigenvalues $\lambda = n^2$ $(n = 0, 1, 2, \ldots)$ and the eigenfunctions

$$\Theta_n(\theta) = A_n \cos n\theta + B_n \sin n\theta \qquad (n = 0, 1, 2, 3, \ldots),$$

where each value of n gives a different solution.

Knowing the eigenvalues, we can now return and solve the first equation, which becomes

$$r^2R'' + rR' - n^2R = 0.$$

This is an Euler equation, which upon substituting $r = e^t$ becomes

$$(D^2 - n^2)R = 0,$$

so that the solutions are e^{nt} and e^{-nt}, or, in terms of r, r^n and r^{-n}. Because $R(r)$ is required to be continuous throughout the unit disk, we must reject the solution r^{-n}. Thus we conclude that the solutions of the differential equation which we were seeking must be of the form

$$u_n(r, \theta) = R_n(r)\Theta_n(\theta) = r^n(A_n \cos n\theta + B_n \sin n\theta), \qquad n = 0, 1, 2, \ldots .$$

In elementary courses in partial differential equations one learns how to superpose a sequence of solutions like this in such a manner that the boundary condition is satisfied as well.

Problems

1. $2x^2D^2y + xDy - y = 0.$
2. $x^2D^2y - 2xDy + 2y = 0.$
3. $x^2D^2y + xDy - 9y = 0.$
4. $x^2D^2y + 6xDy + 6y = x^{-3}.$
5. $x^2D^2y - 6y = 3 \ln x.$
6. $x^2D^2y - 3xDy + 4y = 0.$
7. $9x^2D^2y + 15xDy + y = 0.$
8. $x^2D^2y + xDy + y = 0.$
9. $x^2D^2y - 3xDy + 8y = 0.$
10. $2x^2D^2y + 2xDy + y = x - x^2.$
11. $x^3D^3y + 6x^2D^2y + 7xDy + y = 0.$
12. $x^3D^3y + 4x^2D^2y + xDy - y = 0.$
13. $x^3D^3y + x^2D^2y - 6xDy + y = 0.$
14. $x^3D^3y + 3x^2D^2y + xDy = 2 \ln x.$
15. $x^3D^3y + x^2Dy - 2y = 0.$

3.8 Linear Systems with Constant Coefficients: Matrix Methods

As we have seen previously, linear systems with constant coefficients arise from physical problems involving coupled mechanical systems and electrical circuits. Furthermore, every such problem can always be reduced to a first-order system by renaming some of the dependent variables. It is possible to solve linear systems by operator methods, as we have done with single equations earlier in this chapter. However, matrix methods are considerably neater and more satisfactory, so we shall use them instead, in conjunction with many of the facts that we have proved for single equations, which also turn out to be true for systems. (In the next chapter, we shall consider Laplace transform methods for systems.)

For simplicity of exposition, we shall consider only first-order systems with constant coefficients in two unknowns, that is, systems that can be put into the form

$$\begin{aligned} Dx &= ax + by + f(t), \\ Dy &= cx + dy + g(t), \end{aligned} \tag{3.14}$$

where $x(t)$ and $y(t)$ are the unknown functions and t is the independent variable. Very little generality is lost by treating systems with only two unknowns, since the behavior of systems with more unknowns is quite analogous; meanwhile, our computations will be considerably simpler.

Before we attempt the solution of the system (3.14), we shall establish our notation and review a few simple facts about matrices. We shall use the notations

$$\boldsymbol{x} = \begin{pmatrix} x \\ y \end{pmatrix}, \quad D\boldsymbol{x} = \begin{pmatrix} Dx \\ Dy \end{pmatrix}, \quad \boldsymbol{f} = \begin{pmatrix} f \\ g \end{pmatrix}$$

to represent the unknown *column vector* and its derivative and the known inhomogeneous term, respectively. Likewise we shall use

$$\boldsymbol{A} = \begin{pmatrix} a & b \\ c & d \end{pmatrix}$$

to represent the 2×2 *square matrix* of coefficients of (3.14).

We recall that two column vectors are equal if and only if the corresponding components are equal; likewise, two matrices are equal providing that all corresponding entries are equal. Furthermore, let us recall that a matrix and a column vector, or two matrices, are *multiplied* as follows:

$$\begin{pmatrix} a & b \\ c & d \end{pmatrix}\begin{pmatrix} x \\ y \end{pmatrix} = \begin{pmatrix} ax + by \\ cx + dy \end{pmatrix}; \quad \begin{pmatrix} a & b \\ c & d \end{pmatrix}\begin{pmatrix} e & f \\ g & h \end{pmatrix} = \begin{pmatrix} ae + bg & af + bh \\ ce + dg & cf + dh \end{pmatrix}.$$

Thus equation (3.14) can be written in matrix form as follows:

$$D\boldsymbol{x} = \boldsymbol{A}\boldsymbol{x} + \boldsymbol{f}(t), \quad \text{or} \quad (D - \boldsymbol{A})\boldsymbol{x} = \boldsymbol{f}(t). \tag{3.15}$$

The *associated homogeneous equation* is defined to be

$$D\boldsymbol{x} = \boldsymbol{A}\boldsymbol{x}, \quad \text{or} \quad (D - \boldsymbol{A})\boldsymbol{x} = \boldsymbol{0}. \tag{3.16}$$

It is a simple matter to verify that the *principles of superposition hold* for linear systems; that is,

(1) if $\boldsymbol{x}_1$ and $\boldsymbol{x}_2$ are any two solutions of (3.16), and if c_1 and c_2 are any two constants, then $c_1\boldsymbol{x}_1 + c_2\boldsymbol{x}_2$ is likewise a solution;

(2) if $\boldsymbol{x}_h$ is a solution of (3.16) and $\boldsymbol{x}_p$ is a solution of (3.15), then $\boldsymbol{x}_h + \boldsymbol{x}_p$ is likewise a solution of (3.15).

We leave the proof of these facts to the reader. It can also be shown that (3.16) has precisely two linearly independent solutions, and hence that two arbitrary constants enter into the solution, as in the case of a single second-order equation.

Let us now note the following:

Definition
The special matrix

$$\boldsymbol{I} = \begin{pmatrix} 1 & 0 \\ 0 & 1 \end{pmatrix}$$

is called *the identity matrix.*

$\boldsymbol{I}$ has a role analogous to that of 1 in the real numbers, since for any matrix $\boldsymbol{A}$ it is true that $\boldsymbol{IA} = \boldsymbol{AI} = \boldsymbol{A}$.

Definition
A matrix

$$\boldsymbol{A} = \begin{pmatrix} a & b \\ c & d \end{pmatrix}$$

is said to be *nonsingular* if and only if its determinant (defined to be the number $ad - bc$ and denoted by $\det \boldsymbol{A}$) is nonzero.

Proposition
If $\boldsymbol{A}$ is nonsingular, then it possesses an *inverse* $\boldsymbol{A}^{-1}$; that is, $\boldsymbol{A}^{-1}\boldsymbol{A} = \boldsymbol{A}\boldsymbol{A}^{-1} = \boldsymbol{I}$. In fact, the inverse of $\boldsymbol{A}$ is

$$\boldsymbol{A}^{-1} = \begin{pmatrix} \dfrac{d}{ad - bc} & \dfrac{-b}{ad - bc} \\[2ex] \dfrac{-c}{ad - bc} & \dfrac{a}{ad - bc} \end{pmatrix}.$$

The reader can easily verify this statement.

We shall need these facts in order to answer for the reader a question which no doubt has occurred to him. That is, why are we considering systems in the form (3.14) or (3.15) when the most general first-order system is

$$\begin{aligned} pDx + qDy &= ax + by + f(t), \\ rDx + sDy &= cx + dy + g(t). \end{aligned} \tag{3.17}$$

We now have the necessary ammunition to answer this question. Let us write (3.17) in the matrix form

$$\boldsymbol{P}D\boldsymbol{x} = \boldsymbol{A}\,\boldsymbol{x} + \boldsymbol{f}, \tag{3.18}$$

where

$$\boldsymbol{P} = \begin{pmatrix} p & q \\ r & s \end{pmatrix}, \quad \boldsymbol{A} = \begin{pmatrix} a & b \\ c & d \end{pmatrix}.$$

Let us assume that $\boldsymbol{P}$ is nonsingular; if $\boldsymbol{P}$ is singular, then the system (3.18) is said to be *degenerate* and is of little interest to us, as we shall see at the end of this section.

If $\boldsymbol{P}$ is nonsingular, then $\boldsymbol{P}^{-1}$ exists and we can multiply (3.18) on the left by $\boldsymbol{P}^{-1}$, obtaining a new equation

$$\boldsymbol{P}^{-1}(\boldsymbol{P}D\boldsymbol{x}) = \boldsymbol{P}^{-1}(\boldsymbol{A}\boldsymbol{x}) + \boldsymbol{P}^{-1}\boldsymbol{f}.$$

However, because matrix multiplication is *associative*,* we can write this as

$$(\boldsymbol{P}^{-1}\boldsymbol{P})D\boldsymbol{x} = (\boldsymbol{P}^{-1}\boldsymbol{A})\boldsymbol{x} + \boldsymbol{P}^{-1}\boldsymbol{f}.$$

We note that $\boldsymbol{P}^{-1}\boldsymbol{P} = \boldsymbol{I}$; and if we denote $\boldsymbol{P}^{-1}\boldsymbol{A}$ by $\boldsymbol{B}$ and $\boldsymbol{P}^{-1}\boldsymbol{f}$ by $\boldsymbol{g}$, then (3.18) has been reduced to

$$D\boldsymbol{x} = \boldsymbol{B}\boldsymbol{x} + \boldsymbol{g},$$

which is in the standard form (3.15) Hence every nondegenerate first-order system can be reduced to this standard form.

Example

Let us reduce the system

$$\begin{pmatrix} 1 & 2 \\ 5 & 3 \end{pmatrix} D\boldsymbol{x} = \begin{pmatrix} 2 & 1 \\ 2 & 3 \end{pmatrix} \boldsymbol{x} + \boldsymbol{f}$$

to the form (3.15). Using the result of the proposition, we find that

$$\begin{pmatrix} 1 & 2 \\ 5 & 3 \end{pmatrix}^{-1} = \begin{pmatrix} -\frac{3}{7} & \frac{2}{7} \\ \frac{5}{7} & -\frac{1}{7} \end{pmatrix}.$$

* See problem 28 at the end of this section.

Next we have

$$\begin{pmatrix} -\frac{3}{7} & \frac{2}{7} \\ \frac{5}{7} & -\frac{1}{7} \end{pmatrix}\begin{pmatrix} 2 & 1 \\ 2 & 3 \end{pmatrix} = \begin{pmatrix} -\frac{2}{7} & \frac{3}{7} \\ \frac{8}{7} & \frac{2}{7} \end{pmatrix},$$

and therefore the given system is equivalent to

$$D\boldsymbol{x} = \begin{pmatrix} -\frac{2}{7} & \frac{3}{7} \\ \frac{8}{7} & \frac{2}{7} \end{pmatrix}\boldsymbol{x} + \begin{pmatrix} -\frac{3}{7}f & \frac{2}{7}g \\ \frac{5}{7}f & \frac{1}{7}g \end{pmatrix}.$$

To discover how to find the general solution of the inhomogeneous system (3.15), we shall first consider the associated homogeneous system (3.16)

$$D\boldsymbol{x} = \boldsymbol{A}\boldsymbol{x},$$

exactly as in the case of a single equation. Our study of homogeneous single equations has led us to expect that one might obtain an exponential-type solution, although it may possibly be a complex exponential. Hence let us see whether we can find a solution of (3.16) in the form

$$\boldsymbol{x} = \boldsymbol{u}e^{\lambda t} = \begin{pmatrix} ue^{\lambda t} \\ ve^{\lambda t} \end{pmatrix},$$

where $\boldsymbol{u}$ is a constant vector and λ is an unknown constant. Both $\boldsymbol{u}$ and λ may take on complex values.

Then

$$D\boldsymbol{x} = \lambda \boldsymbol{u}e^{\lambda t};$$

putting this into (3.16) yields

$$\lambda \boldsymbol{u}e^{\lambda t} = \boldsymbol{A}\boldsymbol{u}e^{\lambda t}.$$

Canceling the exponential from both sides, we have

$$\boldsymbol{A}\boldsymbol{u} = \lambda\boldsymbol{u} = \lambda\boldsymbol{I}\boldsymbol{u},$$

or

$$\boldsymbol{A}\boldsymbol{u} - \lambda\boldsymbol{I}\boldsymbol{u} = (\boldsymbol{A} - \lambda\boldsymbol{I})\boldsymbol{u} = \boldsymbol{0}.$$

This matrix equation is equivalent to the two simultaneous, homogeneous linear equations

$$\begin{aligned}(a-\lambda)u+bv&=0,\\ cu+(d-\lambda)v&=0,\end{aligned}\tag{3.19}$$

where

$$\boldsymbol{u}=\begin{pmatrix}u\\ v\end{pmatrix}.$$

According to the principles of algebra, this system has a nontrivial (i.e., nonzero) solution (u, v) only if its determinant vanishes,

$$\det(\boldsymbol{A}-\lambda\boldsymbol{I})=\begin{vmatrix}a-\lambda & b\\ c & d-\lambda\end{vmatrix}=\lambda^2-(a+d)\lambda+(ad-bc)=0.\tag{3.20}$$

We know that (3.20) always has two roots, which may be real and unequal, real and equal, or a complex conjugate pair. We shall consider these three different cases in some detail. In the meantime, the values of λ for which $(\boldsymbol{A}-\lambda\boldsymbol{I})$ is singular are called the *eigenvalues* of $\boldsymbol{A}$, and the corresponding nontrivial solutions $\boldsymbol{u}$ of (3.19) are called the corresponding *eigenvectors* of $\boldsymbol{A}$.

If the roots of (3.20), say λ_1 and λ_2, are unequal (real or complex), and if $\boldsymbol{u}_1$ and $\boldsymbol{u}_2$ are the corresponding eigenvectors, then

$$\boldsymbol{x}_1=\boldsymbol{u}_1e^{\lambda_1 t},\qquad \boldsymbol{x}_2=\boldsymbol{u}_2e^{\lambda_2 t}$$

are the two linearly independent solutions of (3.16), and hence by the first principle of superposition, the general solution is

$$\boldsymbol{x}=c_1\boldsymbol{u}_1e^{\lambda_1 t}+c_2\boldsymbol{u}_2e^{\lambda_2 t}.\tag{3.21}$$

The matrix

$$\boldsymbol{\Psi}=\begin{pmatrix}x_1 & x_2\\ y_1 & y_2\end{pmatrix}=\begin{pmatrix}u_1e^{\lambda_1 t} & u_2e^{\lambda_2 t}\\ v_1e^{\lambda_1 t} & v_2e^{\lambda_2 t}\end{pmatrix},$$

in which the columns are two linearly independent solutions of (3.16), is called the *fundamental* (*solution*) *matrix* of (3.16), and indeed, it is true that

$$D\boldsymbol{\Psi}=\boldsymbol{A}\boldsymbol{\Psi}.$$

Moreover, the general solution $\boldsymbol{x}$ can always be written as $\boldsymbol{x} = \boldsymbol{\Psi c}$, where

$$\boldsymbol{c} = \begin{pmatrix} c_1 \\ c_2 \end{pmatrix},$$

since

$$\boldsymbol{\Psi c} = \begin{pmatrix} x_1 & x_2 \\ y_1 & y_2 \end{pmatrix} \begin{pmatrix} c_1 \\ c_2 \end{pmatrix} = \begin{pmatrix} c_1 x_1 + c_2 x_2 \\ c_1 y_1 + c_2 y_2 \end{pmatrix} = c_1 \boldsymbol{x}_1 + c_2 \boldsymbol{x}_2,$$

which is simply another way of writing (3.21).

Because of the principle of superposition, $\boldsymbol{\Psi}$ can be manipulated in a number of ways to yield different fundamental matrices; for example, either column can be multiplied by a nonzero constant, and any constant times one column can be added to the other. By such manipulations we shall learn how to produce a real solution matrix even in the case when λ_1 and λ_2 are complex.

If λ_1 and λ_2 are equal, then we can obtain only one solution of the form

$$\boldsymbol{x}_1 = \boldsymbol{u}_1 e^{\lambda_1 t}. \tag{3.22}$$

In analogy with the case of a single equation, it is reasonable to expect a second solution of the form

$$\boldsymbol{x}_2 = \boldsymbol{u}_2 t e^{\lambda_1 t}. \tag{3.23}$$

However, since

$$D\boldsymbol{x}_2 = \boldsymbol{u}_2(1 + \lambda_1 t)e^{\lambda_1 t},$$

equation (3.16) yields

$$\boldsymbol{u}_2(1 + \lambda_1 t)e^{\lambda_1 t} = \boldsymbol{A u}_2 t e^{\lambda_1 t},$$

or, canceling the exponential and gathering like powers of t,

$$\boldsymbol{u}_2 + (\boldsymbol{A} - \lambda_1 \boldsymbol{I})\boldsymbol{u}_2 t = \boldsymbol{0}.$$

Since the coefficient of each power of t must vanish, this implies that $\boldsymbol{u}_2 = \boldsymbol{0}$.

Where have we gone wrong? The answer stems from the fact that the coefficient $\boldsymbol{u}_1$ in (3.22) is a vector, not a number; and hence, (3.23) might be expected to contain a term in $e^{\lambda_1 t}$ as well, with a vector coefficient $\boldsymbol{u}_3$ that is *linearly independent* of $\boldsymbol{u}_1$. Indeed, if we assume a solution of the form

$$\boldsymbol{x}_2 = (\boldsymbol{u}_3 + \boldsymbol{u}_2 t)e^{\lambda_1 t},$$

then

$$D\boldsymbol{x}_2 = [(\boldsymbol{u}_3\lambda_1 + \boldsymbol{u}_2) + \boldsymbol{u}_2\lambda_1 t]e^{\lambda_1 t}.$$

Thus (3.16) now gives us

$$(\boldsymbol{u}_3\lambda_1 + \boldsymbol{u}_2) + \boldsymbol{u}_2\lambda_1 t = \boldsymbol{A}\boldsymbol{u}_3 + \boldsymbol{A}\boldsymbol{u}_2 t,$$

after cancellation of the exponential from both sides. Thus

$$(\boldsymbol{u}_3\lambda_1 + \boldsymbol{u}_2 - \boldsymbol{A}\boldsymbol{u}_3) + (\boldsymbol{u}_2\lambda_1 - \boldsymbol{A}\boldsymbol{u}_2)t = \boldsymbol{0}.$$

The individual expressions in parentheses must vanish, and hence

$$(\boldsymbol{A} - \lambda_1\boldsymbol{I})\boldsymbol{u}_2 = \boldsymbol{0},$$
$$(\boldsymbol{A} - \lambda_1\boldsymbol{I})\boldsymbol{u}_3 = \boldsymbol{u}_2.$$

The first of these two equations states that $\boldsymbol{u}_2$ is an eigenvector of $\boldsymbol{A}$ with eigenvalue λ_1; hence $\boldsymbol{u}_2$ can be chosen equal to $\boldsymbol{u}_1$. The second equation is more troublesome because it is inhomogeneous with a vanishing determinant; it can be solved only if one of its two scalar equations is a multiple of the other (i.e., the two straight lines represented by the two equations actually coincide), and this turns out always to be the case. We shall see an example of this later.

Now let us work problems involving each of the three cases described above:

Case 1 Real, unequal eigenvalues
Let us solve

$$D\boldsymbol{x} = \begin{pmatrix} 4 & 1 \\ 5 & -2 \end{pmatrix}\boldsymbol{x}.$$

Then

$$\operatorname{Det}(\boldsymbol{A} - \lambda\boldsymbol{I}) = \begin{vmatrix} 4-\lambda & -1 \\ 5 & -2-\lambda \end{vmatrix} = (4-\lambda)(-2-\lambda) + 5 = \lambda^2 - 2\lambda - 3$$
$$= 0,$$

so that $\lambda_1 = -1$, $\lambda_2 = 3$. Hence two linearly independent solutions will be

$$\boldsymbol{x}_1 = \boldsymbol{u}_1 e^{-t}, \qquad \boldsymbol{x}_2 = \boldsymbol{u}_2 e^{3t},$$

where $\boldsymbol{u}_1$ and $\boldsymbol{u}_2$ are the eigenvectors corresponding to $\lambda_1 = -1$ and $\lambda_2 = 3$, respectively. To find these eigenvectors, we set

$$(\boldsymbol{A} + \boldsymbol{I})\boldsymbol{u}_1 = \boldsymbol{0}, \qquad (\boldsymbol{A} - 3\boldsymbol{I})\boldsymbol{u}_2 = \boldsymbol{0},$$

obtaining

$$\begin{pmatrix} 5 & -1 \\ 5 & -1 \end{pmatrix} \boldsymbol{u}_1 = \boldsymbol{0}, \qquad \begin{pmatrix} 1 & -1 \\ 5 & -5 \end{pmatrix} \boldsymbol{u}_2 = \boldsymbol{0},$$

which have many solutions, among which are

$$\boldsymbol{u}_1 = \begin{pmatrix} 1 \\ 5 \end{pmatrix}, \qquad \boldsymbol{u}_2 = \begin{pmatrix} 1 \\ 1 \end{pmatrix}.$$

Note that $\boldsymbol{u}_1$ and $\boldsymbol{u}_2$ are not unique, since any constant multiple of either one would do as well. Note also that in the two-by-two case that we are considering here there is no difficulty at all in finding $\boldsymbol{u}_1$ and $\boldsymbol{u}_2$; in fact, if our set of equations for $\boldsymbol{u}_1$ is

$$(a - \lambda_1)u_1 + bv_1 = 0,$$
$$cu_1 + (d - \lambda_1)v_1 = 0,$$

then obviously $u_1 = b$, $u_2 = -(a - \lambda_1)$ is a solution. (In the two-by-two case, the second equation is always a multiple of the first.) Thus we have

$$\boldsymbol{x}_1 = \begin{pmatrix} 1 \\ 5 \end{pmatrix} e^{-t}, \quad \boldsymbol{x}_2 = \begin{pmatrix} 1 \\ 1 \end{pmatrix} e^{3t}.$$

Therefore a fundamental matrix is

$$\boldsymbol{\Psi} = \begin{pmatrix} e^{-t} & e^{3t} \\ 5e^{-t} & e^{3t} \end{pmatrix}.$$

The general solution, if desired, can be put in the form

$$\boldsymbol{x} = c_1\boldsymbol{x}_1 = c_2\boldsymbol{x}_2 = c_1 \begin{pmatrix} 1 \\ 5 \end{pmatrix} e^{-t} + c_2 \begin{pmatrix} 1 \\ 1 \end{pmatrix} e^{3t}$$

or in scalar form as

$$x = c_1 e^{-t} + c_2 e^{-3t}, \quad y = 5c_1 e^{-t} + c_2 e^{3t}.$$

Case 2 Real, equal eigenvalues

Here we examine

$$D\boldsymbol{x} = \begin{pmatrix} 1 & 1 \\ -1 & 3 \end{pmatrix} \boldsymbol{x}.$$

In this case,

$$\text{Det}(\boldsymbol{A} - \lambda \boldsymbol{I}) = \begin{vmatrix} 1 - \lambda & 1 \\ -1 & 3 - \lambda \end{vmatrix} = (1 - \lambda)(3 - \lambda) + 1 = \lambda^2 - 4\lambda + 4 = 0,$$

so $\lambda_1 = \lambda_2 = 2$. Hence one solution of the system is of the form $\boldsymbol{x}_1 = \boldsymbol{u}_1 e^{2t}$, $\boldsymbol{u}_1$ being an eigenvector of $\boldsymbol{A}$ for $\lambda = 2$:

$$(\boldsymbol{A} - 2\boldsymbol{I})\boldsymbol{u}_1 = \begin{pmatrix} -1 & 1 \\ -1 & 1 \end{pmatrix} \boldsymbol{u}_1 = \boldsymbol{0},$$

and hence

$$\boldsymbol{u}_1 = \begin{pmatrix} 1 \\ 1 \end{pmatrix}$$

will do. A second solution is given by $\boldsymbol{x}_2 = \boldsymbol{u}_3 e^{2t} + \boldsymbol{u}_2 t e^{2t}$, where, as we showed previously, $\boldsymbol{u}_2$ can be taken equal to $\boldsymbol{u}_1$, and where $\boldsymbol{u}_3$ is a solution of

$$\begin{pmatrix} -1 & 1 \\ -1 & 1 \end{pmatrix} \boldsymbol{u}_3 = \begin{pmatrix} 1 \\ 1 \end{pmatrix};$$

thus, for example, one value for $\boldsymbol{u}_3$ is

$$\boldsymbol{u}_3 = \begin{pmatrix} 0 \\ 1 \end{pmatrix}.$$

At this point, we note that $\boldsymbol{u}_3$ is not unique; for example,

$$\hat{\boldsymbol{u}}_3 = \begin{pmatrix} -1 \\ 0 \end{pmatrix}$$

would do just as well. Does it make any difference which of these many possible values we use? The answer is no, since any two different solutions for $\boldsymbol{u}_3$ differ by a multiple of $\boldsymbol{u}_1$. Because

$$(\boldsymbol{A} - \lambda \boldsymbol{I})\boldsymbol{u}_3 = \boldsymbol{u}_2,$$

$$(\boldsymbol{A} - \lambda \boldsymbol{I})\hat{\boldsymbol{u}}_3 = \boldsymbol{u}_2,$$

we have, subtracting, that

$$(\boldsymbol{A} - \lambda \boldsymbol{I})(\boldsymbol{u}_3 - \hat{\boldsymbol{u}}_3) = \boldsymbol{0},$$

whence $\boldsymbol{u}_3 - \hat{\boldsymbol{u}}_3$ is an eigenvector of $\boldsymbol{A}$ for the eigenvalue λ, and is therefore a scalar multiple of $\boldsymbol{u}_1$. Thus the choice of a different value for $\boldsymbol{u}_3$ causes a change in $\boldsymbol{x}_2 = \boldsymbol{u}_3 e^{2t} + \boldsymbol{u}_2 t e^{2t}$ which is equivalent to adding a multiple of $\boldsymbol{x}_1 = \boldsymbol{u}_1 e^{2t}$ to it, which, by superposition, does no harm and contributes nothing new to $\boldsymbol{x}_2$.

So, since we have that

$$\boldsymbol{u}_1 = \boldsymbol{u}_2 = \begin{pmatrix} 1 \\ 1 \end{pmatrix}, \quad \boldsymbol{u}_3 = \begin{pmatrix} 0 \\ 1 \end{pmatrix},$$

therefore

$$\boldsymbol{x}_1 = \begin{pmatrix} 1 \\ 1 \end{pmatrix} e^{2t}, \quad \boldsymbol{x}_2 = \begin{pmatrix} 0 \\ 1 \end{pmatrix} e^{2t} + \begin{pmatrix} 1 \\ 1 \end{pmatrix} t e^{2t}$$

are two linearly independent solutions, and a fundamental matrix is

$$\boldsymbol{\Psi} = \begin{pmatrix} e^{2t} & te^{2t} \\ e^{2t} & (1 + t)e^{2t} \end{pmatrix}.$$

The general solution in scalar form is

$$x = c_1 e^{2t} + c_2 t e^{2t}, \quad y = (c_1 + c_2)e^{2t} + c_2 t e^{2t}.$$

Case 3 Complex conjugate eigenvalues

In the case

$$D\boldsymbol{x} = \begin{pmatrix} 4 & -1 \\ 5 & 2 \end{pmatrix} \boldsymbol{x},$$

we have

$$\operatorname{Det}(\boldsymbol{A} - \lambda \boldsymbol{I}) = \begin{vmatrix} 4-\lambda & -1 \\ 5 & 2-\lambda \end{vmatrix} = (4-\lambda)(2-\lambda) + 5 = \lambda^2 - 6\lambda + 13 = 0,$$

which has roots $\lambda_1 = 3 + 2i$, $\lambda_1 = 3 - 2i$. Hence two complex solutions are $\boldsymbol{x}_1 = \boldsymbol{u}_1 \exp[(3 + 2i)t]$, $\boldsymbol{x}_2 = \boldsymbol{u}_2 \exp[(3 - 2i)t]$, where $\boldsymbol{u}_1$ and $\boldsymbol{u}_2$ are the complex eigenvectors corresponding to λ_1 and λ_2, respectively:

$$\begin{pmatrix} 1-2i & -1 \\ 5 & -1-2i \end{pmatrix} \boldsymbol{u}_1 = \boldsymbol{0}, \quad \begin{pmatrix} 1+2i & -1 \\ 5 & -1+2i \end{pmatrix} \boldsymbol{u}_2 = \boldsymbol{0};$$

by inspection, two possible solutions are

$$\boldsymbol{u}_1 = \begin{pmatrix} 1 \\ 1-2i \end{pmatrix}, \quad \boldsymbol{u}_2 = \begin{pmatrix} 1 \\ 1+2i \end{pmatrix}.$$

Therefore two independent solutions are

$$\boldsymbol{x}_1 = \begin{pmatrix} 1 \\ 1-2i \end{pmatrix} e^{(3+2i)t}, \quad \boldsymbol{x}_2 = \begin{pmatrix} 1 \\ 1+2i \end{pmatrix} e^{(3-2i)t},$$

and a fundamental matrix is

$$\hat{\boldsymbol{\Psi}} = \begin{pmatrix} e^{(3+2i)t} & e^{(3-2i)t} \\ (1-2i)e^{(3+2i)t} & (1+2i)e^{(3-2i)t} \end{pmatrix} = e^{3t} \begin{pmatrix} e^{2it} & e^{-2it} \\ (1-2i)e^{2it} & (1+2i)e^{-2it} \end{pmatrix}.$$

We can find a fundamental matrix in real form by utilizing the fact that it remains a fundamental matrix if we replace the second column by the sum of the two columns and the first column by their difference, thus:

$$e^{3t} \begin{pmatrix} e^{2it} - e^{-2it} & e^{2it} + e^{-2it} \\ e^{2it} - e^{-2it} - 2i(e^{2it} + e^{-2it}) & e^{2it} + e^{-2it} - 2i(e^{2it} - e^{-2it}) \end{pmatrix}$$

$$= e^{3t} \begin{pmatrix} 2i \sin 2t & 2\cos 2t \\ 2i \sin 2t - 4i \cos 2t & 2\cos 2t + 4\sin 2t \end{pmatrix},$$

which we obtained by using the facts that

$$e^{\alpha it} + e^{-\alpha it} = 2\cos \alpha t, \quad e^{\alpha it} - e^{-\alpha it} = 2i \sin \alpha t.$$

Finally, we can multiply the first column by $1/2i$ and the second by $\frac{1}{2}$, and what remains is now a real fundamental matrix

$$\boldsymbol{\Psi} = e^{3t}\begin{pmatrix} \sin 2t & \cos 2t \\ \sin 2t - 2\cos 2t & \cos 2t + 2\sin 2t \end{pmatrix}.$$

In scalar form, the general solution is

$$x = c_1 \sin 2t + c_2 \cos 2t,$$
$$y = (c_1 + c_2)\sin 2t + (c_2 - 2c_1)\cos 2t.$$

And now, let us consider the problem of solving the inhomogeneous equation (3.15),

$$D\boldsymbol{x} = \boldsymbol{A}\boldsymbol{x} + \boldsymbol{f}(t),$$

assuming that we know the fundamental matrix $\boldsymbol{\Psi}$ of the associated homogeneous equation. We attempt a technique similar to the variation of parameters; namely, since $\boldsymbol{\Psi}(t)\boldsymbol{c}$ (where $\boldsymbol{c}$ is a constant vector) is the general solution of the homogeneous equation, we seek a particular solution of the inhomogeneous equation in the form $\boldsymbol{\Psi}(t)\boldsymbol{c}(t)$, where $\boldsymbol{c}(t)$ is a variable vector. Then $D(\boldsymbol{\Psi c}) = (D\boldsymbol{\Psi})\boldsymbol{c} + \boldsymbol{\Psi}D\boldsymbol{c}$, and therefore, from (3.15),

$$(D\boldsymbol{\Psi})\boldsymbol{c} + \boldsymbol{\Psi}D\boldsymbol{c} = \boldsymbol{A\Psi c} + \boldsymbol{f}.$$

Now because $\boldsymbol{\Psi c}$ is the general solution of the homogeneous equation, we have $(D\boldsymbol{\Psi})\boldsymbol{c} = \boldsymbol{A\Psi c}$, and hence

$$\boldsymbol{\Psi}D\boldsymbol{c} = \boldsymbol{f}.$$

Because $\boldsymbol{\Psi}$ is a fundamental solution matrix, its columns are linearly independent vectors and hence it is nonsingular.* Therefore $\boldsymbol{\Psi}^{-1}$ exists and

$$D\boldsymbol{c} = \boldsymbol{\Psi}^{-1}\boldsymbol{f}.$$

This represents a pair of first-order equations, one for c_1 and the other for c_2, which can be integrated to give the result.

* See problem 30 at the end of the section.

Example

Let us consider

$$Dx = \begin{pmatrix} 4 & 1 \\ 4 & -2 \end{pmatrix} x + \begin{pmatrix} e^t \\ 2 \end{pmatrix}.$$

As we saw under Case 1 previously, the fundamental matrix for this system is

$$\Psi = \begin{pmatrix} e^{-t} & e^{3t} \\ 5e^{-t} & e^{3t} \end{pmatrix}.$$

Since Det $\Psi = e^{2t} - 5e^{2t} = -4e^{2t}$, we have that

$$\Psi^{-1} = \begin{pmatrix} e^{3t}/(-4e^{2t}) & -e^{3t}/(-4e^{2t}) \\ -5e^{-t}/(-4e^{2t}) & e^{-t}/(-4e^{2t}) \end{pmatrix} = \begin{pmatrix} -\frac{1}{4}e^t & \frac{1}{4}e^t \\ \frac{5}{4}e^{-3t} & -\frac{1}{4}e^{-3t} \end{pmatrix}.$$

Therefore the differential equation for c is

$$Dc = \begin{pmatrix} Dc_1 \\ Dc_2 \end{pmatrix} = \begin{pmatrix} -\frac{1}{4}e^t & \frac{1}{4}e^t \\ \frac{5}{4}e^{-3t} & -\frac{1}{4}e^{-3t} \end{pmatrix} \begin{pmatrix} e^t \\ 2 \end{pmatrix},$$

which gives the two equations

$$Dc_1 = -\tfrac{1}{4}e^{2t} + \tfrac{1}{2}e^t,$$
$$Dc_2 = \tfrac{5}{4}e^{-2t} - \tfrac{1}{2}e^{-3t},$$

which have the solutions

$$c_1 = -\tfrac{1}{8}e^{2t} + \tfrac{1}{2}e^t$$
$$c_2 = -\tfrac{5}{8}e^{-2t} + \tfrac{1}{6}e^{-3t},$$

where we have suppressed the constants of integration to yield a particular solution. Then, since the particular solution is $x_p = \Psi(t)c(t)$, we have

$$x_p = \begin{pmatrix} e^{-t} & e^{3t} \\ 5e^{-t} & e^{3t} \end{pmatrix} \begin{pmatrix} -\frac{1}{8}e^{2t} + \frac{1}{2}e^t \\ -\frac{5}{8}e^{-2t} + \frac{1}{6}e^{-3t} \end{pmatrix} = \begin{pmatrix} -\frac{3}{4}e^t + \frac{2}{3} \\ -\frac{5}{4}e^t + \frac{8}{3} \end{pmatrix}.$$

The only remaining question which we have to answer is what happens when the linear system (3.18)

$$PDx = Ax + f$$

is degenerate, that is, when Det $\boldsymbol{P} = 0$. The answer is similar to what happens when the determinant of a system of algebraic equations is zero; there may be either no solutions or infinitely many. Let us examine two examples:

Example 1. No solutions
The system

$$Dx - Dy = t,$$
$$4Dx - 4Dy = t^2,$$

has no solutions at all, since subtracting four times the first equation from the second yields

$$t^2 - t \equiv 0,$$

which is patently untrue.

Example 2. Infinitely many solutions

$$Dx - Dy = t,$$
$$2Dx - 2Dy = 2t.$$

Then here, $D(x - y) = t$, which implies that $x - y = \frac{1}{2}t^2 + c$, and therefore $x = y + \frac{1}{2}t^2 + c$. One can choose any function whatever for y and get a corresponding solution for x.

Problems

Find a fundamental matrix for each of the following systems; in the case of complex eigenvalues, find a real fundamental matrix.

1. $D\boldsymbol{x} = \begin{pmatrix} 2 & 2 \\ -1 & -1 \end{pmatrix} \boldsymbol{x}.$

2. $D\boldsymbol{x} = \begin{pmatrix} 5 & -3 \\ 4 & -2 \end{pmatrix} \boldsymbol{x}.$

3. $D\boldsymbol{x} = \begin{pmatrix} -2 & 3 \\ -1 & 2 \end{pmatrix} \boldsymbol{x}.$

4. $D\boldsymbol{x} = \begin{pmatrix} -1 & 2 \\ -2 & -6 \end{pmatrix} \boldsymbol{x}.$

5. $D\boldsymbol{x} = \begin{pmatrix} -1 & 5 \\ -4 & 8 \end{pmatrix}\boldsymbol{x}.$

6. $D\boldsymbol{x} = \begin{pmatrix} -3 & 4 \\ -1 & 1 \end{pmatrix}\boldsymbol{x}.$

7. $D\boldsymbol{x} = \begin{pmatrix} 0 & 1 \\ -1 & -2 \end{pmatrix}\boldsymbol{x}.$

8. $D\boldsymbol{x} = \begin{pmatrix} -1 & 1 \\ -1 & -3 \end{pmatrix}\boldsymbol{x}.$

9. $D\boldsymbol{x} = \begin{pmatrix} -1 & 2 \\ -2 & 3 \end{pmatrix}\boldsymbol{x}.$

10. $D\boldsymbol{x} = \begin{pmatrix} -2 & -1 \\ 1 & -4 \end{pmatrix}\boldsymbol{x}.$

11. $D\boldsymbol{x} = \begin{pmatrix} 0 & 1 \\ -1 & 0 \end{pmatrix}\boldsymbol{x}.$

12. $D\boldsymbol{x} = \begin{pmatrix} 1 & 5 \\ -1 & -1 \end{pmatrix}\boldsymbol{x}.$

13. $D\boldsymbol{x} = \begin{pmatrix} -1 & -5 \\ 2 & 1 \end{pmatrix}\boldsymbol{x}.$

14. $D\boldsymbol{x} = \begin{pmatrix} 1 & 1 \\ -1 & 1 \end{pmatrix}\boldsymbol{x}.$

15. $D\boldsymbol{x} = \begin{pmatrix} 1 & -2 \\ 1 & 3 \end{pmatrix}\boldsymbol{x}.$

16. $\begin{pmatrix} 2 & 3 \\ 1 & 2 \end{pmatrix}D\boldsymbol{x} = \begin{pmatrix} 5 & 4 \\ 3 & 2 \end{pmatrix}\boldsymbol{x}.$

17. $\begin{pmatrix} 1 & 0 \\ 1 & 3 \end{pmatrix}D\boldsymbol{x} = \begin{pmatrix} -4 & 2 \\ -7 & -1 \end{pmatrix}\boldsymbol{x}.$

18. $\begin{pmatrix} 1 & -1 \\ 0 & 1 \end{pmatrix}D\boldsymbol{x} = \begin{pmatrix} 2 & 6 \\ -2 & -4 \end{pmatrix}\boldsymbol{x}.$

19. $\begin{pmatrix} 2 & -1 \\ -1 & 3 \end{pmatrix}D\boldsymbol{x} = \begin{pmatrix} -1 & 3 \\ -2 & 1 \end{pmatrix}\boldsymbol{x}.$

20. $\begin{pmatrix} 7 & 4 \\ 2 & 5 \end{pmatrix}D\boldsymbol{x} = \begin{pmatrix} 5 & 10 \\ 14 & -1 \end{pmatrix}\boldsymbol{x}.$

Find particular solutions of each of the following:

21. $Dx = \begin{pmatrix} 2 & 2 \\ -1 & -1 \end{pmatrix} x + \begin{pmatrix} -6 - 3e^{-t} \\ 3 + e^{-t} \end{pmatrix}.$

22. $Dx = \begin{pmatrix} -2 & 3 \\ -1 & 2 \end{pmatrix} x + \begin{pmatrix} -1 \\ 0 \end{pmatrix}.$

23. $Dx = \begin{pmatrix} -1 & 5 \\ -4 & 8 \end{pmatrix} x + \begin{pmatrix} 2e^{t} + 5e^{-t} \\ 4e^{t} - 9e^{-t} \end{pmatrix}.$

24. $Dx = \begin{pmatrix} 0 & 1 \\ -1 & 0 \end{pmatrix} x + \begin{pmatrix} -3 \sin 2t \\ 3 \cos 2t \end{pmatrix}.$

25. $Dx = \begin{pmatrix} 1 & 1 \\ -1 & 1 \end{pmatrix} x + \begin{pmatrix} 1 - t - t^2 \\ 3t - t^2 \end{pmatrix}.$

26. Prove the principles of superposition for first-order systems.

27. Prove the formula for the inverse of a nonsingular matrix.

28. Prove that matrix multiplication is associative; i.e., if A, B, and C are three matrices, then $A(BC) = (AB)C$.

29. Is matrix multiplication commutative; i.e., is it always true that $AB = BA$?

30. Prove that, if the columns of a matrix are linearly independent vectors, then the matrix is nonsingular, and conversely.

4

Solving Linear Equations with Constant Coefficients (Laplace Transform Methods)

We have just seen one method for reducing the solution of linear equations with constant coefficients to a purely algebraic process. In this chapter we shall consider another method which uses a device called the Laplace transform. The basic purpose of the Laplace transform is to alter the equation under consideration in such a way that the operations of calculus are reduced to algebraic operations: differentiation to multiplication, and integration to division. Then, once we have solved the resulting algebraic problem, we reverse the transformation to arrive at the solution of the original problem.

4.1 Properties of the Laplace Transform; the Gamma Function

Let us ponder for a moment the problem which we are trying to solve. We shall be working with the initial value problem

$$P(D)y = (a_n D^n + a_{n-1} D^{n-1} + \cdots + a_1 D + a_0)y = f(t),$$
$$y(0) = y_0, \quad Dy(0) = y_1, \ldots, D^{n-1}y(0) = y_{n-1},$$

using t as the independent variable to emphasize that in an initial value problem we are often concerned with time. We shall for this reason restrict ourselves to functions f which are defined for all t, but which are zero for $t < 0$. We shall develop a method for transforming the differential equation $P(D)y = f(t)$ into a new equation in a new variable s in which it will become an algebraic equation, which we can then easily solve.

Given a function f of the variable t defined for all t but vanishing for $t < 0$, then, we wish to perform an operation $\mathscr{L}$ on $f(t)$ which will produce a new function $f^*(s)$; one way we can do this is to multiply $f(t)$ by some function $k(s, t)$ and integrate over t between two limits a and b:

$$\mathscr{L}\{f\} = f^*(s) = \int_a^b f(t)k(s, t)\, dt.$$

Since we have integrated over t, the one remaining variable is s. So far, however, this is very indefinite. We do not know what $k(s, t)$ is or what the values of the limits of integration a and b should be. Let us see whether we can determine reasonable possibilities.

One thing we wish to do is to transform the calculus operation D on f into an algebraic operation on $f^*(s)$. What operation is more natural than multiplication by s? Thus, in some way, $\mathscr{L}\{Df\}$ should give us $sf^*(s)$ plus perhaps a known function $F(s)$. By examining our formula, we see that

$$\mathscr{L}\{Df\} = \int_a^b Df(t)k(s, t)\, dt;$$

let us integrate by parts:

$$\mathscr{L}\{Df\} = -\int_a^b f(t)\frac{\partial k(s, t)}{\partial t}\, dt + f(t)k(s, t)\Big|_a^b. \tag{4.1}$$

The integrated term will give us a function of the parameter s, to which we shall return in a minute; in the meantime, we have asked that

$$\mathscr{L}\{Df\} = sf^*(s) + F(s) = \int_a^b f(t)sk(s, t)\, dt + F(s). \tag{4.2}$$

We want the integrals in (4.1) and (4.2) to be equal; i.e.,

$$\int_a^b f(t)sk(s, t)\, dt = -\int_a^b f(t)\frac{\partial k(s, t)}{\partial t}\, dt,$$

which will certainly be true if

$$-\frac{\partial k(s, t)}{\partial t} = sk(s, t).$$

This is a first-order differential equation for $k(s, t)$ in the independent variable t, if we look at s as being a constant parameter in the equation. It can be solved by separation of variables:

$$\frac{\partial k(s, t)/\partial t}{k(s, t)} = -s,$$

so $\ln(k(s, t)) = -st$, and hence $k(s, t) = e^{-st}$ is one possibility which will work as desired. Let us use this in our expressions and see what we have:

$$\mathscr{L}\{f\} = f^*(s) = \int_a^b e^{-st} f(t)\, dt;$$

$$\mathscr{L}\{Df\} = sf^*(s) + e^{-st} f(t)\Big|_a^b .$$

In the second of the above expressions, if we were to choose $b = \infty$, then the value of $e^{-st} f(t)$ at the upper limit would vanish if $f(t)$ were a function such that

$$\lim_{t \to \infty} e^{-st} f(t) = 0$$

for at least some value of s. And, since the initial value is given for $t = 0$, it makes sense to set the lower limit at $a = 0$. Then the term

$$e^{-st}\Big|_a^b$$

becomes simply $-f(0)$, the negative of the known initial value of $f(t)$. If we do this, we immediately have the following definition and theorem:

Definition

Let $f(t)$ be a given function defined for all t and vanishing for $t < 0$, such that the improper integral

$$\mathscr{L}\{f\} = f^*(s) = \int_0^\infty e^{-st} f(t)\, dt$$

converges; then $\mathscr{L}\{f\} = f^*(s)$ is called the *Laplace transform* of $f(t)$.

Theorem 4.1

If $f(t)$ is a differentiable function with the above properties, and if

$$\lim_{t \to \infty} e^{-st} f(t) = 0,$$

then $\mathscr{L}\{Df\}$ exists and equals $sf^*(s) - f(0)$.

We have not yet specified exactly what properties $f(t)$ must have in order for its Laplace transform to exist. One rather large class of functions for which it does exist is the collection defined as follows:

Definition
A function $f(t)$ defined for all t is said to be *piecewise continuous* if it is continuous everywhere except for a finite number of discontinuities (Figure 4.1);

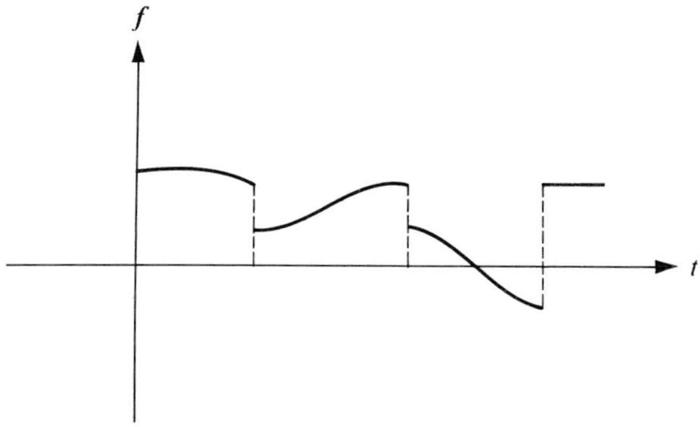

Figure 4.1 Graph of a piecewise continuous function.

it is said to be of *exponential order* if there exist two constants A and α with

$$|f(t)| \leqq Ae^{\alpha t}$$

for all $t \geqq 0$. (Remember, we are assuming f is zero for negative t.)
Then we immediately have

Theorem 4.2
Every piecewise continuous function f which is of exponential order has a Laplace transform.
Proof
According to the calculus, for each h, the integral

$$\int_0^h e^{-st} f(t)\, dt$$

exists and is finite; the Laplace transform $f^*(s)$ (if it exists) will be the limit of this expression as $h \to \infty$. Now again, according to the calculus,*

* See the Appendix.

$$\left| \int_0^h e^{-st} f(t)\, dt \right| \leqq \int_0^h e^{-st} |f(t)|\, dt \leqq \int_0^h e^{-st} A e^{\alpha t}\, dt = A \int_0^h e^{-(s-\alpha)t}\, dt$$

$$= -\frac{A}{s-\alpha} e^{-(s-\alpha)t} \Big|_0^h = -\frac{A}{s-\alpha}(e^{-(s-\alpha)h} - 1).$$

This last expression has a finite limit as $h \to \infty$ provided that $s > \alpha$, which proves that the integral

$$f^*(s) = \int_0^\infty e^{-st} f(t)\, dt$$

exists and is finite, completing the proof.

Virtually every function of interest to us in this course is of exponential order. Naturally, every exponential function e^{mt} is of exponential order, since of course we can take $A = 1$ and $\alpha = m$. Furthermore, every power of t, every sine and cosine function, and various combinations of these, are also of exponential order. For instance, $|t| \leqq e^t$, as can be seen from the graphs of the functions (Figure 4.2), and so t is of exponential order. From the graph

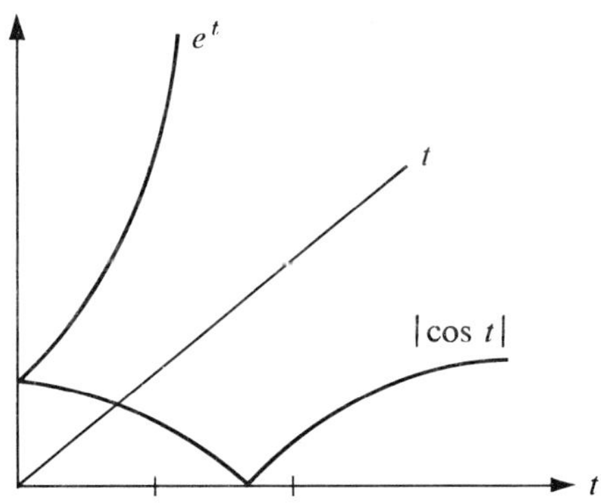

Figure 4.2

we can also see that $|\cos t| \leqq e^t$. Another interesting result is that if $f(t)$ is of exponential order, then so is $tf(t)$. For, we know that $|f(t)| \leqq Ae^{\alpha t}$ and $|t| \leqq e^t$; so, multiplying these two inequalities together, we obtain $|tf(t)| \leqq Ae^{(\alpha+1)t}$. Thus $tf(t)$ is of exponential order with the constants A and $\alpha + 1$. Likewise, by induction, so is $t^n f(t)$.

We have now successfully devised a means of transforming piecewise continuous functions of exponential order in the variable t into functions of the variable s so that differentiation with respect to t corresponds to an algebraic operation with respect to s. Let us now proceed to find the Laplace transform of the second derivative of f, assuming that D^2f exists:

$$\mathscr{L}\{D^2f\} = \mathscr{L}\{D(Df)\} = s\mathscr{L}\{Df\} - Df(0),$$

treating $Df(t)$ as a new function, $Df(0)$ being its known initial value; then, applying the theorem to the right side, we have

$$\mathscr{L}\{D^2f\} = s^2f^*(s) - sf(0) - Df(0). \tag{4.3}$$

A pattern is beginning to emerge! We might suspect that

$$\mathscr{L}\{D^nf\} = s^nf^*(s) - s^{n-1}f(0) - s^{n-2}Df(0) - \cdots - D^{n-1}f(0). \tag{4.4}$$

Let us prove (4.4) by induction. Assuming its truth for $n = k$,

$$\mathscr{L}\{D^kf\} = s^kf^*(s) - s^{k-1}f(0) - s^{k-2}Df(0) - \cdots - D^{k-1}f(0),$$

we attempt from this to prove it for $n = k + 1$:

$$\begin{aligned}\mathscr{L}\{D^{k+1}f\} &= \mathscr{L}\{D(D^kf)\} = s\mathscr{L}\{D^kf\} - D^kf(0)\\ &= s[s^kf^*(s) - s^{k-1}f(0) - s^{k-2}Df(0) - \cdots - D^{k-1}f(0)] - D^kf(0)\\ &= s^{k+1}f^*(s) - s^kf(0) - s^{k-1}Df(0) - \cdots - sD^{k-1}f(0) - D^kf(0),\end{aligned}$$

which completes the induction proof. We have thus verified that operator "multiplication" of f by the operator D^n with respect to the variable t is equivalent to actual algebraic multiplication of f^* by s^n and subtraction of a known polynomial of degree $n - 1$ in the new variable s.

Furthermore, Laplace transforms preserve sums; if a and b are constants, and $f(t)$ and $g(t)$ are two functions which have Laplace transforms, then

$$\begin{aligned}\mathscr{L}\{af(t) + bg(t)\} &= \int_0^\infty e^{-st}[af(t) + bg(t)]\,dt\\ &= a\int_0^\infty e^{-st}f(t)\,dt + b\int_0^\infty e^{-st}g(t)\,dt = a\mathscr{L}\{f\} + b\mathscr{L}\{g\}.\end{aligned}$$

Linearity of the Laplace transform is a direct consequence of the linearity of the integral. It is clear from this that Laplace transformation of a linear differential equation with constant coefficients is an algebraic equation for $f^*(s)$, which can easily be solved for $f^*(s)$. If we can then reverse the process and go back to the variable t, we shall have solved the differential equation by purely algebraic means.

Before we can do this, we shall have to compute the Laplace transforms of a few basic functions—the ones which have proved so important up to now: powers of t, exponentials, and sines, and cosines. We begin with the simplest power of t, namely $t^0 = 1$:

$$\mathscr{L}\{1\} = \int_0^\infty e^{-st} \cdot 1 \, dt = -\frac{1}{s} e^{-st} \Big|_0^\infty = \frac{1}{s},$$

provided that $s > 0$ so that the integral converges; thus

$$\mathscr{L}\{1\} = \frac{1}{s} \qquad \text{for } s > 0. \tag{4.5}$$

It turns out that we can determine the Laplace transform of all integral powers of t from (4.5) by a simple device. In fact, suppose that $f(t)$ is *any* piecewise continuous function of exponential order. Since

$$f^*(s) = \int_0^\infty e^{-st} f(t)\, dt,$$

we can differentiate both sides *with respect to* s and obtain

$$D_s f^*(s) = D_s \int_0^\infty e^{-st} f(t)\, dt = \int_0^\infty D_s(e^{-st}) f(t)\, dt = \int_0^\infty (-t) e^{-st} f(t)\, dt.^*$$

We can interchange the order of differentiation and integration at the second equals sign because of the absolute convergence of the integral (see the Appendix). The final integral converges, because, as we saw earlier, $tf(t)$ is also of exponential order. It is an easy matter to continue the process to find that

$$D_s^n f^*(s) = (-1)^n \int_0^\infty t^n e^{-st} f(t)\, dt;$$

the proof is by induction, and is left to the reader as an exercise. This formula can also be written in the form

$$\mathscr{L}\{t^n f(t)\} = (-1)^n D_s^n f^*(s).$$

* The notation D_s simply means "the derivative with respect to s of what follows."

This is a highly interesting fact! It tells us that nth-order differentiation with respect to the s variable corresponds to multiplication by $(-t)^n$ in the t variable. Thus we have an unusual situation: calculus in t corresponds to algebra in s, and conversely, calculus in s corresponds to much the same sort of algebra in t!

Now, in applying this to the case at hand where $f(t) = 1$, we have

$$\mathscr{L}\{t^n \cdot 1\} = (-1)^n D_s^n \mathscr{L}\{1\} = (-1)^n D_s^n(1/s).$$

Since

$$D_s^n(1/s) = (-1)^n n!/s^{n+1},$$

we have immediately that

$$\mathscr{L}\{t^n\} = \frac{n!}{s^{n+1}} \qquad \text{for } s > 0. \tag{4.6}$$

Another device enables us to determine another large class of transforms, those involving exponentials and powers of t. If, for some function of t, we have computed

$$\mathscr{L}\{f\} = f^*(s) = \int_0^\infty e^{-st} f(t)\, dt,$$

then by simple algebra,

$$\mathscr{L}\{e^{at}f\} = \int_0^\infty e^{-st} e^{at} f(t)\, dt = \int_0^\infty e^{-(s-a)t} f(t)\, dt = f^*(s-a).$$

Thus, in the case at hand, if $f(t) = t^n$, we have

$$\mathscr{L}\{t^n e^{at}\} = \frac{n!}{(s-a)^{n+1}} \qquad \text{for } s > a. \tag{4.7}$$

In particular, for $n = 0$, we have

$$\mathscr{L}\{e^{at}\} = \frac{1}{s-a} \qquad \text{for } s > a. \tag{4.8}$$

This last formula holds also for complex values of s and a provided that the real part of $s - a$ is positive. All of these results could as well have been obtained directly from the definition by integration. However, this would have been considerably more work; and mathematicians are the laziest people on earth when it comes to integration. Any devices which can mitigate this onerous task are eagerly seized upon.

We shall have computed the Laplace transforms of all "interesting" functions once we have computed the transforms of the trigonometric functions $\cos \omega t$ and $\sin \omega t$. Let us proceed first with the latter; by definition

$$\mathscr{L}\{\sin \omega t\} = \int_0^\infty e^{-st} \sin \omega t \, dt,$$

an integral which obviously converges for $s > 0$. However, we shall not actually have to perform the integration to evaluate this Laplace transform; we simply use the formula for the transform of the second derivative

$$\mathscr{L}\{D^2 f\} = s^2 \mathscr{L}\{f\} - sf(0) - Df(0),$$

and choose our function f carefully: because

$$\sin \omega t - D^2\left(-\frac{1}{\omega^2}\sin \omega t\right),$$

we have

$$\mathscr{L}\left\{D^2\left(-\frac{1}{\omega^2}\sin \omega t\right)\right\} = s^2 \mathscr{L}\left\{-\frac{1}{\omega^2}\sin \omega t\right\} - s\left(-\frac{1}{\omega^2}\sin \omega t\right)\bigg|_{t=0} - D\left(-\frac{1}{\omega^2}\sin \omega t\right)_{t=0},$$

or

$$\mathscr{L}\{\sin \omega t\} = -\frac{s^2}{\omega^2}\mathscr{L}\{\sin \omega t\} + \frac{1}{\omega}.$$

The factor $-1/\omega^2$ can be brought outside the $\mathscr{L}$ sign because of the linearity of the Laplace transform, which we proved earlier. We can now solve this equation for $\mathscr{L}\{\sin \omega t\}$:

$$\left(1 + \frac{s^2}{\omega^2}\right)\mathscr{L}\{\sin \omega t\} = \frac{\omega^2 + s^2}{\omega^2}\mathscr{L}\{\sin \omega t\} = \frac{1}{\omega},$$

and therefore

$$\mathscr{L}\{\sin \omega t\} = \frac{\omega}{s^2 + \omega^2} \qquad \text{for } s > 0. \tag{4.9}$$

A very similar calculation now gives

$$\mathscr{L}\{\cos \omega t\} = \frac{s}{s^2 + \omega^2} \qquad \text{for } s > 0. \tag{4.10}$$

We also have immediately from our previous results that

$$\begin{aligned} \mathscr{L}\{e^{at} \sin \omega t\} &= \frac{\omega}{(s-a)^2 + \omega^2} \qquad \text{for } s > a, \\ \mathscr{L}\{e^{at} \cos \omega t\} &= \frac{s-a}{(s-a)^2 + \omega^2} \qquad \text{for } s > a. \end{aligned} \tag{4.11}$$

We can also compute the Laplace transforms of powers of t times the trigonometric functions, an exercise which is left to the reader.

One could also have calculated the Laplace transform of $\cos \omega t$ and $\sin \omega t$ by first finding the transform of $e^{i\omega t}$, and then taking its real and imaginary parts, respectively. Namely, since

$$\mathscr{L}\{e^{i\omega t}\} = \frac{1}{s - i\omega} = \frac{s + i\omega}{s^2 + \omega^2},$$

we immediately have

$$\mathscr{L}\{\cos \omega t\} = \frac{s}{s^2 + \omega^2}, \quad \mathscr{L}\{\sin \omega t\} = \frac{\omega}{s^2 + \omega^2}.$$

Once again, the complex exponential form is admirable in its simplicity and contains in one simple statement all the information in both of the above real formulas.

The reader should note that we have obtained the Laplace transforms of all "interesting" functions, namely those of the form $t^n e^{at} \cos \omega t$ and $t^n e^{at} \sin \omega t$ (for $n = 0, 1, 2, \ldots$, and for any real values of a and ω, including zero); and, although the Laplace transform was defined by means of an integral, we integrated only *once* to find the transforms of all these functions, namely in the calculation of $\mathscr{L}\{1\}$. All of the others were found by using the basic properties of the transform.

Thus, we now have all the necessary information to transform linear differential equations with constant coefficients into algebraic equations in the variable s. However, before we leave this section, there remains one further topic to be discussed. This arises when we attempt to find the Laplace transform of a general power of t, say t^α, where α is no longer restricted to integer values. Since $f(t) = t^\alpha$ is now no longer (in general) a single-valued function, we restrict ourselves to considering only the positive real value of t^α. We have

$$\mathscr{L}\{t^\alpha\} = \int_0^\infty e^{-st} t^\alpha \, dt;$$

the parameter s can be removed from underneath the integral sign by making the substitution $st = v$, so that $dt = dv/s$. The limits of integration remain the same, and we have

$$\mathscr{L}\{t^\alpha\} = \frac{1}{s^{\alpha+1}} \int_0^\infty e^{-v} v^\alpha \, dv.$$

Thus, when the number α is such that the integral converges, the Laplace transform of t^α is the reciprocal of $s^{\alpha+1}$ times the constant value of this integral. We also know that when α is an integer n, we have

$$\int_0^\infty e^{-v} v^n \, dv = n!$$

(This is just the Laplace transform of v^n evaluated at $s = 1$.) This integral, which assumes the values of the factorials when α is an integer, will bear further study. Let us first determine the values of α for which the integral converges. We shall have to test its behavior at the lower limit as well as at the upper limit, since $\alpha < 0$ would give us an improper integral at $v = 0$ as well. Thus we shall have to consider

$$\lim_{\substack{\delta \to 0 \\ \Delta \to \infty}} \int_\delta^\Delta e^{-v} v^\alpha \, dv = \lim_{\delta \to 0} \int_\delta^1 e^{-v} v^\alpha \, dv + \lim_{\Delta \to \infty} \int_1^\Delta e^{-v} v^\alpha \, dv, \text{ assuming } \delta < 1 < \Delta.$$

For the first integral, we have

$$\int_\delta^1 e^{-v} v^\alpha \, dv \le \int_\delta^1 v^\alpha \, dv = \left. \frac{v^{\alpha+1}}{\alpha + 1} \right|_\delta^1 = \frac{2}{\alpha + 1} - \frac{\delta^{\alpha+1}}{\alpha + 1} \qquad \text{if } \alpha \neq -1,$$

$$\int_\delta^1 e^{-v} v^\alpha \, dv \le \int_\delta^1 v^{-1} \, dv = \ln v \Big|_\delta^1 = -\ln \delta \qquad \text{if } \alpha = -1.$$

Clearly when α is -1 or less, the integral does not converge at the lower limit, so we must limit ourselves to $\alpha > -1$. At the upper limit, there will be no difficulty, since we know from the calculus that e^{-v} overpowers any power of v as $v \to \infty$. Hence the integral

$$\int_0^\infty e^{-v} v^\alpha \, dv$$

converges to a definite number for all $\alpha > -1$, and, in particular, to $n!$ if $\alpha = n$. Thus for $\alpha > -1$ this integral is a well-defined function of α which provides a continuous interpolation between the factorials.* This function is called the *gamma function*, which is written

$$\Gamma(\alpha) = \int_0^\infty e^{-v} v^{\alpha - 1} \, dv \qquad \text{for } \alpha > 0. \tag{4.12}$$

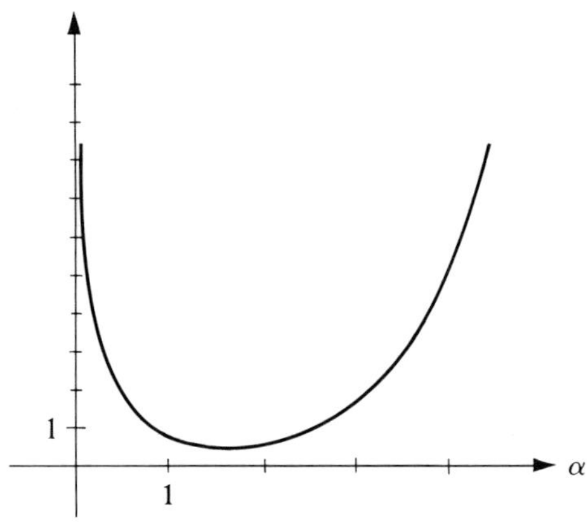

Figure 4.3

(The graph of the gamma function is illustrated in Figure 4.3.) The exponent $\alpha - 1$ is used in (4.12) for historical reasons; our integral

$$\int_0^\infty e^{-v} v^\alpha \, dv$$

is thus $\Gamma(\alpha + 1)$, and hence

$$\mathscr{L}\{t^\alpha\} = \frac{\Gamma(\alpha + 1)}{s^{\alpha + 1}} \qquad \text{for } s > 0 \text{ and } \alpha > -1. \tag{4.13}$$

* We have not proved that this is a *continuous* function of α. This is left to the reader in the problems.

Formula (4.13) is perfectly general and is valid for all α, integer or not, since $\Gamma(n + 1) = n!$

The analogy between the gamma function and the factorial is strengthened by another interesting property. If we write

$$\Gamma(\alpha + 1) = \int_0^\infty e^{-v}v^\alpha \, dv$$

and integrate, by parts, we have

$$\Gamma(\alpha + 1) = -v^\alpha e^{-v}\Big|_0^\infty - \int_0^\infty (-e^{-v})\alpha v^{\alpha-1} \, dv$$

$$= \alpha \int_0^\infty e^{-v}v^{\alpha-1} \, dv = \alpha\Gamma(\alpha)$$

for $\alpha > 0$, so that

$$\Gamma(\alpha + 1) = \alpha\Gamma(\alpha). \tag{4.14}$$

This equation states a sort of generalized factorial property for the gamma function, which holds for any $\alpha > 0$. If we replace α by $\alpha + 1$ in (4.14), we obtain

$$\Gamma(\alpha + 2) = (\alpha + 1)\Gamma(\alpha + 1) = \alpha(\alpha + 1)\Gamma(\alpha).$$

The general formula

$$\Gamma(\alpha + n) = \alpha(\alpha + 1)\cdots(\alpha + n - 1)\Gamma(\alpha) \tag{4.15}$$

can be verified easily by induction. It is instructive to note that if we know the values of the gamma function for $0 < \alpha \leq 1$, then we can obtain the value of Γ at any other value of α by means of formula (4.15). Indeed, in books of mathematical tables, one almost always finds the gamma function tabulated only for $0 < \alpha \leq 1$, since any further values can easily be calculated from (4.15), and tabulating them would be a waste of time.

Generally, values of the gamma function must be calculated using numerical methods. However, it is possible to calculate $\Gamma(\frac{1}{2})$ explicitly, and this we shall do.

$$\Gamma(\tfrac{1}{2}) = \int_0^\infty e^{-v}v^{-\frac{1}{2}} \, dv.$$

We make the change of variables $t^2 = v$, so that $dv = 2t\,dt = 2\sqrt{v}\,dt$, and we have simply that

$$\Gamma(\tfrac{1}{2}) = 2\int_0^\infty e^{-t^2}\,dt.$$

The integral $\int_0^\infty e^{-t^2}\,dt$ is a standard one which can be found in a book of tables or evaluated by means of a simple substitution (see problem 13 at the end of this section); its value is $\sqrt{\pi}/2$. Thus $\Gamma(\frac{1}{2}) = \sqrt{\pi}$. Using (4.15), we can now obtain that

$$\Gamma(\tfrac{3}{2}) = \tfrac{1}{2}\sqrt{\pi};\quad \Gamma(\tfrac{5}{2}) = \tfrac{3}{4}\sqrt{\pi};\quad \Gamma(\tfrac{7}{2}) = \tfrac{15}{8}\sqrt{\pi};$$

and so on.

Our study of the gamma function may have seemed a bit of a diversion. However, we shall meet it again later.

Problems

Find Laplace transforms of the following functions directly, by integration. Use Table 4.1 to check your results.

1. $f(t) = 3t^2$.
2. $f(t) = \cos 4t$.
3. $f(t) = te^{-t}$.
4. $f(t) = 2\cos^2 t$.
5. $f(t) = t^2 + 1$.
6. $f(t) = t \sin 3(t + 2)$.
7. $f(t) = e^{2t} - 3\sin t + t$.
8. $f(t) = (1 + 2t - 3t^2)e^t$.
9. $f(t) = (t + 1)\sin 2t - \cos(t - 1)$.
10. $f(t) = 3t^3 - 2t + 1$.
11. Complete the induction proof that $D_s^n f^*(s) = (-1)^n\int_0^\infty t^n e^{-st} f(t)\,dt$.
12. Complete the induction proof that $\Gamma(\alpha + n) = \alpha(\alpha + 1)\cdots(\alpha + n - 1)\Gamma(\alpha)$.
13. Evaluate $I = \int_0^\infty e^{-x^2}\,dx$ by multiplying it by $\int_0^\infty e^{-y^2}\,dy$ (which has the same value, of course), then interpreting the product $I^2 = \int_0^\infty e^{-x^2}\,dx \int_0^\infty e^{-y^2}\,dy$ as a double integral extended over the first quadrant; evaluate I^2 and hence I by changing to polar coordinates.

Evaluate the following expressions:

14. $\Gamma(7)/2\Gamma(3)$.
15. $3\Gamma(\frac{7}{3})/\Gamma(\frac{1}{3})$.

Table 4.1 Short Table of Laplace Transforms*

Function	Transform
1	$\frac{1}{s}, s > 0$
t	$\frac{1}{s^2}, s > 0$
$t^n (n \geqq 0, \text{integer})$	$\frac{n!}{s^{n+1}}, s > 0$
e^{at}	$\frac{1}{s - a}, s > a$
te^{at}	$\frac{1}{(s - a)^2}, s > a$
$t^n e^{at} (n \geqq 0, \text{integer})$	$\frac{1}{(s - a)^{n+1}}, s > a$
$\cos \omega t$	$\frac{s}{s^2 + \omega^2}, s > 0$
$\sin \omega t$	$\frac{\omega}{s^2 + \omega^2}, s > 0$
$e^{at} \cos \omega t$	$\frac{(s - a)}{(s - a)^2 + \omega^2}, s > a$
$e^{at} \sin \omega t$	$\frac{\omega}{(s - a)^2 + \omega^2}, s > a$
$t^n \cos \omega t (n \geqq 0, \text{integer})$	$(-1)^n D_s^n \left(\frac{s}{s^2 + \omega^2} \right), s > 0$
$t^n \sin \omega t (n \geqq 0, \text{integer})$	$(-1)^n D_s^n \left(\frac{\omega}{s^2 + \omega^2} \right), s > 0$
$t^n e^{at} \cos \omega t (n \geqq 0, \text{integer})$	$(-1)^n D_s^n \left[\frac{s - a}{(s - a)^2 + \omega^2} \right], s > a$
$t^n e^{at} \sin \omega t (n \geqq 0, \text{integer})$	$(-1)^n D_s^n \left[\frac{\omega}{(s - a)^2 + \omega^2} \right], s > a$
$\delta(t - \alpha)$	$e^{-\alpha s}$
$\chi_\alpha(t)$	$e^{-\alpha s}/s$

* More extensive tables of Laplace transforms are available, for instance, in recent editions of *Standard Mathematical Tables*, 19th ed., edited by Samuel M. Selby (The Chemical Rubber Company, Cleveland, Ohio, 1971). See also the Bibliography.

Table 4.2 Table of Formulas for Laplace Transforms

Function	Transform
$f(t)$	$f^*(s)$
$Df(t)$	$sf^*(s) - f(0)$
$D^2f(t)$	$s^2f^*(s) - sf(0) - Df(0)$
$D^nf(t)$	$s^nf^*(s) - s^{n-1}f(0) - s^{n-2}Df(0) - \cdots - D^{n-1}f(0)$
$e^{at}f(t)$	$f^*(s-a)$
$\chi_\alpha(t)f(t-\alpha)$	$e^{-\alpha s}f^*(s)$
$t^nf(t)$	$(-1)^nD_s^nf^*(s)$
$af(t) + bg(t)$	$af^*(s) + bg^*(s)$
$t^ne^{at}f(t)$	$(-1)^nD_s^nf^*(s-a)$

16. $\Gamma(\frac{9}{2})$.

17. $\Gamma(4)\Gamma(\frac{3}{2})/\Gamma(\frac{9}{2})$.

18. Assuming that $\Gamma(\alpha)$ can be defined for negative values of α (except negative integers) by means of the formulas

$$\Gamma(\alpha - 1) = \Gamma(\alpha)/(\alpha - 1)$$
$$\Gamma(\alpha - 2) = \Gamma(\alpha)/(\alpha - 1)(\alpha - 2),$$

etc., evaluate $\Gamma(-\frac{1}{2})$ and $\Gamma(-\frac{3}{2})$.

19. Evaluate $\int_0^\infty x^n e^{-cx^m}\,dx$, where n, m, and c are positive, by means of the substitution $y = cx^m$.

20. Evaluate $\int_0^1 x^n(\log x)^m$, where m is a positive integer and $n > -1$, by means of the substitution $x = e^{-y}$ and then applying the result of problem 19.

21. Using the formula $D_s^n f^*(s) = (-1)^n\int_0^\infty t^n e^{-st}f(t)\,dt$, find the Laplace transforms of $t\cos\omega t$, $t\sin\omega t$, $te^{at}\cos\omega t$, and $te^{at}\sin\omega t$.

22. Do problem 21 by the complex exponential method.

23. Prove that the following functions are of exponential order: t^n (for any integer $n > 0$), t^ne^{mt}, $t^n\cos\omega t$, $t^n\sin\omega t$, $t^ne^{mt}\cos\omega t$, $t^ne^{mt}\sin\omega t$.

24. Prove that e^{t^2} is not of exponential order.

4.2 Inverting the Laplace Transforms

In the previous section we have seen that it is a relatively straightforward, cut-and-dried procedure to determine the Laplace transform of a given function. The inverse problem, however, is much more difficult: given a function $f^*(s)$, how do we find the function $f(t)$ for which $\mathscr{L}\{f(t)\} = f^*(s)$? Two ancillary questions also present themselves at the same time: first, how do we know that $f^*(s)$ is the Laplace transform of anything at all; and second, if $f^*(s)$ is the Laplace transform of $f(t)$, is there some other function $g(t)$ so that $f^*(s)$ is also the Laplace transform of $g(t)$? Unfortunately, the answers to these questions are by no means trivial to determine, and we would have to go far afield in order to do so. We do not have the time to do this, so we shall instead give the answers insofar as they apply to a course at this level, and then proceed about our business.

For the purposes of this section, we shall discuss the inversion only of Laplace transforms of the form

$$f^*(s) = \frac{a_n s^n + a_{n-1}s^{n-1} + \cdots + a_1 s + a_0}{b_m s^m + b_{m-1}s^{m-1} + \cdots + b_1 s + b_0}, \qquad m > n,$$

that is, proper algebraic fractions in s. This is all that we shall need for now, but the reader should bear in mind that there are Laplace transforms which are not rational functions—for example, $\Gamma(\frac{3}{2})/s^{\frac{3}{2}}$ is the Laplace transform of $t^{\frac{1}{2}}$. The reader is also referred to the last section of this chapter.

Next, we shall need the following theorem, which will be stated without proof:

Theorem 4.2

If $f^*(s)$ is the Laplace transform of a function $f(t)$, then there is no other function of t whose Laplace transform is $f^*(s)$.

In other words, the Laplace transform of a function is unique: For example, $\mathscr{L}\{e^t\} = 1/(s-1)$, hence e^t is the only function whose Laplace transform is $1(s-1)$; there is no other. Using this result, we can pose the following definition:

Definition

If $\mathscr{L}\{f(t)\} = f^*(s)$, then $f(t)$ is called the *inverse Laplace transform* of $f^*(s)$, and we write $f(t) = \mathscr{L}^{-1}\{f^*(s)\}$.

It now remains only to determine how, given a rational function $f^*(s)$, we can determine the $f(t)$ of which $f^*(s)$ is the transform. For simple functions such as $1/(s-1)$, the answer is clearly to use the tables.* Since $\mathscr{L}\{e^t\} = 1/(s-1)$, we have $\mathscr{L}^{-1}\{1/(s-1)\} = e^t$. Thus we can find the inverse transforms of rational functions of s when the denominator is of degree one or two simply by finding the function in the right-hand column of the table and reading its inverse transform in the left-hand column.

For example:

$$\mathscr{L}^{-1}\left\{\frac{1}{s^2+4}\right\} = \mathscr{L}^{-1}\left\{\frac{1}{2}\cdot\frac{2}{s^2+4}\right\} = \frac{1}{2}\mathscr{L}^{-1}\left\{\frac{2}{s^2+4}\right\} = \frac{1}{2}\sin 2t.$$

The same idea works if the denominator is an integer power of a linear expression in s and the numerator is a constant. For example,

$$\mathscr{L}^{-1}\left\{\frac{4}{(s-2)^3}\right\} = \mathscr{L}^{-1}\left\{2\cdot\frac{2!}{(s-2)^3}\right\} = 2\mathscr{L}^{-1}\left\{\frac{2!}{(s-2)^3}\right\} = 2t^2e^t.$$

It is apparent, then, that to find the inverse Laplace transform of a more general rational function we must reduce it to a sum of simple terms such as the above, whose inverses we can read off from the table. The way to do this is, of course, by means of partial fractions, the technique used in calculus to evaluate integrals of rational functions. Let us work a few examples illustrating this technique.

Example 1

If

$$f^*(s) = \frac{s}{(s-1)(s-2)},$$

then we know that $f^*(s)$ can be written in the form

$$\frac{s}{(s-1)(s-2)} = \frac{A}{s-1} + \frac{B}{s-2}.$$

To determine A, we multiply both sides by $(s-1)$ and evaluate at $s = 1$:

$$\frac{s}{s-2} = A + \frac{B}{s-2}(s-1), \quad \text{so } A = -1,$$

and similarly for B:

$$\frac{s}{s-1} = \frac{A}{s-1}(s-2) + B, \quad \text{so } B = 2.$$

* Table 4.1 is a short table of Laplace transforms. See the reference there and in the Bibliography for more lengthy tables.

Thus

$$f^*(s) = \frac{-1}{s-1} + \frac{2}{s-1},$$

and hence

$$\mathscr{L}^{-1}\{f^*(s)\} = -e^t + 2e^{2t}.$$

Example 2

If

$$f^*(s) = \frac{2s+1}{(s-2)s^3},$$

in which a linear factor occurs to a higher power, then we can write

$$f^*(s) = \frac{2s+1}{(s-2)s^3} = \frac{A}{s-2} + \frac{B}{s} + \frac{C}{s^2} + \frac{D}{s^3};$$

A can be determined by multiplying through by $s - 2$ and setting $s = 2$:

$$\frac{2s+1}{s^3} = A + \left(\frac{B}{s} + \frac{C}{s^2} + \frac{D}{s^3}\right)(s-2) \Rightarrow A = \tfrac{5}{8};$$

D can then be determined by multiplying through by s^3 and setting $s = 0$:

$$\frac{2s+1}{s-2} = \frac{As^3}{s-2} + Bs^2 + Cs + D \Rightarrow D = -\frac{1}{2}.$$

To find C, we differentiate the last formula with respect to s and set $s = 0$; to facilitate differentiating the left side, we write

$$\frac{2s+1}{s-2} = \frac{2(s-2)+5}{s-2} = 2 + \frac{5}{s-2},$$

whereupon

$$-\frac{5}{(s-2)^2} = \left[\frac{As^3}{(s-2)}\right]' + 2Bs + C \Rightarrow C = -\frac{5}{4}.$$

The first term on the right is zero at $s = 0$ because it will have a factor of at least s^2. To find B, we differentiate once more:

$$\frac{10}{(s-2)^3} = \left[\frac{As^3}{(s-2)}\right]'' + 2B \Rightarrow B = -\frac{5}{8}.$$

Again, the first term on the right has a factor of at least s. Thus we finally have

$$f^*(s) = \frac{\frac{5}{8}}{s-2} + \frac{(-\frac{5}{8})}{s} + \frac{(-\frac{5}{4})}{s^2} + \frac{(-\frac{1}{2})}{s^3},$$

so that

$$\mathscr{L}^{-1}\{f^*(s)\} = \tfrac{5}{8}e^{2t} - \tfrac{5}{8} - \tfrac{5}{4}t - \tfrac{1}{4}t^2.$$

Example 3

If

$$f^*(s) = \frac{2s^2}{(s-1)(s+2)(s^2+1)},$$

then we have a quadratic term in the denominator, and $f^*(s)$ can be put into the form

$$\frac{2s^2}{(s-1)(s+2)(s^2+1)} = \frac{A}{s-1} + \frac{B}{s+2} + \frac{C}{s^2+1} + \frac{Ds}{s^2+1}.$$

By the techniques of the previous examples, we can find that $A = \frac{1}{3}$ and $B = -\frac{8}{15}$. It remains to determine C and D. We can determine C by setting $s = 0$; we then have $^{-}\frac{1}{3} - {}^{-}\frac{4}{15} + C = 0$, or $C = \frac{9}{15} = \frac{3}{5}$. It is then possible to find D by substituting any other value of s; for example, using $s = 2$, we obtain

$$\frac{2}{5} = \frac{1}{3} - \frac{2}{15} + \frac{3}{25} - \frac{2D}{5},$$

which yields $D = \frac{1}{5}$. Hence

$$f^*(s) = \frac{\frac{1}{3}}{s-1} + \frac{-\frac{8}{15}}{s+2} + \frac{\frac{3}{5}}{s^2+1} + \frac{\frac{1}{5}s}{s^2+1}$$

so that

$$\mathscr{L}^{-1}\{f^*(s)\} = \tfrac{1}{3}e^t - \tfrac{8}{15}e^{-2t} + \tfrac{3}{5}\sin t + \tfrac{1}{5}\cos t.$$

Example 4

We can also use the complex number system to simplify the finding of inverse transforms. For example, consider the case

$$f^*(s) = \frac{1}{(s^2 + 4)(s^2 + 1)}.$$

We could attempt to break this up into terms of the form

$$f^*(s) = \frac{As + B}{s^2 + 4} + \frac{Cs + D}{s^2 + 1}.$$

However, it is just as easy to use complex numbers and do this as

$$f^*(s) = \frac{1}{(s^2 + 4)(s^2 + 1)} = \frac{1}{(s + 2i)(s - 2i)(s + i)(s - i)}$$

$$= \frac{A}{s + 2i} + \frac{B}{s - 2i} + \frac{C}{s - i} + \frac{D}{s + i}.$$

Then we evaluate A by multiplying through $(s + 2i)$ and setting $s = -2i$:

$$A = \frac{1}{(-4i)(-i)(-3i)} = -\frac{i}{12}.$$

Similarly,

$$B = \frac{1}{(4i)(3i)i} = \frac{i}{12};\ C = \frac{1}{i(-3i)(-2i)} = \frac{i}{6};\ D = \frac{1}{(3i)(-i)(2i)} = -\frac{i}{6}.$$

Hence,

$$f^*(s) = \frac{-i/12}{s + 2i} + \frac{i/12}{s - 2i} + \frac{i/6}{s - i} + \frac{-i/6}{s + i},$$

and taking inverses, we have

$$f^*(t) = -\frac{ie^{-2it}}{12} + \frac{ie^{2it}}{12} + \frac{ie^{it}}{6} - \frac{ie^{-it}}{6}$$

$$= \frac{i}{12}(e^{2it} - e^{-2it}) + \frac{i}{6}(e^{it} - e^{-it})$$

$$= \frac{i}{12}(2i \sin 2t) + \frac{i}{6}(2i \sin t) = -\frac{1}{6}\sin 2t - \frac{1}{3}\sin t.$$

Problems

Find the inverse Laplace transforms of each of the following functions.

1. $\dfrac{1}{(s-1)(s+2)(s-3)}$.
2. $\dfrac{s-2}{s(s+1)(s+5)}$.
3. $\dfrac{s^2}{(s-6)(s+7)(s+15)(s+16)}$.
4. $\dfrac{s^2}{s(s+1)(s-2)^2}$.
5. $\dfrac{3s+2}{2s(2s-1)(s-1)^2}$.
6. $\dfrac{3}{(s+1)^5(s+2)^2}$.
7. $\dfrac{s-3}{s(s-1)[(s-3)^2+1]}$.
8. $\dfrac{1}{s(3s+2)(s^2+4)}$.
9. $\dfrac{3s}{(s^2+1)(s^2+2)^2}$.
10. $\dfrac{2s-1}{(s^2+1)(s^2+4)}$.
11. $\dfrac{s^2}{s^3+3s^2+4s-8}$.
12. $\dfrac{3}{s^4+5s^3+4s^2}$.

13. $\dfrac{2s + 1}{s^4 - 2s^2 - 3}$.

14. $\dfrac{s^2 + 1}{s^4 - s}$.

15. $\dfrac{s^3 + s + 1}{s^4 + 3s^3 - 3s^2 - 11s - 6}$.

16. $\dfrac{2}{4s^3 + 11s^2 + 5s - 2}$.

17. $\dfrac{2s - 3}{s^4 - 1}$.

18. $\dfrac{s^2 + 2}{s^3 - 7s + 6}$.

19. $\dfrac{2}{s^4 - 2s^3 + 4s^2 - 6s + 3}$.

20. $\dfrac{s^2 - 2s + 2}{s^4 - 4s^3 + s^2 + 8s - 6}$.

4.3 Solving Equations and Systems Using the Laplace Transform

We now have at hand all the tools necessary to solve initial value problems for linear equations (and systems of linear equations) with constant coefficients. The basic technique of solving an initial value problem is as follows: given the problem

$$P(D)y = (a_nD^n + \cdots + a_1D + a_0)y = f(t),$$
$$y(0) = y_0, \quad Dy(0) = y_1, \ldots, D^{n-1}y(0) = y_{n-1},$$

we take the Laplace transform of both sides of the equation and obtain

$$P(s)y^*(s) + Q(s) = f^*(s),$$

where $Q(s)$ is a polynomial of degree $n - 1$ (at most) which contains all the initial data: Thus the initial value problem has been reduced to a *single algebraic equation* which can now easily be solved for $y^*(s)$:

$$y^*(s) = \frac{f^*(s) - Q(s)}{P(s)}.$$

$y^*(s)$ is indeed a Laplace transform: it is certainly a rational function, since $f^*(s)$ is a rational function and $Q(s)$ and $P(s)$ are polynomials. Furthermore, the degree of the denominator of $f^*(s)/P(s)$ is greater than the degree of the numerator since this is true of $f^*(s)$ alone, and $P(s)$ adds n to the degree of the denominator: and the degree of $Q(s)$ is at least one less than the degree of $P(s)$. Hence the combined expression $[f^*(s) - Q(s)]/P(s)$ is a rational function whose denominator has higher degree than its numerator. Thus we can find $\mathscr{L}^{-1}\{y^*(s)\} = y(t)$, the solution of the original initial value problem.

Let us illustrate this technique by means of a few examples:

Example 1

Let us consider the problem

$$(D^2 - 3D - 4)y = 2e^t + 1,$$
$$y(0) = 0, \quad Dy(0) = 1.$$

Then

$$\mathscr{L}\{y\} = y^*(s), \quad \mathscr{L}\{Dy\} = sy^*(s) - y(0) = sy^*(s),$$

and

$$\mathscr{L}\{D^2y\} = s^2y^*(s) - sy(0) - Dy(0) = s^2y^*(s) - 1.$$

Hence the Laplace transform of the initial value problem is

$$(s^2y^* - 1) - 3sy^*(s) - 4y^*(s) = \frac{2}{s-1} + \frac{1}{s} = \frac{3s-1}{s(s-1)},$$

so that

$$(s^2 - 3s - 4)y^* = \frac{3s-1}{s(s-1)} + 1 = \frac{s^2+2s-1}{s(s-1)}$$

and hence

$$y^*(s) = \frac{s^2+2s-1}{s(s-1)(s^2-3s-4)} = \frac{s^2+2s-1}{s(s-1)(s-4)(s+1)}.$$

expanding $y^*(s)$ into partial fractions as in the previous section, we find that

$$y^*(s) = (-\tfrac{1}{4})/s + (-\tfrac{1}{3})/(s-1) + (\tfrac{23}{60})/(s-4) + (+\tfrac{1}{5})/(s+1)$$

and hence

$$\mathscr{L}^{-1}\{y^*(s)\} = y(t) = -\tfrac{1}{4} - \tfrac{1}{3}e^t + \tfrac{23}{60}e^{4t} + \tfrac{1}{5}e^{-t}$$

which is the solution of the initial value problem, as can be easily verified by substituting back into the original problem.

Example 2

Let us consider

$$(D^3 - 2D^2 + D - 2)y = 0,$$
$$y(0) = 0, \quad Dy(0) = 1, \quad D^2y(0) = 2.$$

Then

$$\mathscr{L}\{y\} = y^*, \quad \mathscr{L}\{Dy\} = sy^* - y(0) = sy^*. \quad \mathscr{L}\{D^2y\} = s^2y^* - sy(0) - Dy(0) = s^2y^* - 1,$$

and

$$\mathscr{L}\{D^3y\} = s^3y^* - s^2y(0) - sDy(0) - D^2y(0) = s^3y^* - s - 2.$$

Hence the Laplace transform of the initial value problem is

$$(s^3y^* - s - 2) - 2(s^2y^* - 1) + (sy^*) - 2y^* = 0,$$

so that

$$(s^3 - 2s^2 + s - 2)y^* - s = 0,$$

giving

$$y^*(s) = \frac{s}{s^3 - 2s^2 + s - 2} = \frac{s}{(s-2)(s^2+1)} = \frac{(\frac{2}{5})}{s-2} + \frac{(-\frac{2}{5})s + (\frac{1}{5})}{s^2+1}$$

and hence the solution,

$$\mathscr{L}^{-1}\{y^*\} = y(t) = \tfrac{2}{5}e^{2t} - \tfrac{2}{5}\cos t + \tfrac{1}{5}\sin t.$$

Example 3

As a final example, we examine

$$(D^3 - 3D - 2)y = 3e^{2t},$$

$$y(0) = 1, \quad Dy(0) = -1, \quad D^2y(0) = 0.$$

Then if $\mathscr{L}\{y\} = y^*$, we have $\mathscr{L}\{Dy\} = sy^* - 1$, $\mathscr{L}\{D^2y\} = s^2y^* - s + 1$, and $\mathscr{L}\{D^3y\} = s^3y^* - s^2 + s$. The Laplace transform of the initial value problem is therefore

$$(s^3y^* - s^2 + s) - 3(sy^* - 1) - 2y^* = 3/(s - 2),$$

or

$$(s^3 - 3s - 2)y^* = 3/(s - 2) + s^2 - s - 3 = \frac{s^3 - 3s^2 - s + 9}{(s - 2)},$$

yielding

$$y^*(s) = \frac{s^3 - 3s^2 - s + 9}{(s^3 - 3s - 2)(s - 2)} = \frac{s^3 - 3s^2 - 2 + 9}{(s + 1)^2(s - 2)^2}.$$

A little bit of hard work with partial fractions gives us

$$y^*(s) = \frac{(\frac{4}{3})}{s + 1} + \frac{(\frac{2}{3})}{(s + 1)^2} + \frac{(-\frac{1}{3})}{(s - 2)} + \frac{(\frac{1}{3})}{(s - 2)^2},$$

so that

$$\mathscr{L}^{-1}\{y^*(s)\} = y(t) = \tfrac{4}{3}e^{-t} + \tfrac{2}{3}te^{-t} - \tfrac{1}{3}e^{2t} + \tfrac{1}{3}te^{2t}.$$

Systems of linear equations with constant coefficients, we recall, arise from physical problems involving electrical networks or the motion of several masses, among others. The Laplace transform provides a handy technique for solving such systems. The technique is basically the same as for single equations, although the Laplace transform will produce a system of algebraic equations in several unknowns, which must then be solved simultaneously. For example:

$$Dx + Dy + 2y = 3e^t + 2,$$

$$2(D - 1)x - 3D(D + 2)y = 0,$$

$$x(0) = 3, \quad y(0) = 5, \quad Dy(0) = -8$$

is a system of two equations in two unknowns $x(t)$ and $y(t)$. To solve this system, we take the Laplace transforms of both equations and obtain

$$sx^* - 3 + sy^* - 5 + 2y^* = \frac{3}{s-1} + \frac{2}{s},$$

$$2(s-1)x^* - 6 - 3s(s+2)y^* + 15s + 6 = 0,$$

which simplify to

$$sx^* + (s+2)y^* = \frac{3}{s-1} + \frac{2}{s} + 8 = \frac{8s^2 - 3s - 2}{s(s-1)},$$

$$2(s-1)x^* - 3s(s+2)y^* = -15s.$$

This system can be solved as an algebraic system of two linear equations in two unknowns. The determinant of the system is

$$\begin{vmatrix} s & (s+2) \\ 2(s-1) & -3s(s+2) \end{vmatrix} = (s+2)\begin{vmatrix} s & 1 \\ 2(s-1) & -3s \end{vmatrix}$$
$$= (s+2)(-3s^2 - 2s + 2).$$

Thus

$$x^*(s) = \frac{\begin{vmatrix} \dfrac{8s^2 - 3s - 2}{s(s-1)} & (s+2) \\ -15s & -3s(s+2) \end{vmatrix}}{(s+2)(-3s^2 - 2s + 2)} = \frac{\dfrac{-24s^2 + 9s + 6}{(s-1)} + 15s}{-3s^2 - 2s + 2}$$

$$= \frac{-24s^2 + 9s + 6 + 15s^2 - 15s}{(s-1)(-3s^2 - 2s + 2)} + \frac{-9s^2 - 6s + 6}{(s-1)(-3s^2 - 2s + 2)} - \frac{3}{s-1},$$

so that

$$x(t) = 3e^t.$$

Similarly,

$$y^*(s) = \frac{\begin{vmatrix} s & \dfrac{8s^2 - 3s - 2}{s(s-1)} \\ 2(s-1) & -15s \end{vmatrix}}{s(s+2)(-3s^2 - 2s + 2)} = \frac{-15s^2 + \dfrac{-2(8s^2 - 3s - 2)}{s}}{(s+2)(-3s^2 - 2s + 2)}$$

$$= \frac{-15s^3 - 16s^2 + 6s + 4}{s(s+2)(-3s^2 - 2s + 2)} = \frac{(5s+2)}{s(s+2)} = \frac{1}{s} + \frac{4}{s+2},$$

and thus

$$y(t) = 1 + 4e^{-2t}.$$

The particular system treated in this example was especially sensitive in that the troublesome factor $(-3s^2 - 2s + 2)$ in the denominator would not have canceled if the least algebraic error had been made anywhere along the line. Not every system will involve such horrendous calculations; but systems are certainly more complicated than single equations, so the student must expect them to require more work and more care than single equations.

Problems

Solve the following initial value problems for linear equations and systems with constant coefficients using the Laplace transform.

1. $(D^2 - 2D + 8)y = 0$, $y(0) = 1$, $Dy(0) = 2$.
2. $(D^2 - 4D - 5)y = 2t$, $y(0) = 0$, $Dy(0) = -1$.
3. $(D^4 - 3D^3 + 2D^2)y = 0$, $y(0) = 1$, $Dy(0) = 0$, $D^2y(0) = 2$, $D^3y(0) = 0$.
4. $(D^3 + 4D^2 + D - 6)y = 0$, $y(0) = 1$, $Dy(0) = 4$, $D^2y(0) = -3$.
5. $(D - 2)y = t - 3te^t$, $y(0) = 0$.
6. $(3D + 1)y = 3 \sin t$, $y(0) = -3$.
7. $(2D + 3)y = 2t^2 - e^{-\frac{1}{2}t}$, $y(0) = 2$.
8. $(D^3 + 3D^2 + 3D + 1)y = 0$, $y(0) = 5$, $Dy(0) = 3$, $D^2y(0) = -2$.
9. $(D - 4)y = 2t^2e^{-t} + 1$, $y(0) = -2$.
10. $(3D + 5)y = 4 - 5 \sin t$, $y(0) = 0$.
11. $(D^4 - 16)y = 0$, $y(0) = 0$, $Dy(0) = 0$, $D^2y(0) = 0$, $D^3y(0) = 5$.
12. $(D^2 + D - 2)y = t + \frac{1}{2}t^2$, $y(0) = 0$, $Dy(0) = -3$.
13. $(D^2 + 6D + 5) = 4e^{-t}$, $y(0) = 1$. $Dy(0) = -4$.
14. $(D^3 - 6D^2 + 12D - 8)y = 1 - e^t$, $y(0) = Dy(0) = D^2y(0) = 0$.
15. $(D^3 + D^2 - 2)y = 3$, $y(0) = 1$, $Dy(0) = -5$, $D^2y(0) = 2$.
16. $Dx + Dy = 4e^{2t} - 6t$, $(D - 2)y - D^2x = 6$, $x(0) = 0$, $Dx(0) = 0$, $y(0) = 2$.
17. $3Dx + y = 1$, $Dy - 3x = 0$, $x(0) = 0$, $y(0) = -2$.
18. $3Dx - Dy = 0$, $(D - 2)y + Dx = 2e^{2t}$, $x(0) = -3$, $y(0) = 3$.
19. $Dx - D^2y = 4t$, $5Dx + 4y = 0$, $x(0) = -1$, $y(0) = 0$, $Dy(0) = -5$.
20. $(D - 1)y + Dx = 15e^{3t}$, $(D - 3)x + Dy = -7e^t$, $y(0) = -7$, $x(0) = 5$.

4.4 Applications to the Double Spring Problem and to Circuits

Let us consider the double spring problem under the following conditions: the mass of spring 1 is m, and that of spring 2 is αm; the constant of spring 1 is

k, and that of spring 2 is βk; and initially, spring 1 is held at rest while spring 2 is held at position γl, and then both are released. Then the initial value problem becomes

$$\ddot{x}_1 + (1 + \beta)\delta x_1 - \beta\delta x_2 = -\beta\delta l,$$
$$\alpha\ddot{x}_2 + \beta\delta x_2 - \beta\delta x_1 = \beta\delta l,$$
$$x_1(0) = 0, \quad x_2(0) = \gamma l, \quad Dx_1(0) = 0, \quad Dx_2(0) = 0,$$

where $\delta = k/m$.

Taking the Laplace transform of this system, we obtain the algebraic system

$$[s^2 + (1 + \beta)\delta]x_1^* - \beta\delta x_2^* = -\beta\delta l/s,$$
$$-\beta\delta x_1^* + (\alpha s^2 + \beta\delta)x_2^* = (\beta\delta l + \alpha\gamma l s^2)/s.$$

Upon calculation, we find that the determinant of this system is $\alpha s^4 + (\alpha + \alpha\beta + \beta)\delta s^2 + \beta\delta^2$, so that x_1^* and x_2^* will be of the form

$$x_1^* = \frac{p(s)}{s(\alpha s^4 + (\alpha + \alpha\beta + \beta)\delta s^2 + \beta\delta^2)},$$
$$x_2^* = \frac{q(s)}{s(\alpha s^4 + (\alpha + \alpha\beta + \beta)\delta s^2 + \beta\delta^2)}.$$

From this much information, without further ado, we can see that $x_1(t)$ and $x_2(t)$ will in general be the superposition of a constant and two cosine waves of different frequency. This is because the above denominator always factors into something of the form $s(as^2 + b)(cs^2 + d)$, where a, b, c, and d are nonnegative. Hence upon splitting x_1^* and x_2^* up into partial fractions we will obtain the transforms of sine and cosine terms. That the denominator has factors of this form follows from the fact that the discriminant of the quartic portion,

$$(\alpha + \alpha\beta + \beta)^2\delta^2 - 4\alpha\beta\delta^2,$$

is always nonnegative; this can be shown as follows. Obviously

$$(\sqrt{\alpha} - \sqrt{\beta})^2 = \alpha - 2\sqrt{\alpha\beta} + \beta \geq 0$$

and hence

$$\alpha + \beta \geq 2\sqrt{\alpha\beta};$$

thus certainly

$$\alpha + \alpha\beta + \beta \geq 2\sqrt{\alpha\beta}.$$

Squaring both sides and multiplying by δ^2, we have

$$(\alpha + \alpha\beta + \beta)^2\delta^2 \geq 4\alpha\beta\delta^2,$$

as required.

Physically, the two superposed cosine waves in this motion represent the fact that each of the two springs vibrate at a different frequency, so that the motion of one always includes a component of motion due to the other.

Beyond this, we shall not discuss this problem, relegating further details to the problems at the end of the section. Instead let us turn now to problems involving electrical circuits.

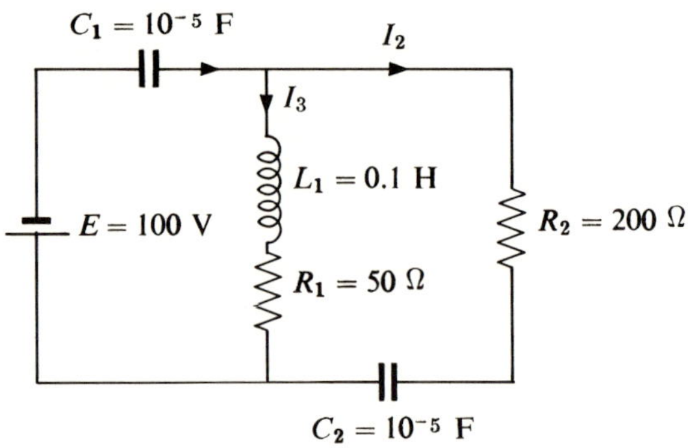

Figure 4.4

Let us consider first the problem delineated in Figure 4.4. The system of units used is such that charge is measured in coulombs, capacitance in farads, inductance in henrys, resistance in ohms, electromotive force in volts, and current in amperes. Then the initial value problem for this circuit is as follows:

$$I_1 = I_2 + I_3,$$

$$50I_3 + 0.1\dot{I}_3 + 10^5Q_1 = 100,$$

$$10^5Q_2 + 200I_2 = 100,$$

$$Q_1(0) = 0, \quad I_1(0) = (0), \quad Q_2(0) = 0, \quad I_2(0) = 0.$$

We may as well use the first equation to eliminate one variable, say I_3; then, rewriting the currents as derivatives of charges, we have

$$0.1D^2Q_1 + 50DQ_1 + 10^5Q_1 - 0.1D^2Q_2 - 50DQ_2 = 100,$$
$$200DQ_2 + 10^5Q_2 = 100,$$
$$Q_1(0) = 0, \quad DQ_1(0) = 0, \quad Q_2(0) = 0, \quad DQ_2(0) = 0.$$

Taking Laplace transforms of this system, we obtain

$$(0.1s^2 + 50s + 10^5)Q_1^* - (0.1s^2 + 50s)Q_2^* = \frac{100}{s},$$

$$(200s + 10^5)Q_2^* = 100.$$

The solution is nearly immediate, since

$$Q_2^* = \frac{1}{2s + 10^3},$$

implying that

$$Q_2 = \tfrac{1}{2}e^{-10^3t/2} = \tfrac{1}{2}e^{-500t}.$$

Then, because

$$(0.1s^2 + 50s + 10)Q_1^* - \frac{(0.1s^2 + 50s)}{2s + 10^3} = \frac{100}{s},$$

we have

$$Q_1^* = \frac{s^2 + 2000}{2s(s^2 + 500s + 10^6)} = \frac{\frac{1}{1000}}{s} + \frac{\frac{499}{1000}(s + 250)}{(s + 250)^2 + 937500} - \frac{\frac{501}{4}}{(s + 250)^2 + 937500}.$$

Upon taking inverse transforms, we have

$$Q_1 = \frac{1}{1000} + e^{-250t}\left[\frac{499}{1000}\cos\sqrt{937500}t - \frac{501}{4\sqrt{937500}}\sin\sqrt{937500}t\right].$$

Thus, the charge Q_2 goes to zero very rapidly with increasing time, as does the current I_2. On the other hand, the charge on capacitor 1 approaches 10^{-3} coulombs as t increases, while I_1 also goes to zero quite rapidly; the effect of the inductor in this circuit is to cause the current to oscillate at a circular frequency of $\sqrt{937500}$ radians per second, about 154 Hz.

Problems

1. Complete the solution of the double spring problem discussed in the text when $\alpha = \beta = 1$, $\delta = \frac{1}{2}$, $\gamma = \frac{11}{10}$, $l = 2$. Calculate the function $x_2 - x_1$ and explain what you find.

2. Show that the same current flows from the generator in both circuits in Figure 4.5, provided that $c = c_1 + c_2$. Hence, capacitors wired in parallel behave the same as a single capacitor whose capacitance is the sum of those of the individual ones.

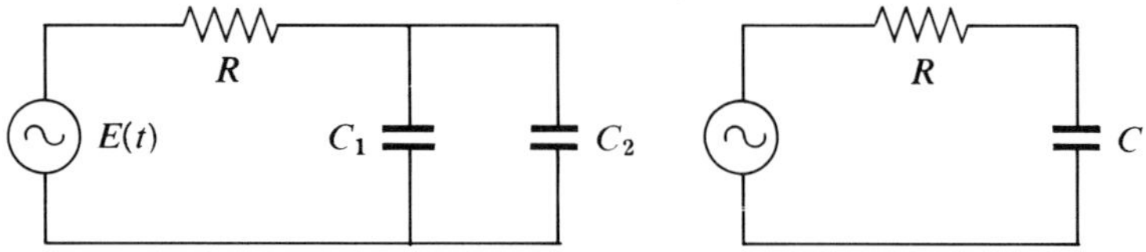

Figure 4.5

3. Set up and solve the systems of differential equations describing the circuits in Figures 4.6 and 4.7. In each case, all charges and currents are initially zero.

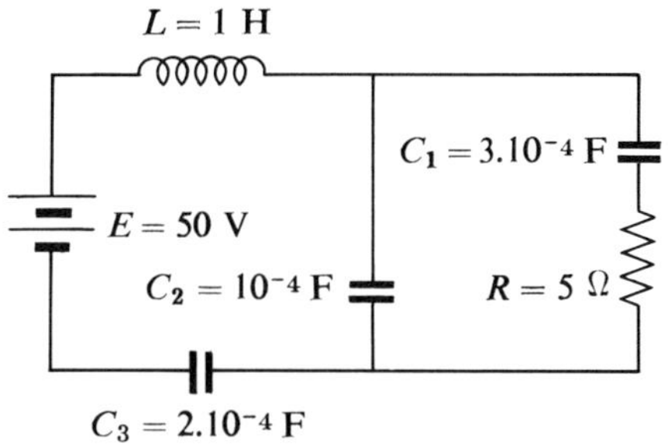

Figure 4.6

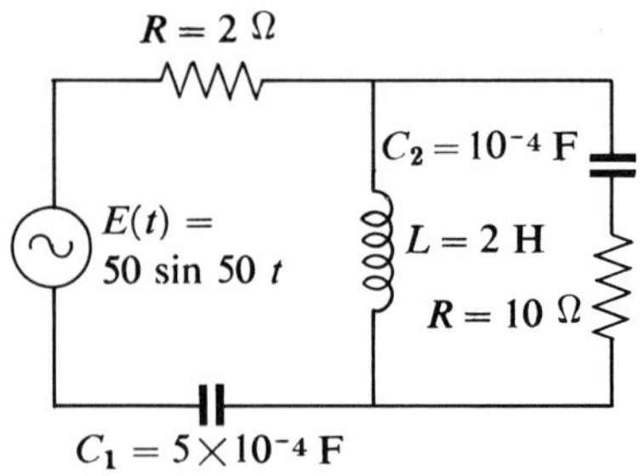

Figure 4.7

4. In Figure 4.8, capacitor 1 initially bears a charge of 10^4 coulombs; capacitor 2, 10^5 coulombs. The switches are closed simultaneously. Find an expression for the current.

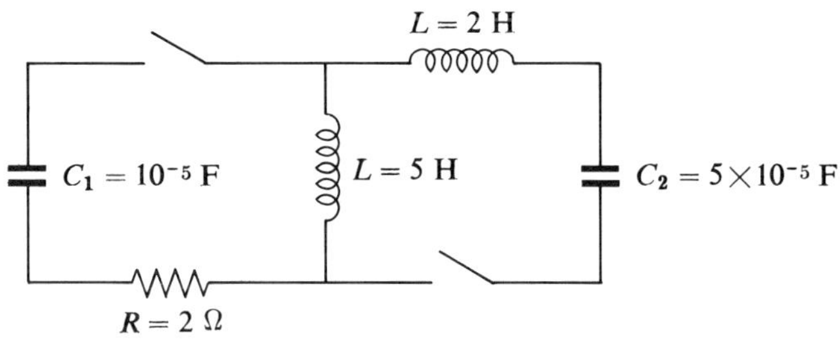

Figure 4.8

4.5 Step Functions and Impulse Functions

In this section we shall confine ourselves to initial value problems for equations with constant coefficients of the form

$$a\ddot{x} + b\dot{x} + cx = f(t),$$
$$x(0) = x_0, \qquad \dot{x}(0) = v_0. \tag{4.16}$$

If x is the displacement from equilibrium of a mass on a vibrating spring, for instance, then $f(t)$ is the external force exerted on the mass at time t; if x represents the *current* (*not* the charge) in a circuit, then $f(t)$ is the time rate of change of the applied electromotive force. On the other hand, if x were used to represent the charge, then $f(t)$ would simply be the electromotive force.

In particular, we are interested in solving the above problem when $f(t)$ exhibits a discontinuity corresponding to one of a number of different physical situations, which we shall consider one by one.

(1) *The step function.* Suppose that x represents a charge in a circuit; $f(t)$ then represents a source of electromotive force. Suppose that this source is not turned on until time α, after which it continues to provide a constant voltage E. Then

$$f(t) = \begin{cases} 0, t < \alpha, \\ E, t \geqq \alpha. \end{cases}$$

(See Figure 4.9.) A function with this type of behavior is called a *step function* for the obvious reason that its graph represents a step.

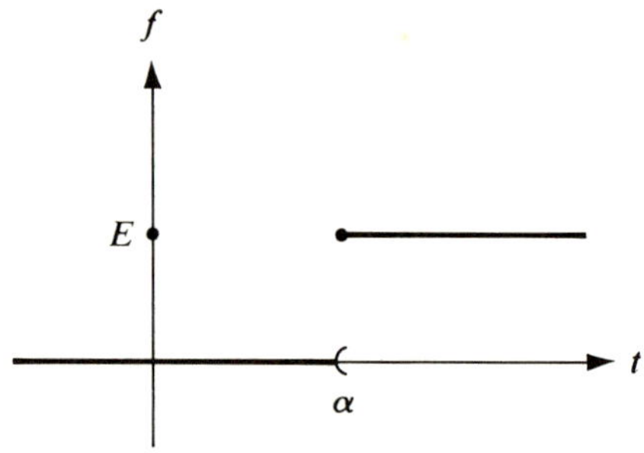

Figure 4.9

It is not at all difficult to find the Laplace transform of f; however, for the purposes of our future studies of more complicated functions, we shall need to pause briefly and define a new notation:

Definition
A function which vanishes for $t < \alpha$ and is equal to unity for $t \geqq \alpha$ is called the characteristic function of the interval $[\alpha, \infty)$; it is denoted by $\chi_\alpha(t)$. In other words,

$$\chi_\alpha(t) = \begin{cases} 0, t < \alpha, \\ 1, t \geqq \alpha. \end{cases}$$

If we find the Laplace transform of χ_α, then we need only multiply it by E to obtain the Laplace transform of our step function $f(t)$. Now

$$\mathscr{L}\{\chi_\alpha\} = \int_0^\infty e^{-st}\chi_\alpha(t)dt = \int_\alpha^\infty e^{-st}\,dt = -\frac{1}{s}e^{-st}\Big|_\alpha^\infty = \frac{e^{-s\alpha}}{s}.$$

Therefore we also have that

$$\mathscr{L}\{f\} = Ee^{-s\alpha}/s.$$

The characteristic function has other interesting properties as well; for example, given any function $g(t)$ defined for all t, then the graph of the function $\chi_\alpha(t)g(t - \alpha)$ is simply the graph of the positive portion of $g(t)$ translated α units to the right (Figure 4.10). [Note that $\chi_\alpha(t) \equiv \chi_\alpha(t) \cdot 1$ represents the translation of the function which is zero for negative t and 1 elsewhere α units to the right.]

To take the Laplace transform of a function of the form $\chi_\alpha(t)g(t - \alpha)$, we consider

$$\mathscr{L}\{\chi_\alpha g(t - \alpha)\} = \int_0^\infty e^{-st}\chi_\alpha(t)g(t - \alpha)dt = \int_\alpha^\infty e^{-st}g(t - \alpha)dt;$$

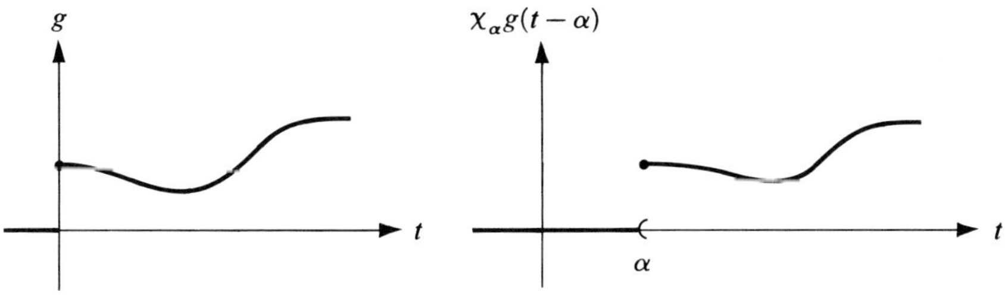

Figure 4.10

then, making the change of variable $\tau = t - \alpha$, we see that

$$\mathscr{L}\{\chi_\alpha g(t - \alpha)\} = \int_0^\infty e^{-s(\tau+\alpha)}g(\tau)d\tau = e^{-s\alpha}\int_0^\infty e^{-s\tau}g(\tau)d\tau,$$

or

$$\mathscr{L}\{\chi_\alpha g(t - \alpha)\} = e^{-s\alpha}\mathscr{L}\{g(t)\} = e^{-s\alpha}g^*(s).$$

Taking inverse transforms of the first and last quantities, we see that

$$\mathscr{L}^{-1}\{e^{-s\alpha}g^*(s)\} = \chi_\alpha(t)g(t - \alpha).$$

Note the similarity between this formula and the one from section 4.1

$$\mathscr{L}\{e^{\alpha t}g(t)\} = g^*(s - \alpha).$$

We have seen enough examples of this sort of "duality" between functions and their Laplace transforms by now to realize that there is an interesting field of study here which we have barely tapped.

Let us solve the following problem using the above methods:

$$\ddot{Q} + Q = \chi_\alpha(t)E, \quad Q(0) = 1, \quad \dot{Q}(0) = 0.$$

This equation represents the charge in an electric circuit containing a capacitor (with initial charge 1) and an inductor, but no resistance; a source of voltage of strength E is turned on at time α. Taking Laplace transforms of both sides,

$$s^2Q^* - s + Q^* = Ee^{-\alpha s}/s.$$

Solving for Q^*, we obtain

$$Q^* = \frac{Ee^{-\alpha s}}{s(s^2 + 1)} + \frac{s}{s^2 + 1}.$$

Now by partial fractions,

$$\frac{1}{s(s^2 + 1)} = \frac{1}{s} - \frac{s}{s^2 + 1},$$

and therefore

$$Q^* = Ee^{-\alpha s}\left(\frac{1}{s} - \frac{s}{s^2 + 1}\right) + \frac{s}{s^2 + 1}.$$

Using the fact that

$$\mathscr{L}^{-1}\{e^{-\alpha s}g^*(s)\} = \chi_\alpha(t)g(t - \alpha),$$

we have

$$Q(t) = E\chi_\alpha(t)[1 - \cos(t - \alpha)] + \cos t.$$

The first term of this solution is due to the electromotive force which is turned on at $t = \alpha$; it has no effect until that time. The second term represents the charge that would be present in the absence of the electromotive force due to the initial conditions $Q(0) = 1$, $\dot{Q}(0) = 0$.

(2) *An impulse over a finite length of time.* Let us now suppose that equation (4.16) represents a vibrating spring problem, where $f(t)$ is an external force of magnitude $E/2\eta$ which is applied between $t = \alpha - \eta$ and $t = \alpha + \eta$, thus,

$$f(t) = \begin{cases} 0 & t < \alpha - \eta, \\ E/2\eta & \alpha - \eta \leqq t \leqq \alpha + \eta, \\ 0 & t > \alpha + \eta. \end{cases}$$

(See Figure 4.11.) For any force, or forcing term in an equation of the form (4.16), the *impulse* I is defined to be

$$I = \int_0^\infty f(t)dt.$$

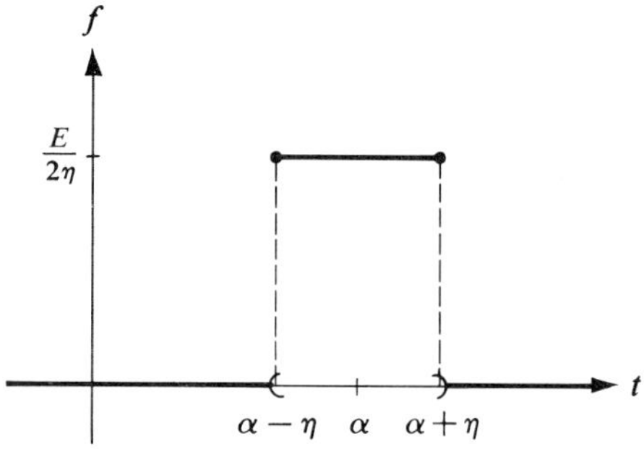

Figure 4.11

In our case, we see that the impulse is precisely E, since

$$I = \int_0^{\infty} f(t)dt = \int_{\alpha-\eta}^{\alpha+\eta}\left(\frac{E}{2\eta}\right)dt = \left(\frac{E}{2\eta}\right)t\bigg|_{\alpha+\eta}^{\alpha+\eta} = \left(\frac{E}{2\eta}\right)\cdot 2\eta = E.$$

This selfsame impulse E could be caused as well by a small force exerted over a considerable length of time as by a large force exerted over a short length of time, depending on the value of η.

Note that if x in equation (4.16) were interpreted to be the *current* through a circuit, then $f(t)$ would be the time rate of change of the electromotive force, meaning that a *total potential* of E volts would be impressed on the circuit over the interval of time $[\alpha - \eta, \alpha + \eta]$.

To find the Laplace transform of an impulse function like the above, we observe that $f(t)$ can be written as

$$f(t) = \frac{E}{2\eta}\left[(\chi_{\alpha-\eta}(t) - \chi_{\alpha+\eta}(t)\right].$$

Why? Because $\chi_{\alpha-\eta}$ is equal to 1 from $t = \alpha - \eta$ on; from $t = \alpha + \eta$ on, $\chi_{\alpha+\eta}$ equals 1 and thus the difference between the two functions is 0. Hence, by the linearity of the Laplace transform,

$$\mathscr{L}\{f\} = \frac{E}{2\eta}\cdot\frac{1}{s}[e^{-(\alpha-\eta)s} - e^{-(\alpha+\eta)s}] = \frac{E}{\eta s}e^{-\alpha s}\sinh \eta s.$$

For instance, consider the problem

$$\ddot{x} + x = \frac{E}{2\eta}[\chi_{\alpha-\eta}(t) - \chi_{\alpha+\eta}(t)],\ x(0) = 0,\ \dot{x}(0) = 0.$$

This represents the displacement of a mass attached to a frictionless vibrating spring subjected to an impulse E over the interval from $t = \alpha - \eta$ to $t = \alpha + \eta$. Taking Laplace transforms,

$$s^2x^* + x^* = \frac{E}{2\eta}\left(\frac{e^{-(\alpha-\eta)s} - e^{-(\alpha+\eta)s}}{s}\right),$$

which leads to

$$x^* = \frac{E}{2\eta}\left[\frac{e^{-(\alpha-\eta)s}}{s(s^2+1)} - \frac{e^{-(\alpha+\eta)s}}{s(s^2+1)}\right],$$

or, by partial fractions,

$$x^* = \frac{E}{2\eta}\left[e^{-(\alpha-\eta)s}\left(\frac{1}{s} - \frac{1}{s^2+1}\right) - e^{-(\alpha+\eta)s}\left(\frac{1}{s} - \frac{1}{s^2+1}\right)\right].$$

Inverting this, we obtain the solution

$$x(t) = \frac{E}{2\eta}(\chi_{\alpha-\eta}(t)\{1 - \cos[t - (\alpha - \eta)]\} - \chi_{\alpha+\eta}(t)\{1 - \cos[t - (\alpha + \eta)]\}).$$

The first term inside the parentheses represents the motion due to a force of constant magnitude applied from $t = \alpha - \eta$ on; from this, we subtract the motion due to a constant force of the same magnitude applied from $t = \alpha + \eta$ on: another manifestation of superposition. Note that the two terms cancel each other, so that there is no motion after $t = \alpha + \eta$, if 2η is equal to one period of the cosine (2π, 4π, etc.).

(3) *An instantaneous impulse of finite size.* We are now going to examine—not very rigorously—what happens as η approaches zero in the previous case. Physically, this would correspond to a larger and larger force being exerted over a shorter and shorter time interval, the impulse remaining constant at E; or (if x represents current) a voltage E being applied to our circuit over a shorter and shorter time interval.

Mathematically, if we examine

$$\lim_{\eta\to 0}\frac{f(t)}{E} = \lim_{\eta\to 0}\frac{\chi_{\alpha-\eta}(t) - \chi_{\alpha+\eta}(t)}{2\eta},$$

we see that the limit function is zero for $t \neq \alpha$, and undefined (or infinite) for $t = \alpha$. This limit function is usually denoted by $\delta(t - \alpha)$ and is called the *Dirac delta function*. Since, as we have seen previously,

$$\int_0^\infty \frac{f(t)}{E}\,dt = 1$$

for all values of η, one might reasonably expect that since

$$\lim_{\eta \to 0} \int_0^\infty \frac{f(t)}{E}\,dt = 1,$$

we must have

$$\int_0^\infty \lim_{\eta \to 0} \frac{f(t)}{E}\,dt = \int_0^\infty \delta(t - \alpha)dt = 1;$$

however, exchanging the limit and the integral sign is unjustified, particularly since the function δ is 0 everywhere except at one point, where its value is infinite. Therefore it is rather difficult to see how its integral could even be defined, let alone equal to 1.

The Dirac delta function is actually an example of what mathematicians call a *generalized function*, and one can develop and study such entities quite rigorously. Such a study would be beyond the scope of this book, however; so we shall simply take it as a convenient *definition* that $\delta(t - \alpha)$ represents an instantaneous force at $t = \alpha$ producing a unit impulse:

$$\delta(t - \alpha) = \lim_{\eta \to 0} \frac{\chi_{\alpha-\eta}(t) - \chi_{\alpha+\eta}(t)}{2\eta}, \quad \int_0^\infty \delta(t - \alpha)dt = 1.$$

To find the Laplace transform of the Dirac delta function, we note that

$$\mathscr{L}\left\{\frac{\chi_{\alpha-\eta}(t) - \chi_{\alpha+\eta}(t)}{2\eta}\right\} = \frac{e^{-\alpha s}\sinh \eta s}{\eta s},$$

and we simply *define* $\mathscr{L}\{\delta(t - \alpha)\}$ to be the limit of this expression as $\eta \to 0$:

$$\mathscr{L}\{\delta(t - \alpha)\} = \lim_{\eta \to 0} \frac{e^{-\alpha s}\sinh \eta s}{\eta s} = e^{-\alpha s} \lim_{\eta \to 0} \frac{\eta \cosh \eta s}{\eta} = e^{-\alpha s}.$$

In particular, for $\alpha = 0$, $\mathscr{L}\{\delta(t)\} = 1$.

Let us use this to solve the problem

$$\ddot{x} + 2\dot{x} + 5x = \delta(t - \alpha),$$
$$x(0) = 0, \quad \dot{x}(0) = 0.$$

This equation represents the damped vibration of a spring which is initially at rest and is struck at $t = \alpha$ with a force producing a unit impulse. Taking Laplace transforms, we obtain

$$s^2x^* + 2sx^* + 5x^* = e^{-\alpha s},$$

which yields

$$x^* = \frac{e^{-\alpha s}}{(s+1)^2 + 2^2},$$

the inverse transform of which is

$$x(t) = \chi_\alpha(t)e^{-(t-\alpha)} \sin 2(t-\alpha).$$

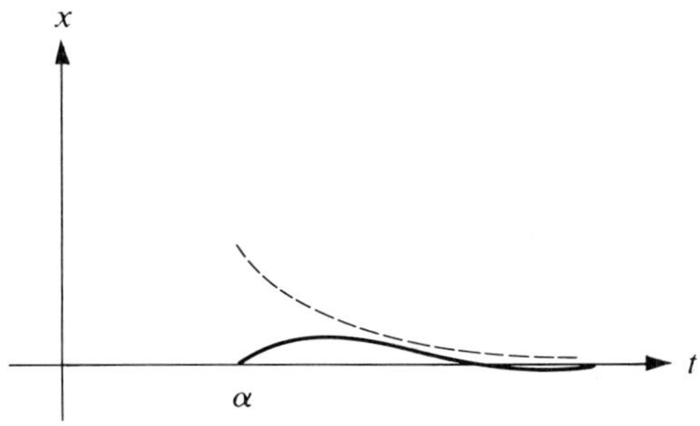

Figure 4.12

Figure 4.12 illustrates this solution. It is simply the translation of a damped sine wave α units to the right. Note that if $\alpha = 0$, then the solution is simply

$$x(t) = e^{-t} \sin 2t;$$

and this function is the solution not only of our original problem, but also of the problem

$$\ddot{x} + 2\dot{x} + 5x = 0,$$
$$x(0) = 0, \quad \dot{x}(0) = 2.$$

These two problems are completely equivalent, and we see that the impulse can also be expressed as in the second problem by imparting an initial velocity to a spring initially at rest. This is in agreement with our intuition; hence, thus pragmatically, we have justified the use of the questionably defined Dirac delta function.

Problems

Solve the following problems by Laplace transform methods.

1. $\ddot{x} + 2\dot{x} + x = \chi_\alpha(t)$; $x(0) = 0$, $\dot{x}(0) = 0$.
2. $\ddot{x} + 2\dot{x} - 3x = \chi_\alpha(t)$; $x(0) = 0$; $\dot{x}(0) = 0$.
3. $\ddot{x} + 2\dot{x} + 5x = \chi_\alpha(t)$; $x(0) = 0$, $\dot{x}(0) = 0$.

4. $\ddot{x} + 2\dot{x} + x = \chi_\alpha(t) - \chi_\beta(t)$; $x(0) = 0$, $\dot{x}(0) = 0$.
5. $\ddot{x} + 4x = \delta(t - \alpha)$; $x(0) = 0$, $\dot{x}(0) = 0$.
6. $\ddot{x} + 2\dot{x} + x = \delta(t - \alpha)$; $x(0) = 0$, $\dot{x}(0) = 0$.
7. $\ddot{x} + 2\dot{x} - 3x = \delta(t - \alpha)$; $x(0) = 0$, $\dot{x}(0) = 0$.
8. $\ddot{x} + 2\dot{x} + 5x = \delta(t - \alpha) - \delta(t - \beta)(\alpha < \beta)$; $x(0) = 0$, $\dot{x}(0) = 0$.
9. $\ddot{x} + 4x = \delta(t - \alpha) - \delta(t - \beta)(\alpha < \beta)$; $x(0) = 0$, $\dot{x}(0) = 0$.
10. $\ddot{x} + 2\dot{x} = \delta(t - \alpha)$; $x(0) = 0$, $\dot{x}(0) = 0$.

5

Infinite Series Methods for Linear Equations

A very great many of the differential equations encountered in practice cannot be solved as easily as those we have considered up to now. Indeed, the differential equation whose solution can be found in a simple closed form as a function of the independent variable is rare. This is even true of linear equations, the simplest type of differential equation, as we shall see in the text that follows. Nevertheless, we need not give up all hope. It is quite often possible to secure approximate solutions of differential equations either by numerical methods (which we shall consider in Chapter 6) or by means of techniques involving infinite series (which we are now proposing to do).

In a sense, solution by means of infinite series is an approximation technique, since it is not normally possible to find the exact value of an infinite series for some specific value of x, but only to approximate this sum by using a finite number of terms of the series.

5.1 Ordinary and Singular Points of Linear Equations

In this section we shall be considering the general inhomogeneous linear equation

$$a_n(x)\frac{d^n y}{dx^n} + a_{n-1}(x)\frac{d^{n-1} y}{dx^{n-1}} + \cdots + a_1(x)\frac{dy}{dx} + a_0(x)y = f(x). \qquad (5.1)$$

For the most part we shall confine our attention to problems in which the coefficients are polynomials in x; however, all the theory which we shall develop will be equally valid for equations with more general functions as coefficients.

It should seem reasonably clear that those values of x for which the leading coefficient $a_n(x)$ vanishes assume a particular importance, for at these points equation (5.1) is no longer effectively of order n—the term involving the highest derivative is somehow removed from our ken. Another obvious difficulty is that if we attempt to solve (5.1) for the highest derivative,

$$\frac{d^n y}{dx^n} = -\left[\frac{a_{n-1}(x)}{a_n(x)}\frac{d^{n-1}y}{dx^{n-1}} + \cdots + \frac{a_1(x)}{a_n(x)}\frac{dy}{dx}\right] + \frac{f(x)}{a_n(x)},$$

then we will be dividing by zero at points where $a_n(x)$ vanishes.

At the same time, any difficulty with $a_n(x)$ may be more apparent than real, because for instance the equation

$$xy'' + 2xy' + xy = 0$$

is really the same as the equation

$$y'' + 2y' + y = 0,$$

except at $x = 0$; but the point $x = 0$ can effectively be ignored—the difficulty is removable. Likewise, the equation

$$xy'' + (\sin x)y' + (1 - \cos x)y = 0$$

can be solved for y'' when $x \neq 0$ to yield

$$y'' = -\frac{\sin x}{x}y' - \frac{(1 - \cos x)}{x}y.$$

Again the point $x = 0$ causes no real difficulty because even though the functions

$$\frac{\sin x}{x}, \quad \frac{(1 - \cos x)}{x}$$

are not defined at $x = 0$, nevertheless their limits exist as $x \to 0$; hence they can be redefined at that one troublesome point. It makes sense to do this, because differential equations represent the culmination of a limiting process anyhow.

To avoid spurious cases such as the above, we divide equation (5.1) through by the leading coefficient $a_n(x)$, and rename the new coefficients

$$b_k(x) = a_k(x)/a_n(x) \quad (k = 0, 1, \ldots, n - 1),$$

$$g(x) = f(x)/a_n(x),$$

so that we have

$$D^n y + b_{n-1}(x)D^{n-1}y + \cdots + b_1(x)Dy + b_0(x)y = g(x). \qquad (5.2)$$

We then pose the following:

Definition
A point $x = x_0$ is said to a *singular point* of equation (5.2) if there is at least one value of k between 0 and $n - 1$ for which

$$\lim_{x \to x_0} |b_k(x)| = \infty.$$

Any point which is not a singular point is called an *ordinary point.*

For example, in the equation

$$x(x^2 - 1)y'' - (2x + 3)y' + 7x^2 y = e^x,$$

we have, upon division by $x(x^2 - 1)$, that

$$y'' - \frac{2x + 3}{x(x^2 - 1)}y' + \frac{7x}{x^2 - 1}y = \frac{e^x}{x(x^2 - 1)}.$$

By examining the coefficient $b_1 = -(2x + 3)/[x(x^2 - 1)]$, it is clear that $x = 0$ and $x = \pm 1$ are singular points. By examining the other coefficient b_0, we see that there are no others.

In the equation

$$(\sin 2x)y'' - x(x - 1)y' + x^2 y = 0,$$

there are an infinite number of singular points, namely $x = \pm\pi/2, \pm\pi, \pm 3\pi/2, \cdots$. Note that $x = 0$ is *not* a singular point, since

$$\lim_{x \to 0} \left| \frac{-x(x - 1)}{\sin 2x} \right| = \frac{1}{2} < \infty, \quad \lim_{x \to 0} \left| \frac{x^2}{\sin 2x} \right| = 0 < \infty.$$

A differential equation need not have singular points at all, of course; any linear equation with constant coefficients lacks singular points since a_n is by definition a nonzero constant. Another possibility is that an equation might have imaginary or complex singular points. For example

$$(x^2 + 1)y'' - (x + 1)y' + 3y = 0$$

has singular points at $x = i$ and $x = -i$. And the equation

$$(x^2 + 4x + 8)y'' + 3xe^x y' = 0$$

has singular points at $x = -2 + 2i$ and $-2 - 2i$. The reader may well question why we worry about imaginary or complex singular points, since we are only interested in the solutions of equations for real values of x. The answer, as we shall see in the next section, is that the location of *all* singular points has a bearing on the domain of convergence of the series solution.

Euler equations, which we studied in Chapter 3, provide another example of linear equations with singular points. Indeed, every Euler equation can be put in the form

$$D^n y + \frac{b_{n-1}}{x} D^{n-1} y + \frac{b_{n-2}}{x^2} D^{n-2} y + \cdots + \frac{b_1}{x^{n-1}} Dy + \frac{b_0}{x^n} y = g(x),$$

where $b_0, b_1, \ldots, b_{n-1}$ are constants. Clearly, then, each Euler equation has a singular point at $x = 0$ and nowhere else.

In the general case, the coefficients $b_k(x)$ can go to infinity more or less rapidly as x approaches a singular point x_0; for example, $1/(x - x_0)^3$ certainly grows more rapidly than $1/(x - x_0)$ as $x \to x_0$. A *regular* singular point is a singular point for which this rate of growth, for each coefficient, is no more rapid than that of the corresponding coefficient of an Euler equation. To be precise:

Definition
$x = x_0$ is said to be a *regular singular point* of equation (5.2) if it is a singular point and if

$$\lim_{x \to x_0} |(x - x_0)^{n-k} b_k(x)| < \infty$$

for $k = 0, 1, 2, \ldots, n - 1$. If a singular point x_0 is not a regular singular point, then it is called an *irregular singular point*.

In one of our previous examples, we had a second-order equation with

$$b_1(x) = -\frac{2x + 3}{x(x^2 - 1)}, \quad b_0(x) = \frac{7x}{(x^2 - 1)};$$

the factors $x + 1$ and $x - 1$ occur in both denominators to the first power, and in $b_1(x)$, x occurs to the first power and in $b_0(x)$ to the power zero. Thus according to our definition all three singular points are regular.

In the equation

$$x^3(x-1)^2y'' + x(x-1)y' - 2y = 0,$$

we have

$$y'' + \frac{y'}{x^2(x-1)} - \frac{2y}{x^3(x-1)^2} = 0,$$

and in this case, $x = 0$ is an irregular singular point because x occurs squared in the denominator of b_1, and cubed in the denominator of b_0, so that in both cases

$$\lim_{x \to 0} |x^{n-k} b_k(x)| = \infty.$$

On the other hand, $x = 1$ is a regular singular point because $(x - 1)$ occurs to the first power in the denominator of b_1, and squared in the denominator of b_0, so that in either case

$$\lim_{x \to 1} |(x-1)^{n-k} b_k(x)|$$

is finite.

Most of our problems will concern equations with polynomial coefficients. For these equations there is a relatively simple rule for distinguishing between the two types of singular points: After dividing through by $a_n(x)$, examine the denominators of the remaining coefficients (after canceling common factors). $(x - x_0)$ should not occur to a degree higher than the first in the first coefficient, to a degree higher than the second in the second coefficient, and so on, for x_0 to be a regular singular point. If $(x - x_0)$ occurs in *even one* of the coefficients raised to too high a power, then x_0 must be an irregular singular point.

Most of our work in this chapter and in Chapter 7 will be concerned with second-order equations. For that reason, it will be useful to summarize the definitions of this section for the special case of second-order equations. We consider those second-order equations which have been put into the following form:

$$y'' + b_1(x)y' + b_0(x)y = 0.$$

Then

(1) $x = x_0$ is an *ordinary point* if both of the following limits exist and

are finite:

$$\lim_{x \to x_0} b_1(x), \quad \lim_{x \to x_0} b_0(x).$$

(2) $x = x_0$ is a *singular point* if either

$$\lim_{x \to x_0} |b_1(x)| = \infty, \quad \lim_{x \to x_0} |b_0(x)| = \infty.$$

(a) $x = x_0$ is a *regular singular point* if it is a singular point and if both of the following limits exist and are finite:

$$\lim_{x \to x_0} |(x - x_0)b_1(x)|, \quad \lim_{x \to x_0} |(x - x_0)^2 b_0(x)|.$$

(b) If either (or both) of the above limits is infinite, then $x = x_0$ is an *irregular singular point.*

Problems

Determine and classify the singular points (if any) of each of the following equations:

1. $xy'' + y' + xy = 0.$
2. $x(x - 1)^3 y'' + 2y' - (x - 1)y = 2 \cos x.$
3. $(x^3 + 7x + 8)y'' - (x + 3)y' + xy = 0.$
4. $(x^4 - 16)y'' + (x - 2)y' + (x + 2)y = 0.$
5. $(x^4 - 4x^2 + 4)y'' + xy' = 0.$
6. $\sin 2xy'' - (x - 3)y' + e^x y = 0.$
7. $x^3(x - 1)D^3 y + x^2 e^x y'' - 3x(x - 1)y' + 14 \cos xy = 0.$
8. $3x^2(x + 2)^2 y'' - 2x(x - 1)y' + (x^2 + 2)y = -xe^x.$
9. $(x^2 + x + 1)y'' + (\sin 2x)y' - 4x^2 y = \frac{1}{2} - \sin 2x.$
10. $(1 - \cos^2 x)y'' - 3xy' + 4y = 0.$
11. $(e^x - 1)y'' + 14y' - 12e^x \cos xy = 0.$
12. $(\sin 3x^2)y'' - x(x - 2)y' + 5xe^x y = 2e^{2x}.$
13. $(x - 2)^2(x - 3)^3 y'' + (x^2 - 3x + 2)y' - 2(\sin 2x)y = 0.$
14. $(\frac{1}{2} - \cos^2 2x)xy'' + 2x(\sin x)y' - 7y = \cos 2x.$
15. $(2 - \sin x)y'' - 4x^2(x - 1)y' = -10e^x + 1.$
16. $(x^2 - 2x + 2)(x + 13)y''6xy' - 17x^2 y = 0.$
17. $x^2(x - 1)y'' + 12y = \sin x - 2.$
18. $(5 \cosh x + 1)y'' - 7(\sinh x)y' + 4e^{5x} y = 0.$

19. $(e^x - e^{-2x})y'' + 13(e^x \cos x)y = 2x^2 - 3.$

20. $x(\sinh x)y'' - 3(\sin x)y' + 14e^x y = 0.$

5.2 Solutions of Homogeneous Linear Differential Equations near an Ordinary Point

Let us consider a homogeneous linear differential equation

$$a_n(x)D^n y + \cdots + a_1(x)Dy + a_0(x)y = 0,$$

which has an ordinary point at $x = x_0$. Then, as we have seen, this equation can be put into the form

$$D^n y + b_{n-1}(x)D^{n-1}y + \cdots + b_1(x)Dy + b_0(x)y = 0$$

by dividing through by $a_n(x)$. Throughout this section we shall make the further assumption that the functions $b_0(x), b_1(x), \ldots, b_{n-1}(x)$ all have infinitely many derivatives near $x = x_0$, which is certainly true if the a_k's are all polynomials and if x_0 is not a singular point. Then it turns out that the above equation has a solution of the form

$$y(x) = \sum_{n=0}^{\infty} c_n(x - x_0)^n,$$

where $c_0, c_1, \ldots, c_{n-1}$ are arbitrary constants. The series converges in an interval around x_0, the radius of which turns out to be the distance between x_0 and the nearest singular point of the equation to x_0.

Let us prove the first part of this statement, that an infinite series solution $y(x)$ can be found, using for simplicity the origin $x_0 = 0$ as the ordinary point under consideration, and considering only a second-order equation. Then our equation is of the form

$$y'' + b_1(x)y' + b_0(x)y = 0.$$

We certainly know that if $y(x)$ can be expanded in a series around $x = 0$ at all, then this series must be the Maclaurin series:

$$y(x) = \sum_{n=0}^{\infty} \frac{D^n y(0)}{n!} x^n = y(0) + y'(0)x + \frac{y''(0)}{2!}x^2 + \cdots.$$

If initial data are given, then $y(0)$ and $y'(0)$ are known constants; if not, then they are arbitrary, say $y(0) = c_0$ and $y'(0) = c_1$. To find $y''(0)$, we appeal to the differential equation, which can be written

$$y''(x) = -\, b_1(x)y'(x) - b_0(x)y(x), \tag{5.3}$$

which, at $x = 0$, gives

$$y''(0) = -b_1(0)Dy(0) - b_0(0)y(0) = -b_1(0)c_1 - b_0(0)c_0 = 2c_2.$$

We can now find c_3 by differentiating (5.3) with respect to x, which gives

$$y'''(x) = -b_1(x)y''(x) - [b_1'(x) + b_0(x)]y'(x) - b_0(x)y(x),$$

and then putting $x = 0$, which yields

$$y'''(0) = 6c_3 = -b_1(0)2c_2 - [b_1'(0) + b_0(0)]c_1 - b_0'(0)c_0;$$

this process can be continued indefinitely, giving us all the constants c_n, since we have assumed that $b_0(x)$ and $b_1(x)$ can be differentiated infinitely many times.

It turns out that the series solution obtained in this manner converges absolutely for all real and complex values of x closer to the origin than the nearest singular point. This distance is measured in the complex sense, where the complex number $(a + ib)$ is identified with the point whose coordinates are (a, b). Thus the distance from the origin to $a + ib$ is $\sqrt{a^2 + b^2}$. For example, the series solution of

$$(x^2 + 1)y'' - 2xy' + y = 0$$

around the origin will converge up to a distance of one unit away, i.e., for $-1 < x < 1$, because the singular points $x = \pm i$ are each one unit away from the origin. The series must be examined for convergence at the end-points $x = \pm 1$ separately.

We shall not prove this contention here, because in any event one can always determine the radius of convergence of a power series by means of the ratio test or some similar device. (See the Appendix for a summary of facts about infinite series).

In practice, finding the series solution of a linear differential equation is much simpler than the cumbersome process used in the above proof. The actual technique used involves substituting $\sum c_n x^n$ for y in the differential equation, gathering the coefficients of like powers of x together, and equating them to the coefficients of the right side (zero in the case of the homogeneous equation). This step is valid because, as we recall, the powers of x are linearly independent. We then obtain a formula for $c_2, c_3, \ldots, c_n, \ldots$ in terms of c_0 and c_1. This process is best illustrated by some examples.

Example 1

Normally we shall use the $\sum$ notation for sums; however, we shall work one simple example without this notation in order that the method might be easier to see. This is the equation

$$y'' - y = 0,$$

which by the methods of Chapter 3 has as its general solution

$$y = a_1 e^x + a_2 e^{-x}.$$

If there is any justice in the universe, then the series method should give us the same solution. Let us see if it does. We write

$$y = c_0 + c_1 x + c_2 x^2 + c_3 x^3 + c_4 x^4 + \cdots;$$

then

$$y' = c_1 + 2c_2 x + 3c_3 x^2 + 4c_4 x^3 + 5c_5 x^4 + \cdots$$
$$y'' = 2c_2 + 2 \cdot 3c_3 x + 3 \cdot 4c_4 x^2 + 4 \cdot 5c_5 x^3 + 5 \cdot 6c_6 x^4 + \cdots.$$

Putting the expressions for y'' and y into the differential equation and gathering like powers of x, we have

$$(2c_2 - c_0) + (2 \cdot 3c_3 - c_1)x + (3 \cdot 4c_4 - c_2)x^2 + (4 \cdot 5c_5 - c_3)x^3 + (5 \cdot 6c_6 - c_4)x^4 + \cdots = 0.$$

Because the powers of x are linearly independent, we have

$$\begin{aligned} 2c_2 - c_0 &= 0, & 2 \cdot 3c_3 - c_1 &= 0, \\ 3 \cdot 4c_4 - c_2 &= 0, & 4 \cdot 5c_5 - c_3 &= 0, \\ 5 \cdot 6c_6 - c_4 &= 0, & 6 \cdot 7c_7 - c_5 &= 0, \\ &\vdots & &\vdots \end{aligned}$$

where we have placed the coefficients with even subscripts in one column and those with odd subscripts in the other. From the first column we obtain

$$c_2 = \frac{c_0}{2}, \quad c_4 = \frac{c_2}{3 \cdot 4} = \frac{c_0}{2 \cdot 3 \cdot 4}, \quad c_6 = \frac{c_4}{5 \cdot 6} = \frac{c_0}{2 \cdot 3 \cdot 4 \cdot 5 \cdot 6}, \ldots$$

and from the second,

$$c_3 = \frac{c_1}{2 \cdot 3}, \quad c_5 = \frac{c_3}{4 \cdot 5} = \frac{c_1}{2 \cdot 3 \cdot 4 \cdot 5}, \quad c_7 = \frac{c_1}{2 \cdot 3 \cdot 4 \cdot 5 \cdot 6 \cdot 7}, \ldots$$

and hence we have

$$y = c_0 + c_1 x + \frac{c_0}{2}x^2 + \frac{c_1}{3!}x^3 + \frac{c_0}{4!}x^4 + \frac{c_1}{5!}x^5 + \frac{c_0}{6!}x^6 + \frac{c_1}{7!}x^7 + \cdots$$

$$= c_0\left(1 + \frac{x^2}{2!} + \frac{x^4}{4!} + \frac{x^6}{6!} + \cdots\right) + c_1\left(x + \frac{x^3}{3!} + \frac{x^5}{5!} + \frac{x^7}{7!} + \cdots\right).$$

We recognize the first series to be that for $\cosh x$, and the second, $\sinh x$; thus

$$y = c_0 \cosh x + c_1 \sinh x = c_0\left(\frac{e^x + e^{-x}}{2}\right) + c_1\left(\frac{e^x - e^{-x}}{2}\right)$$

$$= \frac{c_0 + c_1}{2}e^x + \frac{c_0 - c_1}{2}e^{-x} = a_1 e^x + a_2 e^{-x},$$

and there is indeed justice in the universe!

Example 2
Let us solve

$$(x^2 + 1)y'' - 3xy' + 3y = 0$$

around the ordinary point at $x = 0$. If we put

$$y = \sum_{n=0}^{\infty} c_n x^n,$$

then we have that

$$y' = \sum_{n=0}^{\infty} n c_n x^{n-1}, \quad y'' = \sum_{n=0}^{\infty} n(n-1)c_n x^{n-2},$$

assuming that we can differentiate an infinite series term by term.* We now substitute these series expressions for y and its derivatives directly into the

* See the Appendix.

equation and obtain

$$(x^2 + 1) \sum_{n=0}^{\infty} n(n-1)c_n x^{n-2} - 3x \sum_{n=0}^{\infty} nc_n x^{n-1} + 3 \sum_{n=0}^{\infty} c_n x^n = 0.$$

Next, we multiply under the summation signs by the coefficients of the equation and gather like powers of x together under one summation sign:

$$\sum_{n=0}^{\infty} n(n-1)c_n x^n + \sum_{n=0}^{\infty} (-3n)c_n x^n + \sum_{n=0}^{\infty} 3c_n x^n + \sum_{n=0}^{\infty} n(n-1)c_n x^{n-2} = 0$$

or

$$\sum_{n=0}^{\infty} [n(n-1) - 3n + 3]c_n x^n + \sum_{n=0}^{\infty} n(n-1)c_n x^{n-2} = 0,$$

finally yielding

$$\sum_{n=0}^{\infty} (n-1)(n-3)c_n x^n + \sum_{n=0}^{\infty} n(n-1)c_n x^{n-2} = 0.$$

At this point, we need to change one of the indices of summation so that x will occur with the same exponent in both series. The usual technique is to take the series with the higher powers of x and decrease its summation index. Thus, in the first series, we replace n by $n-2$; hence $n-1$ becomes $n-3$ and $n-3$ becomes $n-5$. To preserve the terms of the series, the sum must now run from 2 to ∞; thus we have

$$\sum_{n=2}^{\infty} (n-3)(n-5)c_{n-2} x^{n-2} + \sum_{n=0}^{\infty} n(n-1)c_n x^{n-2} = 0.$$

One further step is now necessary before we can gather the two series under the same summation sign: We must remove the first two terms from the second series so that both summations start with $n = 2$:

$$0 \cdot c_0 x^{-2} + 0 \cdot c_1 x^{-1} + \sum_{n=2}^{\infty} [n(n-1)c_n + (n-3)(n-5)c_{n-2}]x^{n-2} = 0.$$

For this expression to vanish, the coefficients of like powers of x must all vanish, because the powers of x are linearly independent. Thus

$$0 \cdot c_0 = 0, \quad 0 \cdot c_1 = 0, \quad \text{and} \quad n(n-1)c_n + (n-3)(n-5)c_{n-2} = 0,$$

the latter for $n = 2, 3, 4, \ldots$. The first two equations simply tell us that c_0 and c_1 are arbitrary. The third implies that

$$c_n = \frac{-(n-3)(n-5)}{n(n-1)} c_{n-2}, \qquad n = 2, 3, 4, \ldots.$$

This formula, called a *recursion formula*, gives each coefficient in terms of the one whose subscript is two less. Thus, ultimately, all the coefficients with even subscripts are given in terms of c_0, and those with odd subscripts, in terms of c_1 (the two arbitrary constants of integration). We can determine a general formula for the coefficients c_n by first evaluating the recursion formula for various values of n in columns, as shown in Table 5.1.

Table 5.1

c_0 arbitrary	c_1 arbitrary
$c_2 = -\frac{(-1)(-3)}{2 \cdot 1} c_0$	$c_3 = -\frac{0 \cdot (-2)}{3 \cdot 2} c_1 = 0$
$c_4 = -\frac{(1)(-1)}{4 \cdot 3} c_2$	$c_5 = -\frac{2 \cdot 0}{5 \cdot 4} c_3 = 0$
$c_6 = -\frac{(3)(1)}{6 \cdot 5} c_4$	$c_7 = -\frac{4 \cdot 2}{7 \cdot 6} c_5 = 0$
$\vdots$	$\vdots$
$c_{2k} = -\frac{(2k-3)(2k-5)}{2k(2k-1)} c_{2k-2}$	$c_{2k+1} = 0 \quad (k \geqq 1)$.

To find c_{2k} in terms of c_0, we multiply all the left sides in the first column together, and equate to the product of the right sides, giving

$$c_2 c_4 c_6 \cdots c_{2k} = (-1)^k \frac{(-1)(-3)(1)(3)(1)\cdots(2k-3)(2k-5)}{1 \cdot 2 \cdot 3 \cdot 4 \cdot 5 \cdot 6 \cdots (2k-1)(2k)} c_0 c_2 c_4 \cdots c_{2k-2}.$$

The common factor $c_2 c_4 \cdots c_{2k-2}$ can be canceled from both sides, and we obtain, after some rearrangement of factors,

$$c_{2k} = (-1)^k \frac{[(-1)(1) \cdot 3 \cdots (2k-3)][(-3)(-1) \cdot 1 \cdots (2k-5)]}{(2k)!} c_0, \qquad k \geqq 1. \tag{5.4}$$

For the coefficients with odd subscripts, we notice that $c_3 = 0$, and hence $c_5 = c_7 = \cdots = c_{2k+1} = \cdots = 0$, since a multiplication of the left and right sides of the equations in the second column yields

$$c_{2k+1} = (-1)^k \frac{(0)(-2)(2)(0)\cdots(2k)(2k-2)}{(2k+3)!} c_1 = 0 \cdot c_1 = 0,$$

and every odd coefficient thus vanishes except for c_1. Now

$$\begin{aligned} y(x) &= \sum_{n=0}^{\infty} c_n x^n = \sum_{k=0}^{\infty} c_{2k} x^{2k} + \sum_{k=0}^{\infty} c_{2k+1} x^{2k+1} \\ &= c_0 + \sum_{k=1}^{\infty} c_{2k} x^{2k} + \sum_{k=0}^{\infty} c_{2k+1} x^{2k+1}. \end{aligned}$$

We have just split the series for $y(x)$ into two series, one containing all the even powers of x, the other with all the odd powers. We removed the $k = 0$ term from the first series because formula (5.4) holds only for $k \geqq 1$. Substituting (5.4), we have

$$y(x) = c_0 \left\{ 1 + \sum_{k=1}^{\infty} (-1)^k \frac{[(-1)\cdot 1 \cdot 3 \cdots (2k-3)][(3)(-1)\cdot 1 \cdots (2k-5)]x^{2k}}{(2k)!} \right\} + c_1 x.$$

Note that y is of the form $c_0 y_0 + c_1 y_1$, i.e., it is a linear combination of two linearly independent particular solutions, which, of course, we have grown to expect of the solution of a second-order homogeneous linear equation.

Where does this series solution converge? Naturally the term $c_1 x$ is defined for all values of x, and according to the ratio test, the series of even powers converges for all values of x for which

$$x^2 < \lim_{k\to\infty} \left| \frac{c_{2k}}{c_{2k+2}} \right|.$$

(Convergence at the endpoints of the interval of convergence must be investigated separately.) Now a little calculation shows that

$$\lim_{k\to\infty} \left| \frac{c_{2k}}{c_{2k+2}} \right| = \lim_{k\to\infty} \left| \frac{(2k+1)(2k+2)}{(2k-1)(2k-3)} \right| = 1;$$

therefore the series of even powers converges for $-1 < x < 1$. The series portion of the solution does not converge at either of the endpoints $x = \pm 1$ because c_{2k} does not even approach zero as $k \to \infty$. Note that the singular points of the equation are $x = \pm i$, each of which is a distance of one unit from the origin. Because of our rule that the infinite series solution of an ordinary differential equation around an ordinary point always converges absolutely inside an interval whose radius is the distance between the ordinary point and the nearest singular point, in this case we can read from the equation that the radius of convergence around the origin must be 1.

Example 3

In example 2, one of the series solutions terminated after a finite number (one) of terms. This does not always happen. As a case in point, let us consider the following equation:

$$2y'' - 3xy' - 6y = 0.$$

Making the substitution $y = \sum_{n=0}^{\infty} c_n x^n$ as before, we obtain

$$\sum_{n=0}^{\infty} 2n(n-1)c_n x^{n-2} + \sum_{n=0}^{\infty} (-3n)c_n x^n + \sum_{n=0}^{\infty} (-6c_n)x^n = 0.$$

Changing the index of summation in the second two series from n to $n-2$, and going through the same process as before, we obtain

$$0 \cdot c_0 x^{-2} + 0 \cdot c_1 x^{-1} + \sum_{n=2}^{\infty} [2n(n-1)c_n - 3nc_{n-2}]x^{n-2} = 0,$$

which implies that

$$0 \cdot c_0 = 0, \qquad 0 \cdot c_1 = 0, \qquad [2n(n-1)c_n - 3nc_{n-2}] = 0 \quad \text{for } n = 2, 3, 4, \ldots,$$

which as before implies that c_0 and c_1 are arbitrary and that

$$c_n = \frac{3n}{2n(n-1)} c_{n-2} = \frac{3}{2(n-1)} c_{n-2} \qquad \text{for } n \geqq 2.$$

Proceeding as in the previous example, we set up separate columns for the coefficients of even and odd powers of x (Table 5.2).

Table 5.2

c_0 arbitrary	c_1 arbitrary
$c_2 = \dfrac{3}{2 \cdot 1} c_0$	$c_3 = \dfrac{3}{2 \cdot 2} c_1$
$c_4 = \dfrac{3}{2 \cdot 3} c_2$	$c_5 = \dfrac{3}{2 \cdot 4} c_3$
$c_6 = \dfrac{3}{2 \cdot 5} c_4$	$c_7 = \dfrac{3}{2 \cdot 6} c_5$
$\vdots$	$\vdots$
$c_{2k} = \dfrac{3}{2(2k-1)} c_{2k-2}$	$c_{2k+1} = \dfrac{3}{2(2k)} c_{2k-1}$

Multiplying together the two sides of the first column in Table 5.2, we have

$$c_2 c_4 c_6 \cdots c_{2k} = \frac{3^k}{2^k \cdot 1 \cdot 3 \cdot 5 \cdots (2k-1)} c_0 c_2 c_4 \cdots c_{2k-2}$$

or

$$c_{2k} = \frac{3^k}{2^k \cdot 1 \cdot 3 \cdot 5 \cdots (2k-1)} c_0, \qquad k = 1, 2, \ldots.$$

This expression can be simplified in appearance by multiplying numerator and denominator by $2 \cdot 4 \cdot 6 \cdots 2k = 2^k k!$, whereupon the denominator becomes $(2k)!$ and we have

$$c_{2k} = \frac{3^k k!}{(2k)!} c_0.$$

To find the coefficients with odd subscripts, we multiply the two sides of column 2 together and obtain

$$c_3 c_5 c_7 \cdots c_{2k+1} = \frac{3^k}{2^k \cdot 2 \cdot 4 \cdot 6 \cdot 8 \cdots (2k)} c_1 c_3 c_5 \cdots c_{2k-1},$$

and since $2 \cdot 4 \cdot 6 \cdot 8 \cdots (2k) = 2^k(1 \cdot 2 \cdot 3 \cdot 4 \cdots k) = 2^k k!$, we have

$$c_{2k+1} = \frac{3^k}{2^{2k} k!} c_1, \qquad k = 1, 2, 3, \ldots.$$

Utilizing the formulas for c_{2k} and c_{2k+1}, and recalling again that

$$\sum_{n=0}^{\infty} c_n y^n = c_0 + \sum_{k=1}^{\infty} c_{2k} x^{2k} + c_1 x + \sum_{k=1}^{\infty} c_{2k+1} x^{2k+1},$$

we have

$$y(x) = c_0\left[1 + \sum_{k=1}^{\infty} \frac{3^k k!}{(2k)!} x^{2k}\right] + c_1\left(x + \sum_{k=1}^{\infty} \frac{3^k}{2^k k!} x^{2k+1}\right).$$

The original recursion formula tells us that

$$\lim_{n\to\infty} \left|\frac{c_n}{c_n + 2}\right| = \lim_{n\to\infty} \frac{2(n+1)}{3} = \infty,$$

so that the two series solutions given above converge for all finite values of x. We could also determine this from the fact that the equation has no singular points at all; thus the distance from the origin to the nearest singular point is infinite.

Example 4
Now let us consider the equation

$$2(x-1)(x^2 - 5x + 7)y'' - (x-2)^2 y' - 9(x-2)y = 0.$$

This time we shall seek a solution around the ordinary point at $x = 2$, that is, we want a series solution of the form $\sum_{n=0}^{\infty} c_n(x-2)^n$. To do this, we shall have to write the coefficients in terms of powers of $(x-2)$. Now

$$(x-1) = [(x-2) + 1] \text{ and}$$
$$x^2 - 5x + 7 = (x-2)^2 - (x-2) + 1.$$

[Recall that a polynomial in powers of x can be expanded in terms of powers of $(x - x_0)$ by using Taylor's formula

$$p(x) = a_n x^n + \cdots + a_1 x + a_0 = \frac{p^{(n)}(x_0)}{n!}(x-x_0)^n + \cdots$$
$$+ \frac{p'(x_0)}{1!}(x-x_0) + p(x_0),$$

which is how we arrived at the above result.] Thus the leading coefficient $2(x-1)(x^2-5x+7)$ can be written

$$2[(x-2)+1][(x-2)^2-(x-2)+1] = 2[(x-2)^3-1].$$

Hence, in terms of powers of $x-2$, our differential equation is

$$2[(x-2)^3-1]y''-(x-2)^2y'-9(x-2)y=0.$$

Making the substitution

$$y = \sum_{n=0}^{\infty} c_n(x-2)^n,$$

$$y' = \sum_{n=0}^{\infty} nc_n(x-2)^{n-1},$$

$$y'' = \sum_{n=0}^{\infty} n(n-1)c_n(x-2)^{n-2},$$

we have

$$\sum_{n=0}^{\infty} 2n(n-1)c_n(x-2)^{n+1} + \sum_{n=0}^{\infty} (-2)n(n-1)c_n(x-2)^{n-2}$$
$$+ \sum_{n=0}^{\infty} (-1)nc_n(x-2)^{n+1} + \sum_{n=0}^{\infty} (-9)c_n(x-2)^{n+1} = 0$$

or

$$\sum_{n=0}^{\infty} (-2)n(n-1)c_n(x-2)^{n-2} + \sum_{n=0}^{\infty} [2n(n-1)-n-9]c_n(x-2)^{n+1} = 0.$$

Changing the index of summation in the second series from n to $n-3$, and simplifying the expression in brackets, we have

$$\sum_{n=0}^{\infty} (-2)n(n-1)c_n(x-2)^{n-2} + \sum_{n=3}^{\infty} (2n-3)(n-6)c_{n-3}(x-2)^{n-2} = 0,$$

giving

$$0\cdot c_0(x-2)^{-2} + 0\cdot c_1(x-2)^{-1} + (-2)(2)(1)c_2$$
$$+ \sum_{n=3}^{\infty} [-2n(n-1)c_n + (2n-3)(n-6)c_{n-3}](x-2)^{n-2} = 0.$$

so that

$$0 \cdot c_0 = 0, \quad 0 \cdot c_1 = 0, \quad -4c_2 = 0, \quad c_n = \frac{(2n-3)(n-6)}{2n(n-1)} c_{n-3}$$
$$(n = 3, 4, 5, \ldots).$$

The first two equations tell us that c_0 and c_1 are arbitrary, and the third, that $c_2 = 0$. To get formulas for the coefficients, three columns are now necessary

Table 5.3

c_0 arbitrary	c_1 arbitrary	$c_2 = 0$
$c_3 = \frac{(3)(-3)}{2 \cdot 3 \cdot 2} c_0 = -\frac{3}{4} c_0$	$c_4 = \frac{5(-2)}{2 \cdot (4)(3)} c_1$	$c_5 = \frac{7 \cdot 2}{2 \cdot 5 \cdot 4} c_2 = 0$
$c_6 = \frac{9 \cdot 0}{2 \cdot 6 \cdot 5} c_3 = 0$	$c_7 = \frac{11(1)}{2 \cdot 7 \cdot 6} c_4$	$c_8 = 0$
$c_9 = \frac{15 \cdot 3}{2 \cdot 9 \cdot 8} c_6 = 0$	$c_{10} = \frac{17(4)}{2 \cdot 10 \cdot 9} c_7$	$c_{11} = 0$
$\vdots$	$\vdots$	$\vdots$
$c_{3k} = 0$	$c_{3k+1} = \frac{(6k-1)(3k-5)}{2(3k+1)3k} c_{3k-2}$	$c_{3k+2} = 0$

(Table 5.3). We note from the last column in the table that because $c_2 = 0$, all coefficients with subscripts of the form $3k + 2 (k = 0, 1, 2, 3, \ldots)$ are also zero. From the first column, we see that $c_{3k} = 0$ for $k = 2, 3, 4, \ldots$; i.e., only the first two coefficients with subscripts of the form $3k$ are not zero. The second column then gives

$$c_4 c_7 c_{10} \cdots c_{3k+1} = \frac{5 \cdot 11 \cdot 17 \cdots (6k-1)(-2) \cdot 1 \cdot 4 \cdots (3k-5)}{2^k \cdot 4 \cdot 7 \cdot 10 \cdots (3k+1) \cdot 3 \cdot 6 \cdot 9 \cdots 3k} c_1 c_4 c_7 \cdots c_{3k-2}$$

or

$$c_{3k+1} = \frac{5 \cdot 11 \cdot 17 \cdots (6k-1)(-2) \cdot 1 \cdot 4 \cdots (3k-5)}{6^k k! 4 \cdot 7 \cdot 10 \cdots (3k+1)} c_1$$

since $3 \cdot 6 \cdot 9 \cdots 3k = 3^k k!$. Therefore, because

$$\sum_{n=0}^{\infty} c_n (x-2)^n = \sum_{k=0}^{\infty} c_{3k} (x-2)^{3k} + \sum_{k=0}^{\infty} c_{3k+1} (x-2)^{3k+1}$$
$$+ \sum_{k=0}^{\infty} c_{3k+2} (x-2)^{3k+2},$$

we have

$$y(x) = c_0[1 - \tfrac{3}{4}(x-2)^3]$$
$$+ c_1\left[(x-2) + \sum_{k=1}^{\infty} \frac{5\cdot 11\cdot 17\cdots(6k-1)(-2)\cdot 1\cdot 4\cdots(3k-5)}{6^k k!\cdot 4\cdot 7\cdot 10\cdots(3k+1)}(x-2)^{3k+1}\right].$$

In this case we have as usual extracted the first term $c_1(x-2)$ from the series and written it separately, since the formula for c_{3k+1} holds only for $k \geqq 1$. The ratio test tells us that the infinite series converges for $-1 < (x-2) < 1$.

Problems

In problems 1–15, find the series solution valid near $x = 0$, after verifying that $x = 0$ is an ordinary point.

1. $(x^2 + 4)y'' - 4xy' + 6y = 0.$
2. $(2x^2 - 1)y'' - 9xy' + 5y = 0.$
3. $3y'' + 2xy' + 4y = 0.$
4. $y'' + 3xy = 0.$
5. $(1 - x^2)y'' + xy' + 3y = 0.$
6. $(3x^2 + 3)y'' + 11xy' + 4y = 0.$
7. $2y'' - xy' + 6y = 0.$
8. $2y'' \quad 5x^2y = 0.$
9. $(x^2 + 3)y'' + 3xy' + y = 0.$
10. $(6x^2 - 1)y'' - 6xy' + 6y = 0.$
11. $(3 - 2x^2)y'' - 6xy + 6y = 0.$
12. $3y'' - 2xy' - 7y = 0.$
13. $(1 - x^2)y'' - 4xy' + 4y = 0.$
14. $y'' - 6xy' - 4y = 0.$
15. $y'' + 4y = 0.$

In problems 16–20, find the series solution around the point indicated.

16. $x(x-2)y'' + 7(x-1)y' + 5y = 0$, around $x = 1$.
17. $y'' - 3(x-3)y = 0$, around $x = 3$.
18. $y'' + 4(x+2)y' - 3y = 0$, around $x = -2$.
19. $y'' + 2(x-1)y = 0$, around $x = 1$.
20. $(2x+2)(5-x)y'' - 4(x-2)y' + 4y = 0$, around $x = 2$.
21. Solve problem 3 by another method, and compare your solution with the one found in the text.

5.3 Solutions of Homogeneous Linear Equations near a Regular Singular Point (the General Case)

If the differential equation which we are trying to solve has a singular point at $x = x_0$, then it is not in general possible to find a Taylor series for the solution $y(x)$ around the point x_0. To see this, let us consider the second-order equation

$$3x^2y'' + 8xy' - 2y = 0,$$

which is an Euler equation. Making the usual substitution $x = e^t$, we obtain

$$3\frac{d^2y}{dt^2} + 5\frac{dy}{dt} - 2y = (3D - 1)(D + 2)y = 0$$

and hence $y = c_1e^{1/3t} + c_2e^{-2t}$. Therefore the general solution is

$$y = c_1x^{1/3} + c_2x^{-2}.$$

Neither of these solutions is a Maclaurin series, or even a truncated one, since x nowhere occurs to a positive integral power.

The same is true of more general equations, for example

$$y'' + a(x)y' + b(x)y = 0; \tag{5.5}$$

let us assume that $x = 0$ is a regular singular point. This means that either or both of $a(x)$ and $b(x)$ approach infinity as $x \to 0$ since

$$0 < \lim_{x\to 0} |xa(x)| < \infty \quad \text{and} \quad 0 < \lim_{x\to 0} |x^2b(x)| < \infty;$$

in fact, we can expand $a(x)$ and $b(x)$ in series in the following form:

$$a(x) = a_0/x + a_1 + a_2x + \cdots +,$$
$$b(x) = b_0/x^2 + b_1/x + b_2 + b_3x + \cdots;$$

this can be done by breaking $a(x)$ and $b(x)$ up into partial factions and by expanding terms not involving x alone in Maclaurin series.* For example:

$$1/[x(1 - x)] = 1/x + 1/(1 - x) = 1/x + 1 + x + x^2 + \cdots.$$

Since $x = 0$ is a regular singular point, no more than two of the three coefficients a_0, b_0, b_1 can vanish.

* Recall that $a(x)$ and $b(x)$ are the quotients of polynomials, so this is easily done.

Now if the solution of (5.5) had a Maclaurin series expansion

$$y(x) = \sum_{n=0}^{\infty} c_n x^n = c_0 + c_1 x + c_2 x^2 + \cdots,$$

then upon substituting in (5.5) we would have

$$\begin{aligned}(2c_2 + \cdots) + (a_0/x + a_1 + \cdots)(c_1 + 2c_2 x + \cdots)\\ + (b_0/x^2 + b_1/x + b_2 + \cdots)(c_0 + c_1 x + c_2 x^2 + \cdots) = 0,\end{aligned}$$

which upon gathering like powers of x, becomes

$$\begin{aligned}\frac{b_0 c_0}{x^2} + \frac{a_0 c_1 + b_0 c_1 + b_1 c_0}{x} + 2c_2 + 2a_0 c_2 + a_1 c_1 + b_0 c_2\\ + b_1 c_1 + b_2 c_0 + \cdots = 0,\end{aligned}$$

or

$$\begin{aligned}\frac{b_0 c_0}{x^2} + \frac{(a_0 + b_0)c_1 + b_1 c_0}{x} + c_2(2 + 2a_0 + b_0) + c_1(a_1 + b_1)\\ + b_2 c_0 + \cdots = 0.\end{aligned}$$

It is not difficult to see that if, for instance, a_0, b_0, and b_1 are all positive and the rest of the a_i's and b_i's vanish, then this would imply that $c_0 = c_1 = c_2 = 0$, and in fact that all the c_i's vanish. Thus we would arrive at the trivial solution $y(x) = 0$. We must conclude, then, that equation (5.5) does not *in general* possess a Maclaurin series solution (although it may in certain special cases).

We are then prompted to ask what form the solution of (5.5) must take. Our clue is in examining the solutions of Euler equations, which are the prototypes of this type of equation. While neither $x^{1/3}$ nor x^{-2} (for example) can be considered Maclaurin series, they do resemble Maclaurin series multiplied by a power of x. Thus, instead of writing $y(x)$ as a Maclaurin series, let us attempt a solution of the form

$$y = x^r \sum_{n=0}^{\infty} c_n x^n = \sum_{n=0}^{\infty} c_n x^{n+r} = c_0 x^r + c_1 x^{r+1} + \cdots.$$

If we substitute this expansion into (5.5), we obtain

$$\begin{aligned}c_0 r(r-1)x^{r-2} + (a_0/x + \cdots)(rc_0 x^{r-1} + \cdots)\\ + (b_0/x^2 + \cdots)(c_0 x^r + \cdots) = 0,\end{aligned}$$

where only the terms of lowest degree have been written. Gathering these terms together, we have

$$[r(r-1)c_0 + a_0c_0r + b_0c_0]x^{r-2} + \cdots = 0$$

or

$$[r^2 + (a_0 - 1)r + b_0]c_0x^{r-2} + \cdots = 0. \tag{5.6}$$

We intend that c_0 should be an arbitrary constant, so if the given expression for $y(x)$ is to be a solution, we must have

$$r^2 + (a_0 - 1)r + b_0 = 0. \tag{5.7}$$

Equation (5.7) is called the *indicial equation* of the differential equation (5.5). We note that the indicial equation is a quadratic equation with known real coefficients, and it in general has two solutions r_1 and r_2. There are a number of possibilities for these roots:

(1) r_1 and r_2 may be complex. We shall not consider this case, except for Euler equations.

(2) r_1 and r_2 may be equal.

(3) r_1 and r_2 may not be equal; their difference may be an integer or a noninteger.

We shall consider in this section the general case in which $r_2 - r_1$ is not an integer. The other cases will be discussed in Chapter 7.

Naturally the fact that the indicial equation vanishes if $r = r_1$ or $r = r_2$ still does not guarantee us a solution of the form $x^r \sum_{n=0}^{\infty} c_nx^n$. Only the lowest power of x has a vanishing coefficient in (5.5) when r satisfies the indicial equation. Nevertheless, we should be able to obtain a recursion formula for the rest of the coefficients by examining the higher powers of x in (5.5). And indeed, this turns out to be the case. From this point on, the technique of solution is quite similar to that used in the last section to find Maclaurin series solutions. Let us consider a few examples to see how the process goes.

Example 1

For our first example we shall attempt to find a solution of

$$3xy'' + (3x + 2)y' - 4y = 0,$$

having the form $y = \sum_{n=0}^{\infty} c_nx^{n+r}$ valid near the regular singular point at $x = 0$. Since

$$y' = \sum_{n=0}^{\infty} (n + r)c_nx^{n+r-1} \quad \text{and} \quad y'' = \sum_{n=0}^{\infty} (n + r)(n + r - 1)c_nx^{n+r-2},$$

we have

$$\sum_{n=0}^{\infty} 3(n + r)(n + r - 1)c_n x^{n+r-1} + \sum_{n=0}^{\infty} 3(n + r)c_n x^{n+r}$$
$$+ \sum_{n=0}^{\infty} 2(n + r)c_n x^{n+r-1} - \sum_{n=0}^{\infty} 4c_n x^{n+r} = 0$$

or

$$\sum_{n=0}^{\infty} [3(n + r)(n + r - 1) + 2(n + r)]c_n x^{n+r-1} + \sum_{n=0}^{\infty} [3(n + r) - 4]c_n x^{n+r} = 0.$$

Simplifying the first summand by removing the common factor $(n + r)$, and changing the index of summation in the second from n to $n - 1$, we obtain

$$\sum_{n=0}^{\infty} (n + r)[3(n + r) - 1]c_n x^{n+r-1} + \sum_{n=1}^{\infty} [3(n + r) - 7]c_{n-1} x^{n+r-1} = 0$$

or finally,

$$r(3r - 1)c_0 x^{r-1} + \sum_{n=1}^{\infty} \{(n + r)[3(n + r) - 1]c_n + [3(n + r) - 7]c_{n-1}\}x^{n+r-1} = 0.$$

Once again the coefficient of each power of x must vanish, so that

$$r(3r - 1)c_0 = 0 \text{ and } (n + r)[3(n + r) - 1]c_n + [3(n + r) - 7]c_{n-1} = 0$$
$$\text{for } n = 1, 2, 3, \ldots.$$

The first of these is the indicial equation, and tells us that $r_1 = 0$ and $r_2 = \frac{1}{3}$ are the two roots, and that c_0 is an arbitrary constant. The second is the recursion formula; since $r_1 - r_2 = \frac{1}{3}$ is not an integer, we will obtain two different recursion formulas when we insert r_1 and r_2 into it, each of which will give us a different solution of the differential equation. For, if $r_1 - r_2$ is an integer, then the two recursion formulas might duplicate each other—one reason why this case must be treated separately.

Using $r = 0$, we have

$$c_n = -\frac{3n - 7}{n(3n - 1)}c_{n-1}, \qquad n = 1, 2, 3, \ldots,$$

so

$$c_1 = -\frac{(-4)}{1\cdot 2}c_0,\quad c_2 = -\frac{(-1)}{2\cdot 5}c_1,\quad c_3 = -\frac{2}{3\cdot 8}c_2,\quad c_4 = -\frac{5}{4\cdot 11}c_3, \ldots$$

and, multiplying the two sides of these equations together and canceling like factors from both sides as is our custom, we get the formula

$$c_n = (-1)^n \frac{(-4)(-1)2\cdot 5 \cdots (3n-7)}{n!2\cdot 5\cdots(3n-1)}c_0;$$

our first solution

$$y = c_0 x^{r_1} + \sum_{n=1}^{\infty} c_n x^{n+r_1}$$

is thus

$$y_1 = 1 + \sum_{n=1}^{\infty} \frac{(-1)^n(-4)(-1)2\cdot 5\cdots(3n-7)}{n!2\cdot 5\cdots(3n-1)}x^n,$$

where we have set $c_0 = 1$ for a particular solution. (Notice that this solution is actually a Maclaurin series.)

To find the second particular solution, we put $r = \frac{1}{3}$ into the recursion formula and proceed as before:

$$c_n = -\frac{3(n+\frac{1}{3})-7}{(n+\frac{1}{3})[3(n+\frac{1}{3})-1]}c_{n-1} = -\frac{3(n-2)}{(3n+1)n}c_{n-1}, \qquad n = 1, 2, 3, \ldots,$$

and hence if $c_0 = 1$ (for a particular solution),

$$c_0 = 1,\quad c_1 = -\frac{3(-1)}{4\cdot 1} = \frac{3}{4},\quad c_2 = 0, \ldots$$

and all coefficients from c_2 on vanish. Thus our second solution becomes simply

$$y_2 = x^{1/3} + \tfrac{3}{4}x^{4/3}$$

Naturally the general solution of the differential equation is then

$$y = a_1 y_1 + a_2 y_2,$$

where a_1 and a_2 are arbitrary constants.

Example 2
Let us now consider the equation

$$x^2y'' + xy' + (x^2 - \nu^2)y = 0,$$

where ν is a positive constant such that 2ν is not an integer. This equation, of which we shall see more in future sections, is called *Bessel's equation of index* ν. Making the substitution $y = \sum_{n=0}^{\infty} c_n x^{n+r}$ in order to find a solution valid near the regular singular point at $x = 0$, we see that

$$\sum_{n=0}^{\infty} (n + r)(n + r - 1)c_n x^{n+r} + \sum_{n=0}^{\infty} (n + r)c_n x^{n+r} + \sum_{n=0}^{\infty} c_n x^{n+r+2} + \sum_{n=0}^{\infty} (-\nu^2)c_n x^{n+r} = 0,$$

which leads to

$$\sum_{n=0}^{\infty} [(n + r)(n + r - 1) + (n + r) - \nu^2]c_n x^{n+r} + \sum_{n=0}^{\infty} c_n x^{n+r+2} = 0.$$

Simplifying the first sum* and replacing n by $n - 2$ in the second, we obtain

$$(r + \nu)(r - \nu)c_0 x^r + (r + \nu + 1)(r - \nu + 1)c_1 x^{r+1} + \sum_{n=2}^{\infty} [(r + n - \nu)(r + n + \nu)c_n + c_{n-2}]x^{n+2} = 0.$$

We thus obtain

$$(r + \nu)(r - \nu)c_0 = 0,$$
$$(r + \nu + 1)(r - \nu + 1)c_1 = 0,$$
$$(r + n - \nu)(r + n + \nu)c_n + c_{n-2} = 0, \quad \text{for } n = 2, 3, 4, \ldots.$$

The first of these is the indicial equation, and its roots are $r_1 = -\nu$ and $r_2 = \nu$, and again $r_2 - r_1 = 2\nu$ is not an integer (by hypothesis). The second of these gives one of the following:

$$(-2\nu + 1)c_1 = 0, \quad (2\nu + 1)c_1 = 0$$

* We always do this by treating $(n + r)$ as an algebraically inseparable unit. Thus, here, $(n + r)(n + r - 1) + (n + r) - \nu^2 = (n + r)^2 - (n + r) + (n + r) - \nu^2 = (n + r)^2 - \nu^2 = (n + r + \nu)(n + r - \nu)$.

according as r is $-\nu$ or ν, and in either case we must have $c_1 = 0$ since $-2\nu + 1$ and $2\nu + 1$ cannot vanish.

Once again let us obtain two particular solutions by choosing $c_0 = 1$. In the case $r_1 = -\nu$, we have

$$c_n = -\frac{c_{n-2}}{n(n-2\nu)}, \qquad n = 2, 3, 4, \ldots,$$

so that

$$c_2 = -\frac{1}{2(2-2\nu)}, \quad c_4 = -\frac{1}{4(4-2\nu)}c_2, \quad c_6 = -\frac{1}{6(6-2\nu)}c_4, \ldots,$$

$$c_{2k} = -\frac{1}{2k(2k-2\nu)}c_{2k-2}.$$

Note that $c_3, c_5, \ldots$ all vanish because c_1 does. Thus

$$\begin{aligned} c_{2k} &= \frac{(-1)^k}{2\cdot 4\cdot 6\cdots 2k(2-2\nu)(4-2\nu)\cdots(2k-2\nu)} \\ &= \frac{(-1)^k}{2^{2k}k!(1-\nu)(2-\nu)\ldots(k-\nu)}. \end{aligned}$$

Therefore, the first solution is

$$\begin{aligned} y_1(x) &= c_0 x^{r_1} + \sum_{k=1}^{\infty} c_{2k} x^{2k+r_1} \\ &= x^{-\nu} + \sum_{k=1}^{\infty} \frac{(-1)^k x^{2k-\nu}}{2^{2k}k!(1-\nu)(2-\nu)\cdots(k-\nu)}, \end{aligned}$$

because all the coefficients with odd subscripts vanish. Now, in the case $r_2 = \nu$, we have

$$c_n = -\frac{c_{n-2}}{n(n+2\nu)}.$$

The only difference between this and the previous calculation is that $-\nu$ has been replaced by ν. Thus our second solution will be

$$y_2(x) = x^{\nu} + \sum_{k=1}^{\infty} \frac{(-1)^k x^{2k+\nu}}{2^{2k}k!(1+\nu)(2+\nu)\cdots(k+\nu)}.$$

The general solution of the original equation will be $y = a_1 y_1 + a_2 y_2$, where a_1 and a_2 are arbitrary constants.

The solutions of Bessel's equation are most often seen in another form, which we can obtain from the previous representations by using the gamma function. We recall that the gamma function has the factorial property

$$t\Gamma(t) = \Gamma(t+1).$$

This can be used to simplify the coefficients in the series for y_1 and y_2, for we have

$$\begin{aligned}(1+v)\Gamma(1+v) &= \Gamma(2+v),\\ (2+v)\Gamma(2+v) &= \Gamma(3+v),\ldots,\\ (k+v)\Gamma(k+v) &= \Gamma(k+v+1),\end{aligned}$$

and therefore

$$\Gamma(1+v)(1+v)(2+v)\cdots(k+v) = \Gamma(k+v+1),$$

and similarly for the terms with $-v$. Hence

$$\frac{1}{2^{-v}\Gamma(1-v)}y_1(x) = \frac{1}{\Gamma(1-v)}\left(\frac{x}{2}\right)^{-v} + \sum_{k=1}^{\infty}\frac{(-1)^k}{k!\Gamma(k-v+1)}\left(\frac{x}{2}\right)^{2k-v}$$

and

$$\frac{1}{2^{v}\Gamma(1+v)}y_2(x) = \frac{1}{\Gamma(1+v)}\left(\frac{x}{2}\right)^{v} + \sum_{k=1}^{\infty}\frac{(-1)^k}{k!\Gamma(k+v+1)}\left(\frac{x}{2}\right)^{2k+v}.$$

The first of these two functions is customarily denoted by $J_{-v}(x)$, and the second by $J_v(x)$, and they are called *Bessel functions of the first kind* of index (or order) $-v$ and v, respectively. Since $J_{-v}(x)$ and $J_v(x)$ are obtained from linearly independent particular solutions y_1 and y_2 by multiplying them by nonzero constants, we conclude that $J_{-v}(x)$ and $J_v(x)$ are themselves a linearly independent set of solutions of Bessel's equation of index v, *provided that $2v$ is not an integer.* We shall see in subsequent sections of Chapter 7 what happens when $2v$ is an integer or zero.

Example 3

Finally, let us consider

$$3(x-1)^2y'' - 2(x-1)y' - 2y = 0$$

near the regular singular point at $x = 1$. In other words, we seek solutions of the form $y = \sum_{n=0}^{\infty} c_n(x-1)^{n+r}$. Substituting this into the equation, we obtain

$$\sum_{n=0}^{\infty} 3(n+r)(n+r-1)c_n(x-1)^{n+r} - \sum_{n=0}^{\infty} 2(n+r)c_n(x-1)^{n+r} - \sum_{n=0}^{\infty} 2c_n(x-1)^{n+r} = 0.$$

Right away we notice that all terms have the same power of $(x-1)$ in them, an occurrence we have not seen previously. This need not, however, be any cause for panic. Let us first gather all terms together under the same summation sign:

$$\sum_{n=0}^{\infty} [3(n+r)(n+r-1) - 2(n+r) - 2]c_n(x-1)^{n+r} = 0,$$

which becomes

$$\sum_{n=0}^{\infty} [3(n+r)^2 - 5(n+r) - 2]c_n(x-1)^{n+r} = \sum_{n=0}^{\infty} [3(n+r)+1][(n+r)-2]c_n(x-1)^{n+r} = 0.$$

Again, as in our previous examples, the coefficients of each individual power of $x-1$ must vanish. Thus we have for $n = 0$ that

$$(3r+1)(r-2)c_0 = 0.$$

As in previous cases, the $n = 0$ term tells us that we must have $r_1 = -\frac{1}{3}$, $r_2 = 2$; and $r_2 - r_1 = \frac{7}{3}$, which is not an integer. For $n \geqq 1$, we have

$$[3(n+r)+1][(n+r)-2]c_n = 0;$$

for $r_1 = -\frac{1}{3}$ this is

$$3n(n - \tfrac{7}{3})c_n = 0$$

and for $r_2 = 2$ it is

$$(3n+7)nc_n = 0;$$

in either case, since $3n(n-\frac{7}{3})$ and $(3n+7)n$ never vanish for $n \geqq 1$, it follows that $c_n = 0$ for all $n \geqq 1$. Therefore both solutions of the differential equation have only a single term:

$$y_1 = (x - 1)^{-1/3}, \quad y_2 = (x - 1)^2.$$

As usual the general solution is a linear combination of these two.

Actually, the reader should not be too surprised that both of the solutions of the previous example degenerate to a single term. The equation is, after all, only a slightly disguised form of Euler's equation. In fact, the change of independent variable $t = x - 1$ will reduce it to an Euler equation in the new variable t. It is indeed comforting to know that the series method gives the same solution as our previous technique!

The technique illustrated in these examples differs only slightly from that used in the last section for equations near an ordinary point: instead of using $\sum c_n x^n$ we use $\sum c_n x^{n+r}$. Then when the time comes to equate coefficients to zero, we always use the $n = 0$ term to find r_1 and r_2, the two roots of the indicial equation. As long as $r_2 - r_1$ is not an integer, we can then obtain two recursion formulas and two linearly independent solutions by the methods of the previous section.

Problems

In 1–16, find the solutions of the form $x^r \sum_{n=0}^{\infty} c_n x^n$ around $x = 0$.

1. $2x^2y'' + x(x - 1)y' + y = 0.$
2. $2x(2x + 1)y'' + y' + y = 0.$
3. $3xy'' + y' - 2y = 0.$
4. $2x^2y'' + 5x(1 - x)y' - 2y = 0.$
5. $2x(x + 1)y'' + y' - 4y = 0.$
6. $5xy'' + 2y' - 2 = 0.$
7. $2x^2y'' + 7x(x + 1)y' + 3y = 0.$
8. $3x(x + 1)y'' + 2y' - 18y = 0.$
9. $x(3x + 5)y'' - 2y' = 0.$
10. $3x^2y'' + (x^2 + 7x)y' + y = 0.$
11. $3x(1 - x)y'' + (1 + x)y' - y = 0.$
12. $4xy'' + 5y' = 0.$
13. $3x^2y'' - 2x(x - 1)y' - 2y = 0.$
14. $x(5x + 2)y'' - y' - 10y = 0.$
15. $6xy'' - y' + 7y = 0.$
16. $3x^2y'' - 5x(x + 1)y' + 5y = 0.$

In 17–20 find two solutions of the form

$$(x - x_0)^r \sum_{n=0}^{\infty} c_n(x - x_0)^n,$$

around the indicated x_0.

17. $2(x - 1)^2 y'' - 1(x - 1)(x - 8)y' - 3y = 0, x_0 = 1.$
18. $2(x + 2)(2x + 7)y'' + 3y' + y = 0, x_0 = -2.$
19. $2(x + 7)y'' + y' + 4, x_0 = -7.$
20. $x(x - 2)y'' + y' - 2y = 0, x_0 = 2.$

5.4 Applications of the Series Method

As in the case of Euler equations, our applications of the series methods come from attempts to find particular solutions of the partial differential equations of mathematical physics. In this section we shall consider a problem that results in an equation which has no singular point at the origin and therefore has Maclaurin series solutions.

This problem is to find the electrical potential throughout the interior of a spherical shell of radius 1, centered at the origin, due to a given charge distribution on the surface of the shell. We shall assume that the charge on the shell depends only on r and φ (the angle measured from the positive z axis to the radius vector) and not on θ [the angle between the positive x axis and the projection of the radius vector on the (x,y) plane]. Then the potential $V = V(r,\varphi)$ is a solution of the boundary value problem*:

$$r\frac{\partial^2}{\partial r^2}(rV) + \frac{1}{\sin\varphi}\frac{\partial}{\partial\varphi}\left(\sin\varphi\frac{\partial V}{\partial\varphi}\right) = 0.$$

$$V(1,\varphi) = F(\varphi).$$

As in Section 3.7, we seek particular solutions of the differential equation of the form $R(r)\Phi(\varphi)$ which are required to be continuous throughout the interior of the spherical shell. Thus, upon substituting $R\Phi$ for V in the differential equation, we obtain

$$r\Phi(rR)'' + \frac{R}{\sin\varphi}(\sin\varphi\Phi')' = 0,$$

* See R. V. Churchill, *Fourier Series and Boundary Value Problems* (McGraw-Hill Book Company, New York, 1941).

and, after dividing both sides by $R\Phi$, we have

$$\frac{r(rR)''}{R} = -\frac{(\sin\varphi\Phi')'}{\sin\varphi\Phi} = \lambda,$$

because the only way a function of r can equal a function of φ is for both to equal the same constant. Hence we have the two eigenvalue problems

$$r^2R'' + 2rR' - \lambda R = 0,$$

$$\sin\varphi\Phi'' + \cos\varphi\Phi' + \lambda\sin\varphi\Phi = 0.$$

As before, we must seek values of λ for which these two equations have nontrivial solutions which are continuous in the interior of the unit sphere.

In the second equation, in order to produce an equation with polynomial coefficients, let us make the substitution $x = \cos\varphi$. Then

$$\Phi' = \frac{d\Phi}{d\varphi} = \frac{d\Phi}{dx}\frac{dx}{d\varphi} = -\sin\varphi\frac{d\Phi}{dx} = -\sqrt{1-x^2}\frac{d\Phi}{dx},$$

$$\Phi'' = -\sqrt{1-x^2}\frac{d}{dx}\left(-\sqrt{1-x^2}\frac{d\Phi}{dx}\right) - (1 \quad x^2)\frac{d^2\Phi}{dx^2} - x\frac{d\Phi}{dx}.$$

Therefore the equation becomes

$$\sqrt{1-x^2}\left[(1-x^2)\frac{d^2\Phi}{dx^2} - x\frac{d\Phi}{dx}\right] + x\left(-\sqrt{1-x^2}\frac{d\Phi}{dx}\right) + \lambda\sqrt{1-x^2} = 0,$$

which simplifies to

$$(1-x^2)\frac{d^2\Phi}{dx^2} - 2x\frac{d\Phi}{dx} + \lambda\Phi = 0.$$

It should be possible to find a Maclaurin series solution of this equation, since $x = 0$ is an ordinary point. However, we should note in advance that $x = \pm 1$ are singular points, so we should beware of solutions that fail to converge at these points, since $x = \pm 1$ correspond to $\varphi = 0$ and $\varphi = \pi$, points on the z axis. Because our solution must be continuous throughout the unit sphere, it must be continuous for values on the z axis just as well as any other values.

Thus forewarned, we make the usual substitution,

$$\Phi = \sum_{k=0}^{\infty} c_k x^k$$

and after a little work obtain

$$0 \cdot c_0 x^{-2} + 0 \cdot c_1 x^{-1} + \sum_{k=2}^{\infty} \{k(k-1)c_k - [(k-1)(k-2) - \lambda]c_{k-2}x^{k-2}\} = 0.$$

As usual the first two terms give us the fact that c_0 and c_1 are arbitrary constants. From the series we have the recursion formula

$$c_k = \frac{(k-1)(k-2) - \lambda}{k(k-1)} c_{k-2} \qquad \text{for } k \geqq 2.$$

Writing the coefficients in columns as usual, we have Table 5.4.

Table 5.4

c_0 arbitrary	c_1 arbitrary
$c_2 = \dfrac{(1 \cdot 0) - \lambda}{2 \cdot 1} c_0$	$c_3 = \dfrac{2 \cdot 1 - \lambda}{3 \cdot 2} c_1$
$c_4 = \dfrac{(3 \cdot 2) - \lambda}{4 \cdot 3} c_2$	$c_5 = \dfrac{4 \cdot 3 - \lambda}{5 \cdot 4} c_3$
$\vdots$	$\vdots$
$c_{2l} = \dfrac{(2l-1)(2l-2) - \lambda}{2l(2l-1)} c_{2l-2}$	$c_{2l+1} = \dfrac{2l(2l-1) - \lambda}{(2l+1)2l} c_{2l-1}$

and therefore

$$c_{2l} = \frac{(1 \cdot 0 - \lambda)(3 \cdot 2 - \lambda) \cdots [(2l-1)(2l-2) - \lambda]}{2l!} c_0,$$

$$c_{2l+1} = \frac{(2 \cdot 1 - \lambda)(4 \cdot 3 - \lambda) \cdots [2l(2l-1) - \lambda]}{(2l+1)!} c_1.$$

Clearly, in general these coefficients do not even approach zero as $l \to \infty$, and therefore in general the series solution of our equation will not converge for $x = \pm 1$. Yet somehow we must meet this requirement. The answer, of course, lies in the particular choice of λ. Obviously, from the formulas for c_{2l} and c_{2l+1}, if λ is chosen properly, then one of the two series solutions will terminate after a finite number of terms and hence will be defined for $x = \pm 1$. It is easy to see that the eigenvalue must be $\lambda = n(n+1)$, where n is an integer. Then, one solution will be a polynomial of degree n—an even polynomial if n

is even, an odd polynomial if n is odd—and the other will be an infinite series which does not converge for $x = \pm 1$ and hence must be discarded.

When $\lambda = n(n+1)$, the equation

$$(1 - x^2)y'' - 2xy' + n(n+1)y = 0$$

is called *Legendre's differential equation of degree n*,* and the polynomial solution with the appropriate choice of the coefficient of lowest subscript, denoted by $P_n(x)$, is called the *Legendre polynomial of degree n*.

Thus, so far we have found the eigenvalues $\lambda = n(n+1)$ for $n = 0, 1, 2, \ldots$ and the eigenfunctions $\Phi_n(\varphi) = P_n(\cos\varphi)$. Our next step is to solve the equation for R,

$$r^2R'' + 2rR' - n(n+1)R = 0.$$

This is an Euler equation and can be solved by means of the substitution $r = e^t$, leading to

$$\frac{d^2R}{dt^2} + \frac{dR}{dt} - n(n+1)R = (D - n)[D + (n+1)]R = 0$$

which has as its solutions e^{nt} and $e^{-(n+1)t}$, or r^n and $r^{-(n+1)}$. Once again we must discard the second of these because it fails to be continuous at the origin. Hence we are left with

$$V_n(r,\varphi) = A_n r^n P_n(\cos\varphi), \qquad n = 0, 1, 2, \ldots,$$

where A_n is an arbitrary constant, as the particular solutions of our partial differential equation.

* Here the word "degree" refers to the degree of the polynomial solution, and not to the degree of the equation, which is of course one.

6

Nonlinear Equations and Numerical Methods

In this chapter we shall consider three rather special nonlinear equations or systems: the pendulum equation, Van der Pol's equation, and an equation of biological equilibrium between two organisms. In each case we shall attack the problems by a combination of thoughtful analysis and numerical methods. Needless to say, numerical methods are also useful for, and can be applied to, solving linear equations as well. But most often, numerical methods supply the only possible way of obtaining a quantitative solution of a nonlinear equation, even though this "solution" is only approximate. We shall confine ourselves to a few "common sense" methods of numerical solution for the equations considered, without worrying—except in the most general way—about the detailed analysis of the error that occurs in approximations.

6.1 The Homogeneous Nonlinear Pendulum Equation, with no Friction Present. Taylor Series Methods

First let us consider a frictionless pendulum 61.25 cm (slightly over two feet) in length, which swings freely with no driving force. Then the boundary value problem whose solution describes the motion of this pendulum is

$$\begin{aligned} \theta''(t) + 16 \sin \theta(t) &= 0, \\ \theta(0) = \theta_0, \quad \theta'(0) &= \omega_0. \end{aligned} \tag{6.1}$$

There is no essentially new information to be gained from considering pendulums of different lengths, so we shall limit all our considerations to this one equation.

Before we attempt to solve (6.1) numerically, let us analyze the equation to see if we can reason out *a priori* what sort of solutions we can expect. Physically it is not difficult to see that if we release the pendulum from rest at some angle θ_0, or even give it a reasonably small push, then it will oscillate back and forth between some maximum and minimum angles. The motion will be periodic. On the other hand, if we give the pendulum a very heavy push, then it will move in a full circle, with minimum angular velocity when it is pointing vertically upward, and maximum angular velocity when it is pointing downward. Its motion will still—in a sense—be periodic, if we agree to set θ back to zero each time the pendulum passes the bottom of its swing. Finally, the mathematician in us now demands that the "principle of betweenness" be satisfied: Since a soft push causes the pendulum to go back and forth, and a hard push causes it to move in a full circle, then there must be a push of intermediate magnitude which is precisely sufficient to cause the pendulum to come to rest in unstable equilibrium, pointing vertically upward. We shall see that mathematically, problem (6.1) admits all three types of solutions, depending on the values θ_0 and ω_0 assigned as the initial data.

It is possible to integrate the differential equation

$$\theta'' + 16 \sin \theta = 0$$

once, in order to get an expression for the first derivative. Namely, we multiply through by θ' and note that $\theta''\theta' = [\frac{1}{2}(\theta')^2]'$ and

$$\sin \theta \cdot \theta' = (-\cos \theta)';$$

hence we have

$$\tfrac{1}{2}(\theta')^2 - 16 \cos \theta = \text{constant.}$$

By applying the initial conditions, we see that

$$\tfrac{1}{2}(\theta')^2 - 16 \cos \theta = \tfrac{1}{2}\omega_0^2 - 16 \cos \theta_0. \tag{6.2}$$

This equation contains sufficient information to enable us to see what the solutions of our original problem look like in general outline. If we graph θ' as a function of θ, that is, ignore t and act as if θ is the abscissa and θ' is the ordinate, then we will obtain a curve in the (θ, θ') plane—called the *phase plane*. For each value of the right side of (6.2) we obtain a different curve. By examining (6.2), we can see that each of these curves has the following characteristics: (1) it is symmetrical with respect to the origin; (2) the θ'

intercepts are $\pm\sqrt{\omega_0^2 + 32(1 - \cos\theta_0)}$—the radicand is never negative and hence the square root is defined—thus a solution curve (or *trajectory*) always crosses the θ' axis; (3) the θ intercepts are $\cos^{-1}(\cos\theta_0 - \frac{1}{32}\omega_0^2)$, so the trajectory crosses the θ axis only if $|\cos\theta_0 - \frac{1}{32}\omega_0^2| \leqq 1$; (4) the slope of a trajectory is always zero where it crosses the θ' axis and infinite where it crosses the θ axis, unless the θ intercept is a multiple of π, in which case the slope is ± 4.

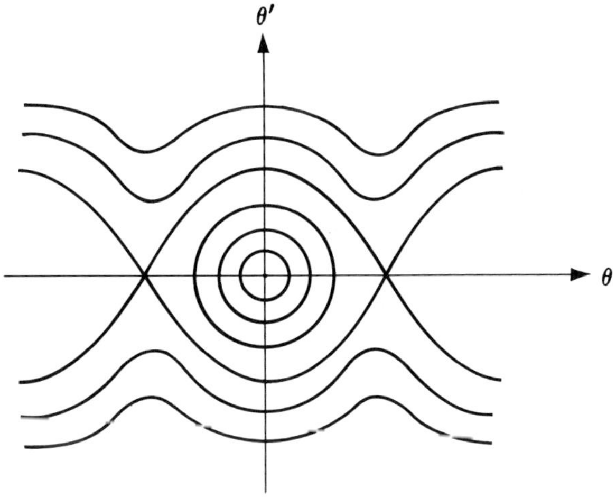

Figure 6.1

Figure 6.1 shows the graphs of a few of the possible trajectories in the phase plane. The closed curves, called *orbits*, represent the case when $|\cos\theta_0 - \frac{1}{32}\omega_0^2| < 1$—in other words, the initial angle and angular velocity are small enough so that the pendulum oscillates back and forth. The wavy trajectories represent the motion when $|\cos\theta_0 - \frac{1}{32}\omega_0^2| > 1$, and the pendulum is revolving in a full circle. The curve in between, where $|\cos\theta_0 - \frac{1}{32}\omega_0^2| = 1$, is the case when precisely the right push is given to cause the pendulum to come to rest in unstable equilibrium pointing upward. Let us examine this case more closely.

We can solve (6.2) for θ', and obtain

$$\theta' = \frac{d\theta}{dt} = \pm\sqrt{\omega_0^2 + 32(\cos\theta - \cos\theta_0)},$$

where the plus sign is chosen in the first quadrant of the phase plane. We can solve this for t as a function of θ as follows:

$$t = \int_{\theta_0}^{\theta(t)} \frac{d\theta}{\sqrt{\omega_0^2 + 32(\cos\theta - \cos\theta_0)}}, \tag{6.3}$$

The integral on the right, which is a so-called *elliptic integral*, cannot be evaluated in closed form. However, one can always determine its value numerically to any desired degree of accuracy. In particular, we can use the integral to tell how much time it takes the pendulum to get to any desired value of θ. Thus, for example, if $\theta_0 = 0$ and $\omega_0^2 = 64$, then the pendulum is pushed just hard enough to achieve its vertical position upward. The amount of time it takes to get to the top is then

$$T = \lim_{\theta \to \pi} \int_0^\theta \frac{d\theta}{\sqrt{32 + 32\cos\theta}},$$

where the limit is used because the integral is improper; the denominator approaches zero as θ approaches π. Now near the point $\theta = \pi$, for example for $|\theta - \pi| \leqq 1$, we know that $\cos\theta \leqq -1 + \frac{1}{2}(\theta - \pi)^2$, because the next term in the Taylor series is a negative one, and the signs of the terms alternate and the terms themselves decrease in value. Hence

$$\begin{aligned} T = \lim_{\theta \to \pi} \int_0^\theta \frac{d\theta}{\sqrt{32 + 32\cos\theta}} &= \int_0^{\pi - 1} \frac{d\theta}{\sqrt{32 + 32\cos\theta}} \\ + \lim_{\theta \to \pi} \int_{\pi - 1}^\theta \frac{d\theta}{\sqrt{32 + 32\cos\theta}} &\geqq \int_0^{\pi - 1} \frac{d\theta}{\sqrt{32 + 32\cos\theta}} \\ + \lim_{\theta \to \pi} \int_{\pi - 1}^\theta \frac{d\theta}{4(\pi - \theta)}; & \end{aligned}$$

the last term is a divergent integral—the limit is infinite! Hence if we were able to push the pendulum precisely hard enough to cause it to approach its vertical unstable equilibrium position, it would take the pendulum *forever* (an infinite length of time) to attain this position!

Equation (6.3) can also be used to compute the period of vibration of the nonlinear pendulum; for example, if $\omega_0 = 0$ and $0 < \theta_0 < \pi$ then

$$T = 4 \int_{\theta_0}^0 \frac{d\theta}{\sqrt{32(\cos\theta - \cos\theta_0)}},$$

that is, the period of vibration is four times the amount of time the pendulum takes to reach the bottom of its swing. The integral is improper, but not divergent, at $\theta = \theta_0$, because the denominator behaves there like $\sqrt{\theta - \theta_0}$.*

* For,

$$\begin{aligned} (\cos\theta - \cos\theta_0) &= \left(1 - \frac{\theta^2}{2!} + \frac{\theta^4}{4!} - \cdots\right) - \left(1 - \frac{\theta_0^2}{2!} + \frac{\theta_0^4}{4!}\right) \\ &= \frac{1}{2!}(\theta_0^2 - \theta^2) + \cdots = \frac{1}{2!}(\theta_0 + \theta)(\theta_0 - \theta)h + \cdots. \end{aligned}$$

Near $\theta = \theta_0$, the behavior is like $(\theta_0 - \theta)$.

Quite clearly, T depends on the initial angle θ_0 and in fact is an increasing function of θ_0. This is in complete contrast with what we learned about the linear pendulum: The period of vibration depended only upon the length. However, in the case of the nonlinear pendulum, the period depends upon the amplitude as well. In fact, as we have seen, the period grows without limit as the amplitude approaches π.

All of this mathematical analysis gives us a good idea of the qualitative behavior of the motion of the pendulum. To find the quantitative behavior, that is, a value of θ for each value of t, we shall need to use numerical methods, since the integral (6.3) cannot be evaluated by the methods of calculus. Even evaluating (6.3) numerically would not be of much use, because it would give t as a function of θ, and we want precisely the reverse. We shall have to return to the differential equation itself in order to seek numerical techniques for solving for θ as a function of t.

To be precise, let us consider the following boundary value problem:

$$\theta'' + 16 \sin \theta = 0,$$
$$\theta(0) = \pi/2, \quad \theta'(0) = 0. \tag{6.4}$$

This problem represents a pendulum which has been drawn up to a horizontal position and released. Because of our previous analysis we expect that the pendulum will swing periodically between the angles of $\pi/2$ and $-\pi/2$; thus, if we find the portion of the solution between $\theta = \pi/2$ and $\theta = 0$ only, we can determine the values of θ for all time as follows (see Figure 6.2), where $\theta(t_0) = 0$:

$$\theta(t) = -\theta(2t_0 - t) \quad \text{for} \quad t_0 \leqq t \leqq 2t_0;$$
$$\theta(t) = -\theta(t - 2t_0) \quad \text{for} \quad 2t_0 \leqq t \leqq 3t_0;$$
$$\theta(t) = \theta(4t_0 - t) \quad \text{for} \quad 3t_0 \leqq t \leqq 4t_0; \text{etc.}$$

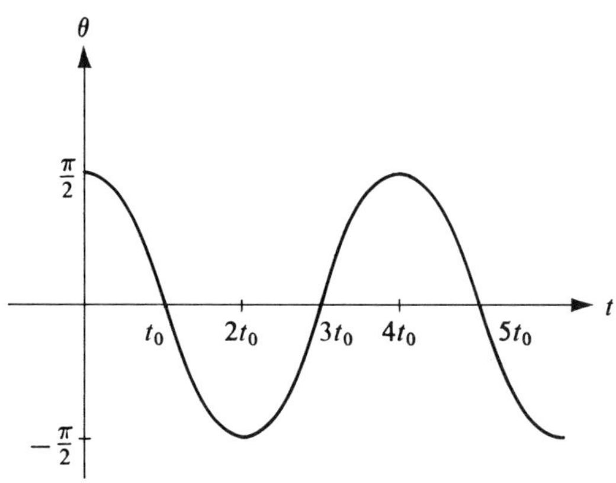

Figure 6.2

The most simple-minded method of determining θ numerically is by means of its Maclaurin series near the origin; however, this method is quite difficult computationally. We first recall that the Maclaurin series for θ is

$$\theta(t) = \theta(0) + t\theta'(0) + \frac{t^2}{2!}\theta''(0) + \frac{t^3}{3!}\theta'''(0) + \cdots.$$

A finite number of terms of this series will approximate θ quite well for very small values of t, but the accuracy of this representation will decrease as t increases. We calculate the coefficients of the series from the initial conditions and the differential equation itself. Thus

$$\theta(0) = \pi/2, \quad \theta'(0) = 0, \quad \theta''(0) = -16 \sin \theta(0) = -16,$$

$$\theta'''(0) = -16\theta'(0) \cos \theta(0) = 0, \quad \theta^{(4)}(0) = -16\theta''(0) \cos \theta(0)$$

$$+ 16[\theta'(0)]^2 \sin \theta(0) = 0,$$

and so on. The process of finding higher derivatives of θ quickly gets extremely involved, as we are computing the derivatives of sums of products of functions of functions. Because of the nice initial conditions in this particular case, most of the coefficients turn out to be zero. In fact, the reader can verify that the first ten terms of the Maclaurin expansion are

$$\theta(t) = \pi/2 - 8t^2 + \tfrac{256}{15}t^6;$$

the other seven coefficients all vanish. Table 6.1 gives the actual values of θ calculated from this series for t between 0 and 1.0 in steps of 0.05. It is easy to see that the solution is not at all accurate for larger values of t; if it were, the minimum value would be closer to -1.5708 and the second maximum would be close to 1.5708, after which θ would again decrease. As it is, the Maclaurin series to ten terms fails to approximate the actual periodic solution even through its first cycle. One solution to this difficulty is to use even more terms of the Maclaurin expansion. It is very gruelling process to calculate further terms, but if we do so, we find that the next nonvanishing term is $-\tfrac{94208}{4725}t^{10}$. Naturally this term has the effect of decreasing the values of θ given in Table 6.1; however, these values overshoot the mark and now the value for 1.00 is clearly too small. The calculation of these values of θ is left to the reader as an exercise.

Table 6.1 Calculated from ten terms of the Maclaurin series; $\theta(0) = \pi/2$, $\theta'(0) = 0$

t	$\theta(t)$
0.00	1.5708
0.05	1.5508
0.10	1.4908
0.15	1.3910
0.20	1.2519
0.25	1.0750
0.30	0.8632
0.35	0.6222
0.40	0.3607
0.45	0.0925
0.50	−0.1625
0.55	−0.3768
0.60	−0.5129
0.65	−0.5221
0.70	−0.3413
0.75	0.1083
0.80	0.9247
0.85	2.2275
0.90	4.1607
0.95	6.8964
1.00	10.6375

Although we have no hard and fast formula available, it seems that the Maclaurin series is not particularly good for values of t equal to or larger than 1.00. For one thing, the calculation of further terms is an extremely onerous process; and for another, the coefficients of the terms of higher degree do not seem to be decreasing very rapidly. Thus, although the Maclaurin series may be quite accurate for values of t close to zero (such as $t = 0.1$), its accuracy and utility are seriously impaired even by the time $t = 1.00$.

Therefore, it makes sense to try the following device: Use a few terms of the Maclaurin series to calculate approximations to $\theta(t)$ and its first few derivatives at some small value of t (such as 0.1 or 0.05); using these values, calculate the first few terms of the Taylor series around this new value of t and use them to approximate $\theta(t)$ at the next value, and so on. Mathematically, we choose a small increment of time Δt, and we write

$$t_0 = 0, \quad t_{n+1} = t_n + \Delta t \qquad (n = 0, 1, 2, 3, \ldots).$$

Then, using the initial data, namely

$$\theta(t_0) = \theta_0, \quad \theta'(t_0) = \omega_0, \quad \theta''(t_0) = -16 \sin \theta_0, \text{etc.},$$

we have the following recursion formula for $\tilde{\theta}(t_{n+1})$, the approximation to the true value $\theta(t_{n+1})$:

$$\tilde{\theta}(t_{n+1}) = \tilde{\theta}(t_n) + \Delta t\tilde{\theta}'(t_n) + \frac{\Delta t^2}{2!}\tilde{\theta}''(t_n) + \cdots + \frac{\Delta t^k \tilde{\theta}^{(k)}}{k!}(t_n).$$

We calculate $\tilde{\theta}'(t_{n+1})$ from the formula $\tilde{\theta}'(t_{n+1}) = \pm 4\sqrt{2\cos\tilde{\theta}(t_{n+1})}$, the sign being chosen appropriately*; $\tilde{\theta}''(t_{n+1}) = -16\sin\tilde{\theta}(t_{n+1})$, and so on. These values $\tilde{\theta}(t_1), \tilde{\theta}(t_2), \ldots, \tilde{\theta}(t_n)$ are only approximate; we use the tilde in order to distinguish them from the true values.

How accurate is such a scheme? We cannot go into too many details here, of course, but there are several sources of error which can accumulate and cause the approximate solution $\tilde{\theta}$ to diverge from the true solution θ for large values of t:

1. We use only $k + 1$ terms of the Taylor series for θ at each step, so there is a "remainder" error caused by this.

2. In moving from t_n to t_{n+1} we go a distance of Δt; for small values of Δt the series is quite accurate, but accuracy decreases as Δt increases.

3. Since $\tilde{\theta}(t_n)$ is not an exact value, $\tilde{\theta}(t_{n+1})$ is computed from the wrong starting point.

4. Finally, we must always recall that this calculation is a numerical one, limited to a certain number of digits. In carrying only this many digits, we cause what is known as roundoff or truncation error, according as we round the numbers or simply drop the extra digits.

The reader may quite correctly conclude that the choice of Δt (the increment in the independent variable) and of k (the order of the highest derivative in the Taylor expansion) are crucial. If Δt is too large, or if k is too small, the accuracy will suffer. On the other hand, if Δt is too small, then excessive roundoff or truncation error may occur before we get to the desired value of t; and if k is too large, the computation may become too time-consuming. Unfortunately we cannot go into a detailed error analysis here to determine optimal values for Δt and k; instead, we shall use the practical means of examining and comparing results to see what visible effects we can observe after changes in these constants.

On the problem at hand, using $k = 1$ will clearly not work, because $\theta'(0) = 0$, so that

$$\theta(t_1) = \theta(0) + \Delta t\theta'(0) = \theta(0) = \pi/2,$$
$$\theta'(t_1) = -4\sqrt{2\cos\theta(t_1)} = 0;$$

* Alternatively, one may calculate $\tilde{\theta}'$ itself from a Taylor series or other approximation, as described later.

that is, because the first derivative is zero, the next value of θ will be the same as the initial value and hence the next value of θ' will be zero, and so it will continue. Physically, at least, this solution is completely spurious; it would mean that the pendulum maintained its initial position forever, a situation which would be possible only with zero gravity or infinite friction. In fact, of course, the solution $\theta \equiv \pi/2$ is mathematically impossible, since $\theta'' = -16 \sin \theta = -16$ if $\theta \equiv \pi/2$; but if θ is a constant, then $\theta'' \equiv 0$, a contradiction of the differential equation.

Thus whenever the initial condition is $\theta'(0) = 0$, we must use at least $k = 2$ in order to have any hope of obtaining a reasonable answer. If $k = 2$, the recursion formulas for the approximate solution are

$$\tilde{\theta}(t_{n+1}) = \tilde{\theta}(t_n) + \Delta t \tilde{\theta}'(t_n) + \frac{\Delta t^2}{2} \tilde{\theta}''(t_n),$$

$$\theta'(t_{n+1}) = \pm 4\sqrt{2 \cos \tilde{\theta}(t_{n+1})},$$

$$\tilde{\theta}''(t_{n+1}) = -16 \sin \tilde{\theta}(t_{n+1}).$$

The sign of $\tilde{\theta}'$ is initially given if $\tilde{\theta}' \neq 0$; otherwise it is taken to be opposite the sign of $\tilde{\theta}$ (since the sign of the velocity will be such as to cause $\tilde{\theta}$ to decrease in absolute value). Then, the sign of $\tilde{\theta}'$ is changed each time that the sign of the correction term

$$\left[\Delta t \tilde{\theta}'(t_n) + \frac{\Delta t^2}{2} \tilde{\theta}''(t_n)\right]$$

changes, i.e., each time the pendulum changes direction. For $k = 3$, the process is the same except that the correction term becomes

$$\left[\Delta t \tilde{\theta}'(t_n) + \frac{\Delta t^2}{2} \tilde{\theta}''(t_n) + \frac{\Delta t^3}{6} \tilde{\theta}'''(t_n)\right],$$

where $\tilde{\theta}'''(t_{n+1}) = -16 \tilde{\theta}'(t_n) \cos \tilde{\theta}(t_n)$.

To calculate the approximate values of the solution, $\tilde{\theta}$, one must be systematic about it. The best way to keep track of calculations is to set up a table with columns headed by the items one wishes to calculate, hopefully in the order in which they will be calculated. Thus, for $k = 2$, for example, we would need columns headed by t, $\tilde{\theta}$, $\sin \tilde{\theta}$, $\cos \tilde{\theta}$, $\tilde{\theta}'$, and $\tilde{\theta}''$, with possibly other columns indicating intermediate stages in the calculation, such as the radicand from which $\tilde{\theta}'$ is computed. If we choose to do the calculation on an electronic computer, then all these intermediate results need not be printed except while the program is being debugged. Unfortunately, hand calculations are very tedious to carry out, if for no other reason than that most commonly available tables of sines and cosines are for angles measured in degrees rather than radians, and conversion between the two is quite time-consuming. In fact, it is in practice not really practical to carry out the solution of even so simple an equation as the pendulum equation on a hand calculator unless the calculator is one of the recent electronic devices which has such functions as sines, cosines, and square roots built in.

However, the first few rows of a table used to solve problem (6.4) with $\Delta t = 0.05$, $k = 2$, might appear as shown in Table 6.2. In the first row, the entries for $\tilde{\theta}$ and $\tilde{\theta}'$ are obtained from the initial data; we compute $\tilde{\theta}''$ from the formula $\tilde{\theta}'' = -16 \sin \tilde{\theta}$. $\Delta\tilde{\theta}$ is given by

$$\Delta\tilde{\theta} = \tilde{\theta}'\Delta t + \tilde{\theta}''\frac{\Delta t^2}{2}.$$

We add this value to $\tilde{\theta}(0)$ to compute the value $\tilde{\theta}(0.05)$; and continue the process. We choose the sign of $\tilde{\theta}'$ to be initially negative, as we explained previously.

Table 6.2

t	$\tilde{\theta}(t)$	$\cos\tilde{\theta}$	$\sin\tilde{\theta}$	$\tilde{\theta}'(t)$	$\tilde{\theta}''(t)$	$\Delta\tilde{\theta}$
0.00	1.5708	0.0000	1.0000	0.0000	−16.0000	−0.0200
0.05	1.5508	0.0200	0.9998	−0.8000	−15.9968	−0.0600
0.10	1.4908	0.0799	0.9968	−1.5991	−15.9488	−0.0999
0.15	1.3909	0.1789	0.9839	−2.3928	−15.7418	−0.1393
0.20	1.2516	0.3138	0.9495	−3.1689	−14.0671	−0.1774
0.25	1.0742	0.4765	0.8792	−3.9047	−12.1395	−0.2128

In what follows, we shall simply tabulate the values of the approximate solution $\tilde{\theta}$. The values of the derivatives and other intermediate results will be omitted. Table 6.3 shows the solutions for various values of Δt, for $k = 2$; Table 6.4 shows the same for $k = 3$. The calculations for these results were actually carried out to fourteen decimal places on an electronic digital computer; these results were rounded to four or five digits to be tabulated. (We show only the values for every 0.05 seconds, even though $\tilde{\theta}$ is calculated more frequently when $\Delta t = 0.01$ or 0.0050.)

From inspecting these tables, we can draw a number of conclusions:

1. Up until approximately $t = 0.95$, from both tables it appears that $\tilde{\theta}$ is converging to a definite value as Δt decreases.

2. For $k = 2$, the convergence appears to be slower than for $k = 3$, which we would expect to be the case; four terms of the Taylor series should fit the solution better than three.

Table 6.3 Approximate solutions of problem (6.4) for various values of Δt, using $k = 2$

t	$\tilde{\theta}: \Delta t = 0.1$	$\Delta t = 0.05$	$\Delta t = 0.01$	$\Delta t = 0.005$
0.0	1.5708	1.5708	1.5708	1.5708
0.05		1.5508	1.5508	1.5508
0.10	1.4908	1.4908	1.4908	1.4908
0.15		1.3909	1.3910	1.3910
0.20	1.2511	1.2516	1.2519	1.2519
0.25		1.0742	1.0749	1.0749
0.30	0.8581	0.8613	0.8628	0.8629
0.35		0.6179	0.6206	0.6207
0.40	0.3401	0.3509	0.3553	0.3554
0.45		0.0700	0.0761	0.0764
0.50	−0.2358	−0.2139	−0.2060	−0.2057
0.55		−0.4893	−0.4800	−0.4797
0.60	−0.7749	−0.7456	−0.7357	−0.7354
0.65		−0.9745	−0.9648	−0.9645
0.70	−1.1971	−1.1699	−1.1611	−1.1608
0.75		−1.3282	−1.3209	−1.3207
0.80	−1.4644	−1.4474	−1.4420	−1.4418
0.85		−1.5268	−1.5235	−1.5234
0.90	−1.5692	−1.5661	−1.5650	−1.5650
0.95		−1.5655	−1.5685	−1.5666
1.00	−1.5118	−1.5248	−1.5350	−1.5282
1.05		−1.4441	−1.4615	−1.4498
1.10	−1.2946	−1.3238	−1.3482	−1.3318
1.15		−1.1645	−1.1961	−1.1751
1.20	−0.9222	−0.9683	−1.0068	−0.9815
1.25		−0.7389	−0.7839	−0.7549
1.30	−0.4188	−0.4823	−0.5330	−0.5011
1.35		−0.2068	−0.2618	−0.2282
1.40	0.1544	0.0772	0.0196	0.0537
1.45		0.3580	0.3002	0.3334
1.50	0.7044	0.6247	0.5691	0.6001

3. The pendulum reaches the other end of its swing ($\theta = -\pi/2$) somewhere between $t = 0.90$ and $t = 0.95$. From this point on, the results within a given table and from one table to the other do not agree at all well.

Point 3 is a direct result of our simple-minded approach to determining the sign of $\tilde{\theta}'$. Naturally, $\tilde{\theta}'$ changes sign as $\tilde{\theta}$ passes through the value $-\pi/2$; we have agreed to change the sign of $\tilde{\theta}'$ when the sign of the correction term changes, but this time may already be too late. It is quite possible that $\tilde{\theta}(t_{n+1})$ may be closer to the value $-\pi/2$ than $\tilde{\theta}(t_n)$, and yet between $t = t_n$

Table 6.4 Approximate solutions of problem (6.4) for various values of Δt, $k = 3$

t	$\tilde{\theta}:\Delta t = 0.1$	$\Delta t = 0.05$	$\Delta t = 0.01$	$\Delta t = 0.005$
0.00	1.5708	1.5708	1.5708	1.5708
0.05		1.5508	1.5508	1.5508
0.10	1.4908	1.4908	1.4908	1.4908
0.15		1.3910	1.3910	1.3910
0.20	1.2515	1.2518	1.2519	1.2519
0.25		1.0748	1.0749	1.0749
0.30	0.8612	0.8626	0.8629	0.8629
0.35		0.6203	0.6207	0.6207
0.40	0.3519	0.3550	0.3555	0.3555
0.45		0.0758	0.0764	0.0764
0.50	−0.2101	−0.2062	−0.2056	−0.2056
0.55		−0.4801	−0.4796	−0.4796
0.60	−0.7382	−0.7357	−0.7353	−0.7353
0.65		−0.9646	−0.9644	−0.9644
0.70	−1.1613	−1.1609	−1.1607	−1.1607
0.75		−1.3206	−1.3206	−1.3206
0.80	−1.4411	−1.4417	−1.4418	−1.4418
0.85		−1.5233	−1.5233	−1.5233
0.90	−1.5645	−1.5649	−1.5649	−1.5649
0.95		−1.5666	−1.5685	−1.5666
1.00	−1.5294	−1.5649	−1.5348	−1.5282
1.05		−1.5233	−1.4613	−1.4499
1.10	−1.3345	−1.4416	−1.3480	−1.3319
1.15		−1.3204	−1.1957	−1.1752
1.20	−0.9847	−1.1605	−1.0064	−0.9817
1.25		−0.9639	−0.7835	−0.7551
1.30	−0.5035	−0.7347	−0.5325	−0.5013
1.35		−0.4789	−0.2614	−0.2284
1.40	0.0521	0.2048	0.0199	0.0534
1.45		0.0773	0.3005	0.3331
1.50	0.5982	0.3563	0.5693	0.5999

and $t = t_{n+1}$ the pendulum could have swung up to $-\pi/2$ and then back again to $\tilde{\theta}(t_{n+1})$. (See Figure 6.3.) We have no way of knowing without further analysis. Because of this problem, some of the results in Tables 6.3 and 6.4 reflect an incorrect choice of the sign for $\tilde{\theta}'(t_n)$. In the next section we shall see one way of ensuring a correct choice of this sign.

Another way of choosing the point where $\tilde{\theta}'$ changes sign is the following: At each step we calculate the nearest point $t > t_n$ where the derivative of

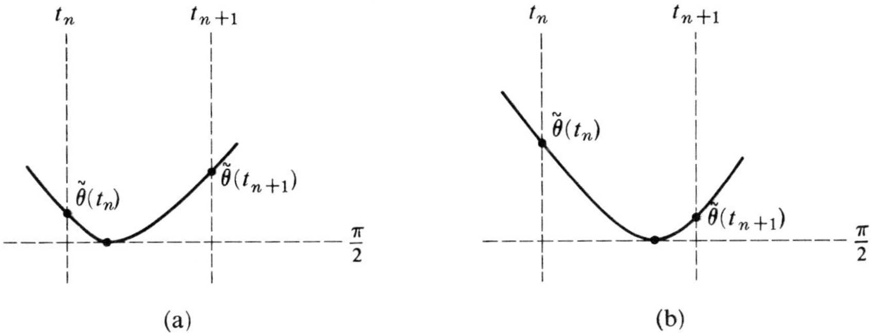

Figure 6.3 In case (b), the sign of θ' is computed incorrectly.

the polynomial

$$\tilde{\theta}(t) = \tilde{\theta}(t_n) + (t - t_n)\tilde{\theta}'(t_n) + (t - t_n)^2\tilde{\theta}''(t_n) + (t - t_n)^3\tilde{\theta}'''(t_n)$$

vanishes. If t turns out to be less than t_{n+1}, then $\tilde{\theta}'(t_{n+1})$ will have the opposite sign from $\tilde{\theta}'(t_n)$. For instance, let us examine $t_n = 0.90$ in Table 6.4, in the column under $\Delta t = 0.01$. Note that the value of $\tilde{\theta}$ at t_n is -1.5649, and this agrees with the values in the two adjacent columns; yet the next value of $\tilde{\theta}$ in this column, -1.5685, disagrees with the value -1.5666 in the two adjacent columns. The polynomial given above in this case turns out to be

$$\tilde{\theta}(t) = -1.5649 - 0.4326(t - 0.9) + \frac{15.9997}{2}(t - 0.9)^2 + \frac{0.0405}{6}(t - 0.9)^3.$$

The derivative of this polynomial is

$$-0.4326 + 15.9997(t - 0.9) + \frac{0.0405}{2}(t - 0.9)^2$$

and its nearest zero to the right of $t_n = 0.90$ is approximately $t = 0.927$. In this case, $t_{n+1} = 0.91$, so $t > t_{n+1}$ and we do nothing. However, when we continue the process to $t_{n+1} = 0.91$ and $t_{n+2} = 0.92$, we find that the pendulum does indeed reach the opposite end of its swing between 0.92 and 0.93, so that $\tilde{\theta}'$ should have a different sign at 0.93 than it does at 0.92.

In the actual calculation, the results of which do not appear in the tables, it turned out that $\tilde{\theta}(0.92) = -1.5704$ and $\tilde{\theta}(0.93) = -1.5707$, so since *the pendulum appeared to be moving toward* $\tilde{\theta} = -\pi/2$ *during the whole interval between* $t = 0.92$ *and* $t = 0.93$, *the sign of the* $\tilde{\theta}'$ *was not changed at* 0.93, *but at* 0.94, *which was too late.* This is the problem with using a formula for $\tilde{\theta}'$ which has an ambiguous sign: It takes a great deal of trouble to make sure that $\tilde{\theta}'$ has the proper sign, and a simple-minded scheme does not always work. Later in the chapter we shall see a somewhat better way to work this.

As a final topic in this section, let us consider the following problem:

$$\tilde{\theta}'' + 16 \sin \theta = 0,$$
$$\theta(0) = 1, \quad \theta'(0) = 0.5. \tag{6.5}$$

Initially the pendulum is displaced 1 radian (about 57 degrees) and is impelled with an angular velocity of 0.5 radians per second in the direction of increasing θ. Table 6.5 illustrates how the results compare, for $\Delta t = 0.005$, using $k = 1$, $k = 2$, and $k = 3$. [Note that it is possible to use $k = 1$ in this case, since $\theta'(0) \neq 0$.] However, the calculation breaks down where the asterisk occurs in the table because the radicand used to compute $\tilde{\theta}'$ came out to be negative, and as a result its square root could not be evaluated. This could not have occurred unless sufficient error had accumulated; thus, clearly $k = 1$ does not supply sufficient accuracy for a practical calculation to be completed. Notice, however, that the results for $k = 2$ and $k = 3$ agree usually right out to the last digit. It thus appears that $k = 2$, $\Delta t = 0.005$ will supply ample accuracy. Both columns show that given this initial impulse, the pendulum first swings up to an angle of 1.0093, reaches its peak, then swings periodically between the angles ± 1.0093, with a period of about 1.680 seconds.

Problems

1. In the problem

$$\theta'' + 16 \sin \theta = 0,$$
$$\theta(0) = \pi, \quad \theta'(0) = -0.5,$$

the pendulum is given a slight push after being raised to a vertical position. Using $k = 2$ and $\Delta t = 0.05$, find $\tilde{\theta}$ for values of t up to 5 seconds. How long does the pendulum take to reach the vertical position again?

2. Solve the problem

$$\theta'' + 16 \sin \theta = 0,$$
$$\theta(0) = 0, \quad \theta'(0) = 5,$$

using $k = 2$ and $\Delta t = 0.01$, for values of t up to $t = 1$ second. Approximately what is the maximum value of $\tilde{\theta}$, and at what value of t does it occur?

3. Solve the problem

$$\theta'' + 16 \sin \theta = 0,$$
$$\theta(0) = 1, \quad \theta'(0) = 0.5,$$

Table 6.5 Approximate solutions of problem (6.5) for various values of k, using $\Delta t = 0.005$

t	$\tilde{\theta}: k = 1$	$k = 2$	$k = 3$
0.000	1.0000	1.0000	1.0000
0.005	1.0025	1.0023	1.0023
0.010	1.0046	1.0043	1.0043
0.015	1.0064	1.0060	1.0060
0.020	1.0078	1.0073	1.0073
0.025	1.0088	1.0083	1.0083
0.030	*	1.0089	1.0089
0.035		1.0092	1.0092
0.040		1.0092	1.0092
0.045		1.0088	1.0088
0.050		1.0081	1.0081
0.100		0.9824	0.9825
0.150		0.9236	0.9236
0.200		0.8329	0.8329
0.250		0.7127	0.7127
0.300		0.5664	0.5664
0.350		0.3988	0.3988
0.400		0.2157	0.2158
0.450		0.0241	0.0242
0.500		−0.1684	−0.1683
0.550		−0.3543	−0.3542
0.600		−0.5264	−0.5263
0.650		−0.6785	−0.6784
0.700		−0.8055	−0.8055
0.750		−0.9039	−0.9038
0.800		−0.9708	−0.9708
0.850		−1.0049	−1.0048
0.870		−1.0091	−1.0091
0.875		−1.0093	−1.0093
0.880		−1.0091	−1.0091
0.900		−1.0052	−1.0052
0.950		−0.9718	−0.9718
1.000		−0.9054	−0.9055
1.290		−0.0182	−0.0185
1.295		0.0011	0.0008
1.705		1.0087	1.0087
1.710		1.0092	1.0091
1.715		1.0092	1.0093
1.720		1.0092	1.0091
1.725		1.0087	1.0087
2.000		0.4984	0.4982

for $k = 3$, $\Delta t = 0.05$, for values of t up to $t = 2$ seconds. Approximately what is the maximum value of θ, and what is the period of the motion?

4. Is there any reason why Δt should always have the same value throughout a calculation? If Δt is to be allowed to vary from one portion of the calculation to another, where should it be chosen smaller and why?

5. Answer the same questions for the value of k.

6.2 The Homogeneous Nonlinear Pendulum Equation, with no Friction Present: the Runge-Kutta Method

The Taylor series method as described in the previous section works reasonably well for the pendulum equation and for some other nonlinear equations which, though nonlinear, are at least not too badly behaved. It is a very simple-minded scheme, however, and it very definitely has its limitations. There are two which most readily come to mind: first, given a typical second-order equation, it may not always be possible to solve for the first derivative (which caused lots of trouble anyhow); and second, not every function is as easily differentiable for Taylor series purposes as is the sine. In fact, very often in real problems, we may not even have an explicit formula for the right side, but only a computational scheme, i.e., a computer subroutine.

To illustrate how the Runge-Kutta method meets these problems, let us first consider how it works for a first-order equation

$$y' = f(x, y);$$

then we shall generalize it to second-order equations in order to solve the nonlinear pendulum equation.

In moving from a point x_0 to a point $x_0 + h$, the first-order Taylor series approximation gives

$$y(x_0 + h) = y(x_0) + hy'(x_0) = y(x_0) + hf(x_0, y_0)$$

where $y_0 = y(x_0)$. The second-order approximation is

$$y(x_0 + h) = y_0 + hf(x_0, y_0) + \frac{h^2}{2!}\left[\frac{\partial f}{\partial x}(x_0, y_0) + \frac{\partial f}{\partial y}(x_0, y_0)f(x_0, y_0)\right]$$

As we pointed out above, the difficulty of calculating the partial derivatives of f may be prohibitive. The purpose of calculating them, of course, is to improve the first-order formula

$$y(x_0 + h) = y(x_0) + hf(x_0, y_0)$$

by including some indication of the change which occurs in f as we proceed from x_0 to $x_0 + h$. The purpose of the Runge-Kutta method is to indicate the change in f without computing its derivatives. One way of doing this would be to calculate the value of f at two different points and then use a weighted average of these two values. Hence it is quite meaningful to ask how we might choose the constants a, b, c, and d in the expression

$$y(x_0 + h) = y(x_0) + h\{af(x_0, y_0) + bf[x_0 + ch, y_0 + dhf(x_0, y_0)]\} \tag{6.6}$$

so that it will give as accurate an approximation to the solution as the second-order Taylor approximation

$$y(x_0 + h) = y(x_0) + hf(x_0, y_0) + \frac{h^2}{2}[f_x(x_0, y_0) + f(x_0, y_0)f_y(x_0, y_0)]. \tag{6.7}$$

Up to terms of order h, we have

$$\begin{aligned} f[x_0 + ch, y_0 + dhf(x_0, y_0)] &= f(x_0, y_0) + chf_x(x_0, y_0) \\ &\quad + dhf(x_0, y_0)f_y(x_0, y_0), \end{aligned}$$

so that (6.6) becomes

$$\begin{aligned} y(x_0 + h) &= y(x_0) + (a + b)hf(x_0, y_0) + h^2[bcf_x(x_0, y_0) \\ &\quad + bdf(x_0, y_0)f_y(x_0, y_0)]. \end{aligned}$$

Comparing this with (6.7), we see that these expressions will be equal provided that the system of equations

$$a + b = 1, \quad bc = \tfrac{1}{2}, \quad bd = \tfrac{1}{2}$$

is satisfied. Since there are three equations and four unknowns, one of the constants can be chosen arbitrarily. Probably the simplest choice is $b = \frac{1}{2}$, giving $a = \frac{1}{2}$, $c = 1$, $d = 1$. Putting these values in (6.6), we have

$$y(x_0 + h) = y(x_0) + h \cdot \tfrac{1}{2}(u_1 + u_2), \tag{6.8}$$

where

$$u_1 = f(x_0, y_0), \quad u_2 = f(x_0 + h, y_0 + hu_1).$$

Formula (6.8) is called the *second-order Runge-Kutta approximation* of $y(x_0 + h)$,* and we see that the correction applied to y_0 is precisely the

* From this point on, we shall omit the "$\sim$" notation for approximate solutions.

average of the values of $f(x, y)$ at the two different points (x_0, y_0) and $(x_0 + h, y_0 + hu_1)$.

One can obtain a third-order scheme by expanding $y(x_0 + h)$ in a Taylor series out to terms containing h^3 and by utilizing a weighted average of the values of f at three different points, and expanding these values out to terms of order h^2. The computations are more complicated, so we shall forego the details here, and simply remark that the third-order Runge-Kutta approximation of $y(x_0 + h)$ is

$$y(x_0 + h) = y(x_0) + h \cdot \tfrac{1}{6}(u_1 + 4u_2 + u_3),$$

where

$$u_1 = f(x_0, y_0), \quad u_2 = f\left(x_0 + \frac{h}{2}, y_0 + \frac{h}{2}u_1\right),$$

$$u_3 = f[x_0 + h, y_0 + h(2u_2 - u_1)].$$

A fourth-order approximation is

$$y(x_0 + h) = y(x_0) + h \cdot \tfrac{1}{6}(u_1 + 2u_2 + 2u_3 + u_4),$$

where u_1 and u_2 are as before, and

$$u_3 = f\left(x_0 + \frac{h}{2}, y_0 + \frac{h}{2}u_2\right), u_4 = f(x_0 + h, y_0 + hu_3).$$

It is easy to see that the Runge-Kutta technique is quite an easy one to apply, given a method for evaluating f, and that it is considerably easier than the Taylor series method, since we need no longer calculate the partial derivatives of f, a process which might be difficult or impossible.

To apply the Runge-Kutta to a second-order equation

$$y'' = f(x, y, y'),$$

we reduce this to a first-order system by writing $y_1 = y$ and $y_2 = y_1'$, so that we have the equivalent first-order system

$$y_1' = y_2,$$

$$y_2' = f(x, y_1, y_2).$$

This is simply a special case of the general first-order system

$$y_1' = f_1(x, y_1, y_2),$$
$$y_2' = f_2(x, y_1, y_2).$$

Clearly a first-order approximation is given by

$$y_1(x_0 + h) = y_1(x_0) + hf_1[x_0, y_1(x_0), y_2(x_0)],$$
$$y_2(x_0 + h) = y_2(x_0) + hf_2[x_0, y_1(x_0), y_2(x_0)].$$

These formulas simply represent the first two terms of the Taylor series for y_1 and y_2.

The second-order Runge-Kutta approximation, we recall, is obtained by averaging the values of f_1 and f_2 at two points, which can be written as follows:

$$y_1(x_0 + h) = y_1(x_0) + h \cdot \tfrac{1}{2}(u_1 + u_2),$$
$$y_2(x_0 + h) = y_2(x_0) + h \cdot \tfrac{1}{2}(v_1 + v_2),$$

where

$$u_1 = f_1[x_0, y_1(x_0), y_2(x_0)], \quad v_1 = f_2[x_0, y_1(x_0), y_2(x_0)],$$
$$u_2 = f_1[x_0 + h, y_1(x_0) + hu_1, y_2(x_0) + hv_1],$$
$$v_2 = f_2[x_0 + h, y_1(x_0) + hu_1, y_2(x_0) + hv_1].$$

Hence, in particular, for our original problem

$$y'' = f(x, y, y')$$

with $y(x_0) = y_0$ and $y'(x_0) = y_1$, we have

$$y(x_0 + h) = y_0 + h \cdot \tfrac{1}{2}(u_1 + u_2),$$
$$y'(x_0 + h) = y_1 + h \cdot \tfrac{1}{2}(v_1 + v_2),$$

where

$$u_1 = y_1, \quad v_1 = f(x_0, y_0, u_1),$$
$$u_2 = y_1 + hv_1, v_2 = f(x_0 + h, y_0 + hu_1, u_2).$$

In quite a similar manner we arrive at the

1. *Third-Order Runge-Kutta Approximation*:

$$y(x_0 + h) = y_0 + h \cdot \tfrac{1}{6}(u_1 + 4u_2 + u_3),$$
$$y'(x_0 + h) = y_1 + h \cdot \tfrac{1}{6}(v_1 + 4v_2 + v_3),$$

where

$$u_1 = y_1, \quad v_1 = f(x_0, y_0, u_1),$$
$$u_2 = y_1 + \tfrac{1}{2}hv_1, \quad v_2 = f(x_0 + \tfrac{1}{2}h, y_0 + \tfrac{1}{2}hu_1, u_2),$$
$$u_3 = y_1 + h(2v_2 - v_1), \quad v_3 = f[x_0 + h, y_0 + h(2u_2 - u_1), u_3].$$

2. *Fourth-Order Runge-Kutta Approximation*:

$$y(x_0 + h) = y_0 + h \cdot \tfrac{1}{6}(u_1 + 2u_2 + 2u_3 + u_4),$$
$$y'(x_0 + h) = y_1 + h \cdot \tfrac{1}{6}(v_1 + 2v_2 + 2v_3 + v_4),$$

where

$$u_1 = y_1, \qquad v_1 = f(x_0, y_0, u_1),$$
$$u_2 = y_1 + \tfrac{1}{2}hv_1, \quad v_2 = f(x_0 + \tfrac{1}{2}h, y_0 + \tfrac{1}{2}hu_1, u_2),$$
$$u_3 = y_1 + \tfrac{1}{2}hv_2, \quad v_3 = f(x_0 + \tfrac{1}{2}h, y_0 + \tfrac{1}{2}hu_2, u_3),$$
$$u_4 = y_1 + hv_3, \quad v_4 = f(x_0 + h, y_0 + hu_3, u_4).$$

Thus in the case of a second-order equation we approximate both the function and its first derivative with a Runge-Kutta type weighted average of the values of the next higher derivative at several different points. In the next section, we shall do the same sort of thing in a case where it is impossible to solve for the first derivative explicitly, as we could have done here.

Let us now specialize these general formulas to the nonlinear pendulum equation,

$$\theta'' = -16 \sin \theta.$$

The second-order Runge-Kutta approximation is

$$\theta(t_0 + h) = \theta(t_0) + h \cdot \tfrac{1}{2}(u_1 + u_2),$$
$$\theta'(t_0 + h) = \theta'(t_0) + h \cdot \tfrac{1}{2}(v_1 + v_2),$$

where $h = \Delta t$ and

$$u_1 = \theta'(t_0), \quad v_1 = -16 \sin \theta(t_0)$$

$$u_2 = \theta'(t_0) + hv_1, \quad v_2 = -16 \sin[\theta(t_0) + hu_1].$$

If we attempt to solve the initial value problem with $\theta(0) = \pi/2$, $\theta'(0) = 0$, then the first few calculations for $\Delta t = 0.05$ are as given in Table 6.6. A similar set of calculations using the third-order process with $\Delta t = 0.01$ are given in Table 6.7.

Table 6.6

t	$\theta(t_0)$	$\theta'(t_0)$	u_1	u_2	v_1	v_2
0.00	1.5707	0.0000	0.0000	−0.800	−16.0000	−16.0000
0.05	1.5507	−0.8000	−0.8000	−1.5998	−15.9967	−15.9712
0.10	1.4908	−1.5992	−1.5992	−2.3966	−15.9488	−15.7957
0.15	1.3909	−2.3928	−2.3928	−3.1799	−15.7418	−15.2875

Table 6.7

t	$\theta(t_0)$	$\theta'(t_0)$	u_1	u_2	u_3	v_1	v_2	v_3
0.00	1.5707	0.0000	0.0000	−0.0799	−0.1599	−16.0000	−16.0000	15.9999
0.01	1.5699	−0.1599	−0.1599	−0.2399	−0.3199	−15.9999	−15.9999	−15.9999
0.02	1.5675	−0.3199	−0.3199	−0.3999	−0.4799	−15.9999	−15.9998	−15.9995
0.03	1.5635	−0.4799	−0.4799	−0.5599	−0.6399	−15.9995	−15.9992	−15.9986

As in the last section, it is interesting to compare the solutions obtained using different values for Δt and the different orders of approximation. In Table 6.8 we see the effect of using differing values of Δt with the third-order approximation scheme. This table should be compared with Table 6.4 of the previous section. Notice in Table 6.8 the obvious trend toward convergence for all values of t as Δt grows smaller. Notice also that in the column for $\Delta t = 0.01$, at $t = 0.95$, the value for $\tilde{\theta}$ is now the same as that for $\Delta t = 0.005$. Recall that at this point the pendulum has swung to the point $\theta = -\pi/2$ and is now beginning to return; the scheme from the previous section gave us trouble here because of the ambiguity in the sign of the expression for θ'. The Runge-Kutta gives us no difficulty on this score, for, although we are using an approximate formula for θ' as well as θ, at least there is no difficulty with the sign.

Table 6.9 is comparable to Table 6.5 and compares the solutions for the second- , third- , and fourth-order Runge-Kutta schemes using a constant value $\Delta t = 0.005$. There are only trifling differences between the three columns, which shows that for this particular problem, the value of Δt used is

Table 6.8 Approximate solutions of problem (6.4) for various values of Δt, using third-order Runge-Kutta

t	$\tilde{\theta}:\Delta t = 0.1$	$\Delta t = 0.05$	$\Delta t = 0.01$	$\Delta t = 0.005$
0.00	1.5707	1.5707	1.5707	1.5707
0.05		1.5507	1.5507	1.5507
0.10	1.4907	1.4908	1.4908	1.4908
0.15		1.3909	1.3909	1.3909
0.20	1.2513	1.2517	1.2518	1.2518
0.25		1.0744	1.0748	1.0749
0.30	0.8599	0.8620	0.8628	0.8629
0.35		0.6191	0.6206	0.6207
0.40	0.3464	0.3529	0.3553	0.3554
0.45		0.0729	0.0762	0.0763
0.50	−0.2229	−0.2101	−0.2058	−0.2056
0.55		−0.4848	−0.4798	−0.4796
0.60	−0.7583	−0.7409	−0.7355	−0.7353
0.65		−0.9699	−0.9645	−0.9644
0.70	−1.1835	−1.1659	−1.1609	−1.1607
0.75		−1.3251	−1.3207	−1.3206
0.80	−1.4594	−1.4454	−1.4418	−1.4417
0.85		−1.5260	−1.5233	−1.5233
0.90	−1.5757	−1.5666	−1.5649	−1.5649
0.95		−1.5672	−1.5665	−1.5665
1.00	−1.5318	−1.5278	−1.5281	−1.5281
1.05		−1.4484	−1.4497	−1.4498
1.10	−1.3281	−1.3294	−1.3317	−1.3318
1.15		−1.1714	−1.1749	−1.1750
1.20	−0.9691	−0.9766	−0.9813	−0.9815
1.25		−0.7486	−0.7547	−0.7549
1.30	−0.4776	−0.4933	−0.5008	−0.5011
1.35		−0.2190	−0.2279	−0.2282
1.40	0.0889	0.0640	0.0539	0.0535
1.45		0.3445	0.3336	0.3332
1.50	0.6428	0.6116	0.6003	0.6000

Table 6.9 Approximate solutions of problem (6.5) for the second-, third-, and fourth-order Runge-Kutta schemes, using $\Delta t = 0.005$

t	$\tilde{\theta}$: 2nd order	3rd order	4th order
0.000	1.0000	1.0000	1.0000
0.005	1.0023	1.0023	1.0023
0.010	1.0043	1.0043	1.0043
0.015	1.0059	1.0059	1.0059

Table 6.9 continued

t	$\tilde{\theta}$: 2nd order	3rd order	4th order
0.020	1.0073	1.0073	1.0073
0.025	1.0082	1.0082	1.0082
0.030	1.0089	1.0089	1.0089
0.035	1.0092	1.0092	1.0092
0.040	1.0091	1.0091	1.0091
0.045	1.0088	1.0088	1.0088
0.050	1.0081	1.0081	1.0081
0.100	0.9824	0.9824	0.9824
0.150	0.9235	0.9235	0.9235
0.200	0.8328	0.8328	0.8328
0.250	0.7126	0.7126	0.7126
0.300	0.5663	0.5664	0.5664
0.350	0.3987	0.3987	0.3988
0.400	0.2156	0.2157	0.2157
0.450	0.0241	0.0241	0.0241
0.500	−0.1684	−0.1683	−0.1683
0.550	−0.3543	−0.3542	−0.3542
0.600	−0.5243	−0.5263	−0.5262
0.650	−0.6784	−0.6784	−0.6783
0.700	−0.8055	−0.8054	−0.8054
0.750	−0.9038	−0.9038	−0.9037
0.800	−0.9708	−0.9708	−0.9707
0.850	−1.0048	−1.0048	−1.0048
0.870	−1.0090	−1.0090	−1.0090
0.875	−1.0092	−1.0092	−1.0092
0.880	−1.0091	−1.0091	−1.0091
0.900	−1.0051	−1.0051	−1.0051
0.950	−0.9717	−0.9718	−0.9718
1.000	−0.9054	−0.9054	−0.9054
1.290	−0.0182	−0.0183	−0.0184
1.295	0.0010	0.0009	0.0008
1.705	1.0086	1.0086	1.0086
1.710	1.0091	1.0091	1.0091
1.715	1.0092	1.0092	1.0092
1.720	1.0089	1.0089	1.0089
1.725	1.0084	1.0084	1.0084
2.000	0.4915	0.4916	0.4918

more important than the order of the Runge-Kutta. Note also that with the Runge-Kutta, there is no danger of obtaining a negative square root and forcing the termination of the process.

Problems

Work the problems at the end of the previous section using the Runge-Kutta technique.

6.3 The Homogeneous Nonlinear Pendulum Equation, with Friction Present

In this section we shall again consider a pendulum approximately two feet long, but we shall assume that a frictional force is present which is proportional to the angular velocity. Thus we are interested in the initial value problem

$$\theta''(t) + \varepsilon\theta'(t) + 16\sin\theta(t) = 0,$$
$$\theta(0) = \theta_0, \quad \theta'(0) = \omega_0,$$

for various values of the frictional coefficient and of the initial data θ_0 and ω_0.

We cannot solve this equation for $\theta'(t)$ as we did in Section 6.1 for $\varepsilon = 0$. However, as we saw in Section 6.2, fewer problems arise if we consider θ' itself as a second unknown. We can do this by one of the two methods we have considered so far: the Taylor series method or the Runge-Kutta. In this section we shall devote ourselves to the latter, both because if its simplicity and its accuracy.

For the second-order Runge-Kutta, our scheme will be

$$\theta(t + \Delta t) = \theta(t) + \tfrac{1}{2}\Delta t(u_1 + u_2),$$
$$\theta'(t + \Delta t) = \theta'(t) + \tfrac{1}{2}\Delta t(v_1 + v_2),$$

where

$$u_1 = \theta'(t), \qquad v = f(\theta(t), u_1),$$
$$u_2 = \theta'(t) + \Delta t v_1, \quad v_2 = f(\theta'(t) + \Delta t u_1, u_2).$$

Here, of course, $f(\theta, \theta') = -\varepsilon\theta' - 16\sin\theta$. The third- and fourth-order schemes are exactly as in Section 6.2 except for the fact that now the right side contains the additional term $-\varepsilon\theta'$. (Note that for $\varepsilon = 0$, we will be solving the same equation as in the previous section.)

Let us first consider the case $\varepsilon > 0$, so that friction is present. Then as in the case of the linear pendulum equation, we would expect the motion of the pendulum to be inhibited or damped by the presence of friction. The amplitude of the arc through which the pendulum swings will decrease—rapidly if ε is large, slowly if it is small—until the pendulum's motion is confined to such a small arc that for all practical purposes the linear equation now describes the motion. Hence we would expect that for ε sufficiently small the motion will eventually appear to be an exponentially damped sine wave, and for ε sufficiently large the motion will be critically damped, that is, no oscillation at all will occur. In fact, because the motion approaches the motion described by the linear equation as t gets large, we would expect that the smallest value of ε for which critical damping takes place should be precisely what it is for the linear equation

$$\theta'' + \varepsilon\theta' + 16\theta = 0,$$

namely, the value where the radicand in $\sqrt{\varepsilon^2 - 4 \cdot 16}$ changes sign, at $\varepsilon = 8$. Our numerical solutions for various values of ε will bear out this conclusion.

First, let us consider the following problem:

$$\begin{aligned} &\theta'' + \theta' + 16 \sin \theta = 0, \\ &\theta(0) = \pi/4, \quad \theta'(0) = 0. \end{aligned} \tag{6.9}$$

The frictional coefficient is 1, a value small enough so that we can see the gradual decrease in amplitude caused by friction. The pendulum is released from rest at an angle of 45 degrees.

Table 6.10

t	$\theta(t)$	$\theta'(t)$	u_1	v_1	u_2	v_2
0.00	0.7853	0.0000	0.0000	−11.3137	−0.5656	−10.7480
0.05	0.7712	−0.5515	−0.5515	−10.6010	−1.0815	− 9.7504
0.10	0.7304	−1.063	−1.063	− 9.6146	−1.5410	8.4873
0.15	0.6653	1.5128	−1.5128	− 8.3650	−1.9311	− 6.9673
0.20	0.5792	−1.8961	−1.8961	− 6.8627	−2.2393	− 5.2126

In the actual calculation of the solution of problem (6.9), as in the previous sections, it is helpful to maintain a table which includes the u_i's and v_i's. For $k = 2$ (the second-order scheme described above) and $\Delta t = 0.05$, the first few entries in the table are shown in Table 6.10. The next value of θ will be

$$0.5792 + \frac{0.05}{2}(-1.8961 - 2.2393) = 0.4759,$$

and so we proceed. Note from the table that there is a sort of "prediction-correction" process going on. Each value of u_2 is a prediction of the next value of u_1, and similarly for the v_i's. We shall see a more direct use of a predictor-corrector scheme in Milne's method, which we shall describe in Section 6.6.

Table 6.11 Approximate solutions of problem (6.9) ($\varepsilon = 1.0$) for $\Delta t = 0.05$ and $k = 2, 3$, and 4

t	θ for $k = 2$	$k = 3$	$k = 4$
0.00	0.7853	0.7853	0.7853
0.10	0.7304	0.7307	0.7311
0.20	0.5792	0.5803	0.5815
0.30	0.3595	0.3616	0.3637
0.40	0.1076	0.1104	0.1133
0.50	−0.1362	−0.1332	−0.1300
0.60	−0.3361	−0.3335	−0.3309
0.70	−0.4666	−0.4652	−0.4639
0.75	−0.5017	−0.5011	−0.5007
0.80	−0.5161	−0.5164	−0.5170
0.85	−0.5105	−0.5117	−0.5133
0.90	−0.4858	−0.4880	−0.4906
1.00	−0.3874	−0.3914	−0.3959
1.10	−0.2410	−0.2464	−0.2522
1.15	−0.1577	−0.1636	−0.1698
1.20	−0.0724	−0.0785	−0.0848
1.25	0.0115	0.0054	−0.0007
1.30	0.0908	0.0851	0.0793
1.40	0.2246	0.2203	0.2160
1.50	0.3117	0.3094	0.3075
1.55	0.3348	0.3338	0.3331
1.60	0.3437	0.3440	0.3448
1.65	0.3390	0.3406	0.3427
1.70	0.3214	0.3242	0.3277
1.80	0.2531	0.2583	0.2639
1.90	0.1534	0.1600	0.1669
2.00	0.0399	0.0470	0.0543
2.05	−0.0159	−0.0091	−0.0021
2.10	−0.0685	−0.0622	−0.0557
2.20	−0.1562	−0.1515	−0.1470
2.30	−0.2188	−0.2095	−0.2076
2.35	−0.2257	−0.2248	−0.2242
2.40	−0.2301	−0.2305	−0.2314
2.45	−0.2252	−0.2271	−0.2293
2.50	−0.2118	−0.2149	−0.2185
2.60	−0.1634	−0.1686	−0.1742

Table 6.11 continued

t	θ for $k = 2$	$k = 3$	$k = 4$
2.70	−0.0948	−0.1013	−0.1081
2.80	−0.0185	−0.0252	−0.0321
2.85	0.0186	−0.0122	0.0056
2.90	0.0533	0.0473	0.0414
3.00	0.1101	0.1059	0.1018

Table 6.12 Solutions of the nonlinear pendulum problem for $k = 4$ and $\Delta t = 0.05$, for various values of the frictional coefficient ε

t	$\varepsilon = 4$	$\varepsilon = 7$	$\varepsilon = 8$	$\varepsilon = 12$
0.00	0.7853	0.7853	0.7853	0.7853
0.05	0.7721	0.7727	0.7729	0.7736
0.10	0.7261	0.7404	0.7417	0.7463
0.15	0.6823	0.6948	0.6985	0.7109
0.20	0.6155	0.6412	0.6484	0.6719
0.25	0.5399	0.5833	0.5950	0.6318
0.30	0.4594	0.5239	0.5407	0.5921
0.35	0.3775	0.4653	0.4874	0.5536
0.40	0.2971	0.4088	0.4363	0.5167
0.45	0.2208	0.3557	0.3882	0.4816
0.50	0.1503	0.3065	0.3435	0.4485
0.55	0.0872	0.2617	0.3024	0.4173
0.60	0.0324	0.2215	0.2651	0.3880
0.65	−0.0135	0.1857	0.2315	0.3606
0.70	−0.0507	0.1534	0.2015	0.3350
0.75	−0.0793	0.1270	0.1747	0.3110
0.80	−0.0999	0.1035	0.1511	0.2887
0.85	−0.1131	0.0834	0.1303	0.2680
0.90	−0.1199	0.0665	0.1121	0.2486
0.95	−0.1211	0.0532	0.0962	0.2306
1.00	−0.1177	0.0406	0.0824	0.2139
1.05	−0.1107	0.0309	0.0704	0.1983
1.10	−0.1010	0.0231	0.0601	0.1839
1.15	−0.0894	0.0168	0.0512	0.1705
1.20	−0.0766	0.0118	0.0436	0.1580
1.25	−0.0634	0.0079	0.0370	0.1465
1.30	−0.0503	0.0048	0.0314	0.1358
1.35	−0.0378	0.0025	0.0266	0.1258
1.40	−0.0262	0.0008	0.0225	0.1166
1.45	−0.0157	−0.0003	0.0190	0.1081

Table 6.12 continued

t	$\varepsilon = 4$	$\varepsilon = 7$	$\varepsilon = 8$	$\varepsilon = 12$
1.50	−0.0065	−0.0012	0.0161	0.1001
1.55	0.0011	−0.0017	0.0135	0.0928
1.60	0.0074	−0.0021	0.0114	0.0860
1.65	0.0122	−0.0022	0.0096	0.0797
1.70	0.0158	−0.0023	0.0081	0.0738
1.75	0.0181	−0.0022	0.0068	0.0684
1.80	0.0194	−0.0021	0.0057	0.0634
1.85	0.0197	−0.0020	0.0048	0.0587
1.90	0.0193	−0.0018	0.0040	0.0544
1.95	0.0182	−0.0016	0.0033	0.0504
2.00	0.0167	−0.0014	0.0028	0.0467

Table 6.11 illustrates the approximate solutions of problem (6.9) for $k = 2, 3,$ and 4 and with $\Delta t = 0.05$. To save space, the results are tabulated every 0.10 second except for maximum, minimum, and near zero values. The differences among the columns are quite noticeable except at the extreme values, where agreement is fairly good. Clearly a smaller value of Δt would be necessary to cause a more obvious convergence. Next, Table 6.12 shows, for $k = 3$ and $\Delta t = 0.05$, the approximate solutions for various values of the frictional coefficient ε [which was equal to 1 in problem (6.9)]. For $\varepsilon = 4$, which is half the critical value, the pendulum is clearly oscillating through an ever-decreasing arc; in fact, from about $t = 1.25$ on, the angle is so small that to four digits it is the same as its sine. Thus for all practical purposes, the linear equation describes the motion from this point on. Even for $\varepsilon = 7$ there is clearly still some oscillatory motion, although the motion is more strongly damped. For $\varepsilon = 8$ (the critical value), the pendulum no longer oscillates, but approaches $\tilde{\theta} = 0$ at an ever-decreasing speed. If one continues the calculation, then at $t = 3.10$ the pendulum is at an angle of 0.0001 radians, and at $t = 3.15$ it is at $\tilde{\theta} = 0.0000$, so to four digits' accuracy it has reached the bottom of its swing.

Finally, for $\varepsilon = 12$ we see an even more retarded swing toward equilibrium; at this point one can almost hear the pendulum creak. Notice how as ε increases, the time the pendulum takes to reach the bottom of its swing (at least to our four digits of accuracy) continually increases. In the case of $\varepsilon = 12$, $\tilde{\theta}$ finally reaches the value 0.0000 when $t = 6.25$.

Problems

1. Solve the problem

$$\theta'' + 0.5\theta' + 16 \sin \theta = 0,$$

$$\theta(0) = \pi/4, \quad \theta'(0) = 0,$$

using $k = 2$ and $t = 0.05$, up to $t = 2.0$ seconds. (a) How long does it take the pendulum to reach the bottom of its swing? (b) How far in the other direction does it swing before it returns?

2. Do the same as problem 1, except use $\varepsilon = 0.1$.
3. Do the same as problem 1, except use $\varepsilon = 0.01$.
4. What happens when one uses $\varepsilon = 0.001$?
5. Solve the problem

$$\theta'' + 0.001\theta' + 16 \sin \theta = 0,$$

$$\theta(0) = \pi/4, \quad \theta'(0) = 0,$$

for t up to 2.00 seconds, using $k = 2$ but letting Δt vary, assuming smaller values when the solution curve is farthest from being linear.

6. Solve the problem

$$\theta'' + 7.5\theta' + 16 \sin \theta = 0,$$

$$\theta(0) = \pi/4, \quad \theta'(0) = 0,$$

for t up to 2.5 seconds, using $k = 3$, $\Delta t = 0.05$. Is the pendulum still oscillating noticeably at the end of this time?

6.4 The Van der Pol Equation

In the previous three sections we have considered the nonlinear pendulum equation, both with and without friction present. In every case, we have found that the behavior of solutions of this equation is not strikingly different from the behavior of solutions of the linearized pendulum equation. That is, lacking friction, we can expect a periodic solution; with friction present we can expect a damped oscillation or critically damped motion, depending on the amount of friction. Naturally the solutions which we obtain differ numerically from those of the linearized equation, but their *qualitative* behavior is the same, and the linear equation is at least a valid predictor of the *type* of solution which we will obtain.

This is emphatically not so with the Van der Pol equation, which we discuss in this section. This equation arises in the theory of electrical circuits containing vacuum tubes, and it presents a highly interesting example of the bizarre phenomena which can occur in the solutions of nonlinear equations. We shall not derive this equation here, as its derivation involves a knowledge of circuit theory beyond that of most students in a differential equations course at this level. So, without further ado, here is Van der Pol's equation:

$$x'' - \varepsilon(1 - x^2)x' + x = 0.$$

The most fruitful way for us to consider this equation is to compare it to the pendulum equation and note similarities. If ε were zero, this equation would be precisely the linearized pendulum equation. The extra term $-\varepsilon(1 - x^2)x'$ can be considered as a frictional term, with the following notable difference: when $|x| < 1$, the "friction" is negative, so that whenever the angle of the pendulum is less than one radian, there is a force acting to increase the amplitude of the motion. This force is zero when $|x| = 1$; then for $|x| < 1$, the force acts to retard the motion.

Looking at the problem in this way, we see that any oscillation which begins with an amplitude of less than one radian will tend to increase in amplitude. And it seems reasonable to expect that a motion through an amplitude sufficiently larger than one radian will tend to decrease in amplitude. All this we deduce from this unusual "frictional" term. But from these conclusions one can now see that it is quite possible that there is some amplitude in the middle somewhere at which periodic oscillation will occur. And, in fact, this is precisely the case! Van der Pol's equation has precisely one periodic solution, and every other solution tends to approach and resemble this periodic solution as time increases indefinitely.

To solve this equation numerically, we shall have to use the method of the previous two sections; that is, to approximate both $x(t)$ and $x'(t)$ by the Runge-Kutta method. We shall use the third-order approximation in this case, so that we shall have

$$x(t + \Delta t) = x(t) + \frac{\Delta t}{6}(u_1 + 4u_2 + u_3),$$

$$x'(t + \Delta t) = x'(t) + \frac{\Delta t}{6}(v_1 + 4v_2 + v_3),$$

where

$$\begin{aligned} u_1 &= x'(t), & v_1 &= f(x(t), u_1), \\ u_2 &= x'(t) + \tfrac{1}{2}\Delta t v_1, & v_2 &= f(x(t) + \tfrac{1}{2}\Delta t u_1, u_2), \\ u_3 &= x'(t) + \Delta t(2v_2 - v_1), & v_3 &= f(x(t) + \Delta t(2u_2 - u_1), u_3). \end{aligned}$$

Naturally $f(x, x') = \varepsilon(1 - x^2)x' - x$.

Since f is algebraic and not transcendental as in the case of the pendulum equation, the solution of the Van der Pol equation by hand calculations is not particularly difficult. For example, for the initial value problem

$$x'' - 0.03(1 - x^2)x' + x = 0,$$
$$x(0) = \pi/2, \quad x'(0) = 0,$$

the first few entries in a table for solving it (with $\Delta t = 0.1$) might look as shown in Table 6.13. The reader can verify that the computations are considerably easier to make here than with the pendulum equation; one does not even need a book of tables.

Table 6.13

t	x	x'	u_1	v_1	u_2	v_2	u_3	v_3
0.0	1.5708	0.0000	0.0000	−1.5707	−0.0785	−1.5673	−0.1567	−1.5561
0.1	1.5629	−0.1566	−0.1566	−1.5561	−0.2344	−1.5451	−0.3111	−1.5267
0.2	1.5395	−0.3109	−0.3109	−1.5267	−0.3873	−1.5086	−0.4618	−1.4834
0.3	1.5008	−0.4617	−0.4617	−1.4834	−0.5359	−1.4587	−0.6076	−1.4272

In Table 6.14 we have tabulated the results of solving the Van der Pol equation numerically under three different initial conditions:

(1) $x(0) = 1.0, x'(0) = 0;$

(2) $x(0) = 2.0, x'(0) = 0;$

(3) $x(0) = 3.0, x'(0) = 0.$

In each case, the results were calculated using $k = 3$ and $\Delta t = 0.1$. The table lists only the values of t and $x(t)$ for which $x(t)$ is either a local maximum or minimum with respect to nearby tabulated values, the values of x analogous to the extreme points of a pendulum's swing.

Notice in the middle column how the extreme values stay quite close to ± 2; considering the error inherent in any numerical approximation process, there is little doubt that the middle column represents the periodic solution. Next note how the extremes of column 1 are monotonically increasing in absolute value (most likely toward the value 2) while those of the last column are monotonically decreasing in absolute value (also apparently toward the value 2). In fact, it is quite reasonable, considering the analogy between Van der Pol's equation and the pendulum, that the amplitude of the periodic solution would be precisely 2, since the driving forces when $-1 < x < 1$ are precisely matched by the retarding forces when $-2 < x < -1$ and $1 < x < 2$.

Table 6.14 Comparisons of extreme values of solutions of Van der Pol's equation for different initial conditions

$x(0) = 1.00$		$x(0) = 2.00$		$x(0) = 3.00$	
t	$x(t)$	t	$x(t)$	t	$x(t)$
0.00	1.000000	0.00	2.000000	0.00	3.000000
3.10	−1.034927	3.10	−1.998563	3.10	−2.840246
6.30	1.071454	6.30	1.999714	6.30	2.720659
9.40	−1.107867	9.40	−1.999961	9.40	−2.620001
12.60	1.143206	12.60	1.998454	12.60	2.540341
15.70	−1.180818	15.70	−2.000388	15.70	−2.472568
18.80	1.216626	18.80	1.999245	18.90	2.414822
22.00	−1.252796	22.00	−1.999848	22.00	−2.367699
25.10	1.289215	25.10	2.000397	25.10	2.323413
28.20	−1.323299	28.30	−1.998349	28.30	−2.289846
31.40	1.359476	31.40	2.000587	31.40	2.257669
34.50	−1.393581	34.50	−1.999753	34.60	−2.230108
37.70	1.426430	37.70	1.999827	37.70	2.206945
40.80	−1.460256	40.80	−2.000682	40.80	−2.183086
43.90	1.491025	43.90	1.998465	44.00	2.166824
47.10	−1.522491	47.10	−2.000658	47.10	−2.149316
50.20	1.552554	50.20	2.000139	50.30	2.134345
53.40	−1.579614	53.40	−1.999681	53.40	−2.122032
56.50	1.608831	56.60	2.000860	56.50	2.107791
59.60	−1.634500	59.60	−1.998962	59.70	−2.099503
62.80	1.659445	62.80	2.000631	62.80	2.089356
65.90	−1.684134	65.90	−2.000434	66.00	−2.080468
69.00	1.705084	69.10	1.999455	69.10	2.073885
72.20	−1.727944	72.20	−2.000956	72.30	−2.064839
75.30	1.747956	75.30	1.999382	75.40	2.060565
78.40	−1.764138	78.50	−2.000531	78.50	−2.054550
81.60	1.785171	81.60	2.000659	81.70	2.048774
84.70	−1.800595	84.80	−1.999163	84.80	−2.045516
87.90	1.816931	87.90	2.000989	87.90	2.039530
91.00	−1.831797	91.00	−1.999742	91.10	−2.037312
94.20	1.843529	94.20	2.000373	94.20	2.033801
97.30	−1.858056	97.30	−2.000828	97.30	−2.029607

Problems

1. Show what the solutions of Van der Pol's equation look like in the phase plane, i.e., in the plane where x is the abscissa and x' is the ordinate. Is the point $(0, 0)$ a solution?

2. Consider the problem

$$x'' - 0.03(1 - x^2)x' + x = 0,$$
$$x(0) = 2.5, \quad x'(0) = 0.$$

Solve it for $k = 3$, $\Delta t = 0.1$ and t up to 100. Is the amplitude nearly 2 at the end of this time?

3. How does changing the value of ε affect the solution? Try $\varepsilon = 1.0$, then $\varepsilon = 0.01$, and solve these problems numerically for $x(0) = 1.00$, $x'(0) = 0$. How do you explain this effect?

4. Suppose that initially the oscillations of a solution of the Van der Pol equation are so small that the x^2 term can be neglected. What does the resulting linear equation approximating Van der Pol's equation look like? What is its solution [under the initial condition $x(0) = \eta$, $x'(0) = 0$]? Does this solution resemble the solution of the nonlinear equation?

6.5 An Equation of Biological Equilibrium

Consider a one-celled plant which nourishes itself by photosynthesis, so that it does not need other living organisms present in order to nourish itself. Then, after a certain amount of time, this plant will divide and form two, which continue photosynthesizing and dividing. Thus the rate of change of the population of these plants is proportional to the number of individuals present, and the population will grow exponentially (at least up to the capacity of the medium in which they live). Hence if $x(t)$ is the number of individuals at time t, then the differential equation describing the growth of this population is

$$x'(t) = \alpha x(t),$$

where α is a positive constant of proportionality.

Next consider the plight of a one-celled animal which is a predator on individuals in the first population, henceforth to be called the prey. Then if a colony of these predators is placed in a medium containing none of their prey, they will all eventually starve to death. The rate of population decrease will be proportional to the number of individuals present; so if $y(t)$ represents the population of predators at time t, the differential equation describing the decline of the population is

$$y'(t) = -\gamma y(t),$$

where γ is a positive constant of proportionality.

Now, if we place colonies of the two organisms together in the same medium, two things will happen: first, the predators will eat some of the prey, thus stemming the prey's population growth; and second, the attrition of the predators by starvation will decline, and in fact, the ones who get the most to eat will multiply, so that the population of predators may actually increase. Thus a term in $y(t)$ will enter the equation for $x(t)$ and vice versa, in the following manner.

First, the number of prey that get eaten must be subtracted from their rate of population growth; this number is proportional both to the number of prey present (the more there are present, the more will be eaten) and to the number of predators present (the more predators there are, the more prey they will consume). Secondly, because the predators are becoming so well fed, their population will increase, again at a rate that is proportional to the number of prey present (every time a predator eats a certain number of prey, he divides) and also to the number of predators present (because of the exponential rate of population growth). Thus the predator-prey relationship can be expressed by the nonlinear first-order *system* of equations

$$x'(t) = \alpha x(t) - \beta x(t)y(t),$$

$$y'(t) = -\gamma y(t) + \delta x(t)y(t),$$

where α, β, γ, and δ are positive constants. This system of equations is *nonlinear* because the product xy appears in each equation; it is also called *autonomous* because t does not appear explicitly. In this respect it is like the homogeneous pendulum equation and Van der Pol's equation.

Let us see what we can determine about the behavior of solutions of these equations without actually solving them. First, because the system is homogeneous, $x = 0$ and $y = 0$ is a solution. This represents the rather uninteresting case when no individuals of either type are present initially and hence for all time—Pasteur's law that spontaneous generation does not occur. Two other rather uninteresting solution pairs are $x = 0$ and $y = y_0e^{-\gamma t}$, $x = x_0e^{\alpha}t$ and $y = 0$. These two solutions represent the cases when one organism is present and the other is absent.

Are there any other instances in which $x \equiv$ constant *and* $y \equiv$ constant are solutions? The answer is yes, for in such a case we would have $x' = 0$ and $y' = 0$, so that

$$0 = \alpha x - \beta xy = x(\alpha - \beta y),$$

$$0 = -\gamma y + \delta xy = y(-\gamma + \delta x),$$

and hence (if β and δ are not zero), $y \equiv \alpha/\beta$ and $x \equiv \gamma/\delta$. Thus there are certain populations of each organism which will remain in perpetual equilibrium;

the number of prey added to the population by division exactly balances the number eaten by the predators, and the number of predators that die of starvation exactly balances the new ones added to the population by division of the well fed.

What does reason tell us might happen if we began with populations of the two organisms which were not in balance like the above? Suppose to start with that there is a "shortage" of the prey. Then the initial tendency will be for the population of predators to decrease by starvation, and the population of prey to increase due to the small number that can be found and eaten by predators. After the population of the prey has increased to the point where they are easier to find and eat, the number of predators will begin to increase; at this point we can say that we have a "surplus" of prey. The population of predators thus well fed will increase until it finally stops the growth of the prey population and causes it to decline. The predators are now in danger, for their numbers are inflated and prey is getting harder to find. Thus will come a time when starving predators outnumber their well-fed colleagues, and the predator population will begin to decline.

This cyclical relationship between prey and predators is well known in nature. For instance, among larger animals, consider the deer (prey) and the cougar (predator). Man virtually eliminated the cougar and as a result deer population increased to the point where there was not adequate forage to feed all deer; as a result, many starved and many of the rest were rather puny specimens. In order to control the deer population and ensure that the surviving deer are healthy and hearty, man has had to increase the length of his hunting season, hunt both sexes, etc; in other words, man himself has had to become a predator to replace the cougar.

Nature also contains many examples of how an animal can multiply in an environment where it lacks predators—witness the rabbit in Australia. Occasionally, too, a predator population gets out of hand, often for reasons nobody fully understands, as in the recent population explosion of a species of coral-eating starfish, which is now threatening the continued existence of live coral reefs in the Western Pacific Ocean.

Because of the apparent cyclical behavior of the prey-predator population, we might ask if $x(t)$ and $y(t)$ are periodic functions of t. It turns out that the answer is in the affirmative when $x(t)$ and $y(t)$ are both nonconstant functions. To see the plausibility of this result, we shall solve the nonlinear system of equations:

$$x'(t) = x(t) - 0.001x(t)y(t),$$

$$y'(t) = -0.3y(t) + 0.000003x(t)y(t);$$

where biological equilibrium occurs when $x = 100{,}000$, $y = 1000$, assuming that t is measured in days. We shall use the initial conditions $x(0) = x_0$,

$y(0) = y_0$, with $t_0 = 0$, $t_{n+1} = t_n + \Delta t (n = 0, 1, 2, \ldots)$, where $\Delta t = \frac{1}{24}$ days; we shall compute $x(t_{n+1})$ and $y(t_{n+1})$ from the formulas

$$x(t_{n+1}) = x(t_n) + \tfrac{1}{6}t(u_1 + 4u_2 + u_3),$$

$$y(t_{n+1}) = y(t_n) + \tfrac{1}{6}t(v_1 + 4v_2 + v_3),$$

where

$$u_1 + f_1(x(t_n), y(t_n)), \quad v_1 = f_2(x(t_n), y(t_n)),$$

$$u_2 = f_1(x(t_n) + \tfrac{1}{2}\Delta t u_1, \quad y(t_n) + \tfrac{1}{2}\Delta t v_1), \quad v_2 = f_2(x(t_n) + \tfrac{1}{2}\Delta t u_1, y(t_n) + \tfrac{1}{2}\Delta t v_1),$$

$$u_3 = f_1(x(t_n) + \Delta t(2u_2 - u_1), y(t_n) + \Delta t(2v_2 - v_1)),$$

$$v_3 = f_2(x(t_n) + \Delta t(2u_2 - u_1), y(t_n) + \Delta t(2v_2 - v_1)),$$

with

$$f_1(x, y) = x - 0.001xy, \quad f_2(x, y) = -0.3y + 0.000003xy.$$

The first few hours' calculations for x and y might look as given in Table 6.15 where the initial values are $x(0) = 101{,}000$ and $y(0) = 1000$ and t is measured in days. Although we show all results as integers, of course we carry out all calculations to as many decimal places as are available. These few calculations show what happens when the initial population contains an excess of prey organisms. The predators devour the prey faster than the prey replaces itself by multiplying. The rate of change of the predator population remains small, however; only after five hours will it increase to 1001.

Table 6.15

t	x	y	u_1	v_1	u_2	v_2	u_3	v_3
0.000	101,000	1000	0	3	−6	3	−12	2
0.041	100,999	1000	−12	2	−18	2	−25	2
0.083	100,998	1000	−25	2	−31	2	−37	2
0.125	100,997	1000	−37	2	−44	2	−50	2
0.166	100,995	1000	−50	2	−56	2	−62	2

Table 6.16 illustrates how the two populations vary when the prey population is initially 150,000 and the predator population is 1000. Time is given in hours. After the first day, we print only every twelve hours, in order to save space. Notice that after 278 hours (about $11\frac{1}{2}$ days) the two populations have returned to pretty close to their original sizes. Note also the very crucial role played by the equilibrium values $x = 100{,}000$ and $y = 1000$.

The prey population will be decreasing whenever $y > 1000$, and increasing whenever $y < 1000$. Likewise, the predator population will be increasing whenever the number of prey exceeds 100,000, and decreasing otherwise.

Table 6.16 Predator and prey populations at selected times, $x(0) = 150{,}000$, $y(0) = 1000$

t	x	y	t	x	y	t	x	y
0	150,000	1000	16	145,005	1101	120	63,486	1058
1	149,980	1006	17	144,374	1108	132	62,578	1000
2	149,922	1013	18	143,710	1114	144	63,434	946
3	149,823	1019	19	143,012	1120	156	65,986	897
4	149,686	1025	20	142,282	1126	168	70,225	855
5	149,508	1032	21	141,521	1132	180	76,174	821
6	149,291	1038	22	140,729	1138	192	83,844	797
7	149,035	1044	23	139,908	1143	204	93,176	783
8	148,740	1051	24	139,059	1149	216	103,958	781
9	148,405	1057	36	127,096	1208	228	115,711	793
10	148,032	1064	48	113,374	1245	240	127,589	819
11	147,620	1070	60	99,882	1257	252	138,321	861
12	147,171	1076	72	87,994	1246	264	146,299	917
13	146,684	1083	84	78,378	1214	276	149,911	987
14	146,160	1089	96	71,184	1169	277	149,974	993
15	145,600	1095	108	66,290	1115	278	149,999	999

Figure 6.4 illustrates a few of the solution curves in the phase plane, i.e., the plane with $x(t)$ as abscissa and $y(t)$ as ordinate. The origin, the point (100,000, 1000), and the x and y axes are each solutions; all other solutions are closed curves (so-called *orbits*) around the point (100,000, 1000). The arrows on the solution curves represent the direction in which time increases.

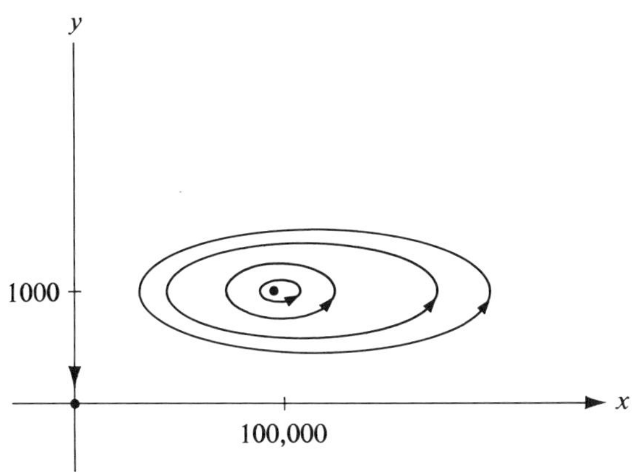

Figure 6.4

It is an interesting fact that all the nontrivial solutions of the equation of biological equilibrium are periodic, and that this equation is easy to solve numerically and approximate solutions are relatively accurate. We have not, of course, proved that the solutions are all periodic, but as we shall see in the problems, the numerical evidence certainly supports this hypothesis.

Problems

1. Solve the equation of biological equilibrium for the initial values $x(0) = 200{,}000$, $y(0) = 2000$. Use $k = 3$, $\Delta t = \frac{1}{24}$ day, as in the test.

(a) How long does it take for the two populations to return to their initial values?

(b) What are the two populations after 30 days?

(c) What are the minimum and maximum populations of prey?

(d) What are the minimum and maximum populations of predators?

2. Answer the same question for $x(0) = 100{,}000$ and $y(0) = 1500$.

3. From the general equations of biological equilibrium $x' = x - xy$ and $y' = -y + xy$, derive a single differential equation for x and one for y.

6.6 Milne's Predictor-Corrector Method

There are many other numerical techniques for solving ordinary differential equations. There is no hard and fast rule for using one or the other; the rule used always depends on the characteristics of the problem under consideration. The Taylor series method, which is certainly the most simple-minded, is very seldom used in practice because it is often impractical or impossible to calculate the derivatives of the right side, and because it requires more calculation than other methods. The Runge-Kutta, one of the most commonly used schemes, is relatively simple and accurate and does not require a knowledge of the derivatives of the right side. However, at each step it requires several evaluations of the right side, the number depending on the order of the scheme.

The Milne predictor-corrector requires only two evaluations of the right side at each step and hence is a more economical calculation. It works as follows: Using the values of the right side at three *previous* points, it gives a *predicted* value for the solution at the current point; then, using this predicted value in the right side, one computes a *corrected* value. In this section we shall derive this scheme and discuss its use.

We shall first derive the scheme for a first-order initial value problem

$$y' = f(x, y), \quad y(x_0) = y_0$$

and then show how it can be generalized to systems and to higher-order equations. We shall for simplicity use the following notation, if h is the length of step we are taking in x:

$$x_i = x_0 + ih, \quad y_i = y(x_i), \quad f_i = f(x_i, y_i).$$

Then we develop the predictor formula by finding a numerical integration scheme for the equation

$$y_{i+1} = y_{i-3} + \int_{x_{i-3}}^{x_{i+1}} y'\,dx. \tag{6.10}$$

The corrector formula comes from a numerical integration scheme for

$$y_{i+1} = y_{i-1} + \int_{x_{i-1}}^{x_{i+1}} y'\,dx. \tag{6.11}$$

These formulas are exact if the solution y is actually known as a function of x. The choice of y_{i-3} for the predictor and y_{i-1} for the corrector is arbitrary, in the sense that other starting points are possible; however, these two choices give a scheme which, as we shall see, is fourth order—that is, the error involved in using the approximation is proportional to h^5.

To find integration schemes for the above equations, it is first necessary to find numerical expressions for the total derivatives of f with respect to x. Using Taylor's expansion [considering $f = f(x, y(x))$ as a function of x alone] we have

$$f_{i+1} = f_i + hf_i' + h^2 f_i'' + \cdots,$$
$$f_{i-1} = f_i - hf'_i + h^2 f_i'' + \cdots.$$

Adding these two expressions, we obtain

$$f_{i+1} + f_{i-1} = 2f_i + h^2 f_i'' + \cdots,$$

and solving for f_i'', we obtain

$$f_i'' = \frac{f_{i+1} - 2f_i + f_{i-1}}{h^2} + \cdots,$$

where the first omitted term has a factor of h^2. We shall use just the term written here to approximate f_i''. Then, to evaluate

$$I = \int_{x_{i-3}}^{x_{i+1}} y'\,dx = \int_{x_{i-3}}^{x_{i+1}} f(x, y(x))\,dx,$$

we first make the change of variable $x = x_{i-1} + \xi$, whereupon

$$I = \int_{-2h}^{2h} f(x_{i-1} + \xi, y(x_{i-1} + \xi))\,dx.$$

Then we expand the integrand in a Taylor series around the point x_{i-1}, obtaining

$$I = \int_{-2h}^{2h} \left[f_{i-1} + \xi f_{i-1} + \frac{\xi^2}{2} f_{i-1}'' + \cdots \right] d\xi = 4hf_{i-1} + \frac{8}{3}h^3 f_{i-1}'' + \cdots,$$

where the first term omitted has a factor h^5. Using the formula which we derived for the second derivative, we have

$$I = 4hf_{i-1} + \tfrac{8}{3}h[f_i - 2f_{i-1} + f_{i-2}] + \cdots = \tfrac{4}{3}h[2f_i - f_{i-1} + 2f_{i-2}] + \cdots.$$

Again the first term omitted has a factor h^5.

Finally, using just the first term of I in (6.10) we have *Milne's predictor formula*

$$y_{i+1} = y_{i-3} + \tfrac{4}{3}h[2f_i - f_{i-1} + 2f_{i-2}]. \tag{6.12}$$

Note that the predictor gives an estimate for y_{i+1} based on previous values of y and f, values which have already been computed; so, provided that these values have been preserved, no new evaluations are necessary to obtain this estimate.

Next, to get the correct formula, we evaluate

$$J = \int_{x_{i-1}}^{x_{i+1}} y'\,dx = \int_{x_{i-1}}^{x_{i+1}} f(x, y(x))\,dx$$

in a similar manner. Letting $x = x_i + \xi$, we have

$$\begin{aligned} J &= \int_{-h}^{h} f(x_i + \xi, y(x_i + \xi))\,d\xi = \int_{-h}^{h} \left(f_i + \xi f_i' + \frac{\xi^2}{2!} f_i'' + \cdots \right) d\xi \\ &= 2hf_i + \tfrac{1}{3}h^3 f_i'' + \cdots, \end{aligned}$$

where as before, the omitted term is of order h^5. Using the approximation for f_i'' as before, we have

$$J = 2hf_i + \tfrac{1}{3}h[f_{i+1} - 2f_i + f_{i-1}] + \cdots = \tfrac{1}{3}h[f_{i+1} + 4f_i + f_{i-1}] + \cdots.$$

Putting this in (6.11) yields *Milne's corrector formula*

$$y_{i+1} = y_{i-1} + \tfrac{1}{3}h[f_{i+1} + 4f_i + f_{i-1}]. \tag{6.13}$$

In this equation, all the values except f_{i+1} have been saved from previous calculations; f_{i+1} represents the value of f at the *predicted* value of y_{i+1} obtained from (6.12). Notice that the corrector formula can be applied repeatedly by using the corrected value y_{i+1} to evaluate f_{i+1} and computing (6.13) again. Normally, one application is sufficient, however. Recall that the omitted term was of order h^5, so for sufficiently small h this scheme will be quite accurate.

We see, then, that Milne's method has the advantage of requiring less calculation than the Runge-Kutta. However, this is countered by the disadvantage that to compute the value y_{i+1} one needs the four previous values $y_{i-3}, y_{i-2}, y_{i-1}$, and y_i. Hence Milne's method cannot be used until later in the computation, after we have used some other method to calculate the first four values of y. The most common technique is to use the Runge-Kutta for these first four values, then switch to Milne's method. (For this reason, Milne's method is said not to be *self-starting*, unlike the Runge-Kutta technique.)

For a first-order system of equations such as

$$y' = f(x, y, z),$$
$$z' = g(x, y, z),$$

the predictor-corrector formulas are simply

$$y_{i+1} = y_{i-3} + \tfrac{4}{3}h[2f_i - f_{i-1} + 2f_{i-2}],$$
$$z_{i+1} = z_{i-3} + \tfrac{4}{3}h[2g_i - g_{i-1} + 2g_{i-2}],$$

and

$$y_{i+1} = y_{i-1} + \tfrac{1}{3}h[f_{i+1} + 4f_i + f_{i-1}],$$
$$z_{i+1} = z_{i-1} + \tfrac{1}{3}h[g_{i+1} + 4g_i + g_{i-1}].$$

From this it is quite simple to obtain a predictor-corrector for a second-order equation

$$y'' = f(x, y, y'),$$

since we can write it as a system in the form

$$y' = z,$$
$$z' = f(x, y, z).$$

Hence the predictor-corrector formulas for a second-order equation are

$$y_{i+1} = y_{i-3} + \tfrac{4}{3}h[2y'_i - y'_{i-1} + 2y'_{i-2}]$$
$$y'_{i+1} = y'_{i-3} + \tfrac{4}{3}h[2f_i - f_{i-1} + 2f_{i-2}]$$

and

$$y_{i+1} = y_{i-1} + \tfrac{1}{3}h[y'_{i+1} + 4y'_i + y'_{i-1}]$$
$$y'_{i+1} = y'_{i-1} + \tfrac{1}{3}h[f_{i+1} + 4f_i + g_{i-1}].$$

As an example of the application of Milne's method, let us consider the Van der Pol equation, with $\varepsilon = 0.03$, given the values of x, y and y' in the first two columns of Table 6.17. The appropriate values of the right side

Table 6.17

t	y	y'	f
1.0	1.098305	−1.648500	−1.088104
1.1	0.928260	−1.749764	−0.935522
1.2	0.748869	−1.835275	−0.773050
1.3	0.561755	−1.904049	−0.600851

$f(y, y') = 0.03(1 - y^2)y' - y$ are given in the fourth column. Using these values, we compute the predictor values

$$\begin{aligned} y_{i+1} &= 1.098305 + \tfrac{4}{3}(0.1)[2(-1.904049) - (-1.835275) + 2(-1.749764)] \\ &= 0.6317012, \\ y'_{i+1} &= -1.648500 + \tfrac{4}{3}(0.1)[2(-1.600851) - (-0.773050) + 2(-1.088104)] \\ &= -1.938661, \end{aligned}$$

and hence the predictor value f_{i+1} is

$$0.03[1 - (0.6310712)^2](-1.938661) - (0.6317012) = -0.666699.$$

Using this in the corrector formulas, we obtain

$$y_{i+1} = 0.748869 + \frac{0.1}{3}[(-1.938661) + 4(-1.904049) + (-1.835275)]$$

$$= 0.43382,$$

$$y'_{i+1} = (-1.835275) + \frac{0.1}{3}[(-0.666699) + 4(-0.600851) + (-0.773050)]$$

$$= -1.941157.$$

To continue the process to $t = 1.5$ it will be necessary to calculate the value of f for the corrected values of y_{i+1} and y'_{i+1}.

Problems

1. Continue the example of this section to $t = 1.5$ and $t = 1.6$.
2. Continue the solution of the problem in Table 6.18 for the next three values of t using Milne's method. This is for the pendulum equation

$$\theta'' + 16 \sin \theta = 0.$$

Table 6.18

t	θ	θ'	f
1.15	−1.1751	3.5116	14.7641
1.20	−0.9816	4.2165	13.3028
1.25	−0.7550	4.8262	10.9657
1.30	−0.5012	5.2973	7.6890

3. Continue the solution of the problem in Table 6.19 for the next three values of t using Milne's method. This is for the pendulum equation

$$\theta'' + 4\theta' + 16 \sin \theta = 0.$$

Table 6.19

t	θ	θ'	f
1.55	0.0011	0.1400	−0.5783
1.60	0.0074	0.1114	−0.5642
1.65	0.0122	0.0839	−0.5323
1.70	0.0158	0.0583	−0.4870

4. Continue the solution of the biological equilibrium problem which was begun in the previous section for three more steps using Milne's method.

7

Further Topics in Series Methods

In this chapter we shall pursue further the study of series solutions of linear equations which we began in Chapter 5. We ended Chapter 5 by considering series solutions in the neighborhood of a regular singular point, where the indicial equation has roots that do not differ by an integer. Now we shall consider the special cases in which the roots are equal or in which they differ by an integer. Finally, after considering an application, we shall consider some problems which are not all sweetness and light.

7.1 Solutions of Homogeneous Linear Equations near a Regular Singular Point (part 2)

In this section we shall consider what happens when the two roots r_1 and r_2 of the indicial equation turn out to be equal. It is clear that in this case the methods of Chapter 5 will produce only one solution. We shall have to devise some method of obtaining a second solution. In order to see what sort of solution to expect, let us first examine an Euler equation with equal roots, for example the equation

$$x^2y'' + 3xy' + y = 0.$$

Under the change of variables $x = e^t$, this equation becomes

$$D^2y + 2Dy + y = (D + 1)^2y = 0,$$

where D represents differentiation with respect to t. The general solution of this equation is

$$y = c_1 e^{-t} + c_2 t e^{-t},$$

or, in terms of the variable x,

$$y = c_1 x^{-1} + c_2 x^{-1} \ln x.$$

Hence the two independent solutions are $y_1 = x^{-1}$ and $y_2 = x^{-1} \ln x = y_1 \ln x$. Thus, at least in this simple case, a repeated root leads to a second solution which contains the first multiplied by a logarithmic term.

For more general equations the situation is more complicated. We shall see that, in general, the second solution is of the form $y_2 = y_1 \ln x + y_3$, where y_3 is another infinite series. The following examples will indicate how one obtains a solution of this form.

Example 1
Let us consider

$$xy'' + y' + xy = 0.$$

This is *Bessel's equation of index zero,* as we can verify by checking example 2 of Section 5.3. To solve it near the regular singular point at $x = 0$, we make the usual substitution

$$y(x) = \sum_{n=0}^{\infty} c_n x^{n+r},$$

and after the usual manipulations obtain

$$\sum_{n=0}^{\infty} [(n + r)(n + r - 1) + (n + r)]c_n x^{n+r-1} + \sum_{n=0}^{\infty} c_n x^{n+r+1} = 0,$$

or

$$\sum_{n=0}^{\infty} (n + r)^2 c_n x^{n+r-1} + \sum_{n=2}^{\infty} c_{n-2} x^{n+r-1} = 0,$$

which leads to

$$r^2 c_0 x^{r-1} + (r + 1)^2 c_1 x^r + \sum_{n=2}^{\infty} [(n + r)^2 c_n + c_{n-2}]x^{n+r-1} = 0 \quad (7.1)$$

We thus obtain, upon equating the coefficients to zero,

$$r^2c_0 = 0, \quad (r+1)^2c_1 = 0, \quad (n+r)^2c_n + c_{n-2} = 0 \qquad \text{for } n = 2, 3, 4, \ldots .$$

The first of these, the indicial equation, tells us that $r_1 = r_2 = 0$, and the second tells us that $c_1 = 0$. Clearly the third equation, the recursion formula, leads to only one solution if we insert the value $r = 0$; so instead we leave r in this expression and solve for c_n as a function of the parameter r. Later, substitution of the value $r = 0$ will lead to the first of the two solutions we seek, but by leaving r in as a parameter, we hope to obtain an insight into the second solution before doing this. Setting $c_0 = 1$ for a particular solution, and using the formula

$$c_n = -\frac{c_{n-2}}{(n+r)^2}, \qquad n = 2, 3, 4, \ldots, \tag{7.2}$$

we have

$$c_2 = -\frac{1}{(2+r)^2}, \quad c_4 = -\frac{c_2}{(4+r)^2}, \ldots, c_{2k} = -\frac{c_{2k-2}}{(2k+r)^2},$$

so that

$$c_{2k}(r) = \frac{(-1)^k}{(2+r)^2(4+r)^2 \cdots (2k+r)^2}.$$

We have written $c_{2k}(r)$ to emphasize the dependence of c_{2k} on the parameter r. Note that in this case as in example 2 of Section 5.3, the coefficients with odd subscripts all vanish because $c_1 = 0$. We now write

$$y(x;r) = x^r + \sum_{k=1}^{\infty} c_{2k}(r)x^{2k+r} = x^r + \sum_{k=1}^{\infty} \frac{(-1)^k x^{2k+r}}{(2+r)^2(4+r)^2 \cdots (2k+r)^2}.$$

If we denote by L the linear differential operator

$$L = xD^2 + D + x,$$

then because of equation (7.1) and because the coefficients $c_{2k}(r)$ satisfy the recursion formula (7.2), we have

$$L[y(x;r)] = r^2c_0x^{r-1}. \tag{7.3}$$

It is now apparent why the choice $r = 0$ makes $y(x;r)$ into a solution of the differential equation: because $r = 0$ makes the right side of (7.3) vanish:

$$L[y(x;0)] = 0 \cdot c_0 x^{-1} = 0.$$

Hence one solution of the differential equation is $y_1(x) = y(x;0)$. How can we go about discovering the second solution? The answer, called *the method of Frobenius*, is obtained by noticing that the right side of (7.3) has a factor of r^2; that is, the zero on the right side when $r = 0$ is a second-order zero. This means that the *derivative* of the right side with respect to r *also* vanishes at $r = 0$. In fact, if we take the partial derivative of both sides of (7.3) with respect to r, we obtain

$$\frac{\partial}{\partial r}\{L[y(x;r)]\} = 2rc_0x^{r-1} + r^2c_0x^{r-1}\ln x.$$

In obtaining this, we have used the product rule, since the right side of (7.3) contains two factors (r^2 and x^{r-1}) depending on r. Furthermore, the reader should recall that the derivative of x^{r-1} *with respect to* r is $x^{r-1}\ln x$, *not* $(r - 1)x^{r-2}$, which is the derivative *with respect to* x. This is the source of the logarithmic term in all such solutions. Now the coefficients of L do not depend on r, so the orders of differentiation can be interchanged and we have

$$L\left[\frac{\partial y}{\partial r}(x;r)\right] = r[2c_0 + rc_0 \ln x]x^{r-1}.$$

Putting $r = 0$ into this equation, we see that

$$L\left[\frac{\partial y}{\partial r}(x;r)\bigg|_{r=0}\right] = 0.$$

Therefore

$$y_2(x) = \frac{\partial y}{\partial r}(x;r)\bigg|_{r=0}$$

is a second solution of the differential equation. Now since

$$y(x;r) = x^r + \sum_{k=1}^{\infty} c_{2k}(r)x^{2k+r}, \tag{7.4}$$

we have

$$\frac{\partial y}{\partial r}(x;r) = \left[x^r + \sum_{k=1}^{\infty} c_{2k}(r)x^{2k+r}\right] \ln x + \sum_{k=1}^{\infty} c'_{2k}(r)x^{2k+r}$$

or

$$\frac{\partial y}{\partial r}(x;r) = y(x;r) \ln x + \sum_{k=1}^{\infty} c'_{2k}(r)x^{2k+r}. \tag{7.5}$$

We need only find an expression for $c'_{2k}(r)$ and we will be done except for substituting $r = 0$ into (7.4) and (7.5).

The expression

$$c_{2k}(r) = \frac{(-1)^k}{(2+r)^2(4+r)^2 \cdots (2k+r)^2}$$

is a rather complicated function of r and does not promise to be easy to differentiate. We shall do this by appealing to the method of logarithmic differentiation; the reader should recall that if a function of r is a product of k other functions of r, that is, if

$$f(r) = f_1(r)f_2(r) \cdots f_k(r),$$

then we can write

$$\ln f(r) = \ln f_1(r) + \ln f_2(r) + \cdots + \ln f_k(r).$$

Differentiating both sides with respect to r, we obtain

$$\frac{f'(r)}{f(r)} = \frac{f_1'(r)}{f_1(r)} + \frac{f_2'(r)}{f_2(r)} + \cdots + \frac{f_k'(r)}{f_k(r)}$$

or, finally,

$$f'(r) = f(r)\left[\frac{f_1'(r)}{f_1(r)} + \frac{f_2'(r)}{f_2(r)} + \cdots + \frac{f_k'(r)}{f_k(r)}\right].$$

Thus to differentiate a product of several functions we simply multiply the original function by the sums of the logarithmic derivatives of each of the factors. If in particular, the factors are powers of a linear function of r,

$$f_i(r) = (a_i r + b_i)^m,$$

as in the case of our function $a_k(r)$, then we have

$$\frac{f_i'(r)}{f_i(r)} = \frac{a_i m(a_i r + b_i)^{m-1}}{(a_i r + b_i)^m} = \frac{a_i m}{(a_i r + b_i)},$$

so that the logarithmic derivative of the factor is the exponent (positive or negative) times the coefficient of r divided by the linear term itself. Thus in our case we obtain

$$c_{2k}'(r) = \frac{(-1)^k}{(2+r)^2(4+r)^2\cdots(2k+r)^2}\left(\frac{-2}{2+r} - \frac{2}{4+r} - \cdots - \frac{2}{2k+r}\right).$$

Evaluating this and $c_{2k}(r)$ at $r = 0$, we have,

$$c_{2k}(0) = \frac{(-1)^k}{2^2 \cdot 4^2 \cdots (2k)^2} = \frac{(-1)^k}{2^{2k}(k!)^2}$$

and

$$c_{2k}'(0) = \frac{(-1)^k}{2^{2k}(k!)}\left(-\frac{2}{2} - \frac{2}{4} - \cdots - \frac{2}{2k}\right) = \frac{(-1)^{k+1}}{2^{2k}(k!)}\left(1 + \tfrac{1}{2} + \tfrac{1}{3} + \cdots + \frac{1}{k}\right).$$

The notation H_k is normally used for the first k terms of the harmonic series

$$\sum_{n=1}^{k} \frac{1}{n} = 1 + \tfrac{1}{2} \cdots + \frac{1}{k};$$

thus, finally,

$$c_{2k}'(0) = \frac{(-1)^{k+1} H_k}{2^{2k}(k!)^2}.$$

Thus, substituting $r = 0$ in (7.4) and (7.5), we obtain

$$y_1(x) = y(x;0) = 1 + \sum_{k=1}^{\infty} \frac{(-1)^k}{(k!)^2}\left(\frac{x}{2}\right)^{2k},$$

$$y_2(x) = y_1(x)\ln x + \sum_{k=1}^{\infty} \frac{(-1)^{k+1} H_k}{(k!)^2}\left(\frac{x}{2}\right)^{2k}.$$

The reader can verify by consulting the last section that $y_1(x)$ is equal to $J_0(x)$, *Bessel's function of the first kind of index 0*. The function $y_2(x)$ is usually denoted by $Y_0(x)$ and is called *Bessel's function of the second kind of index 0*. Clearly $J_0(x)$ and $Y_0(x)$ are a linearly independent pair of particular solutions of Bessel's equation of index 0. There is a close relationship between Bessel functions of the first kind and damped trigonometric functions. In fact, if we write out the first few terms of the power series for the following functions,

$$J_0(x) = 1 - \frac{x^2}{2 \cdot 2} + \frac{x^4}{2 \cdot 2 \cdot 4 \cdot 4} - \frac{x^6}{2 \cdot 2 \cdot 4 \cdot 4 \cdot 6 \cdot 6} + \cdots,$$

$$\cos x = 1 - \frac{x^2}{1 \cdot 2} + \frac{x^4}{1 \cdot 2 \cdot 3 \cdot 4} - \frac{x^6}{1 \cdot 2 \cdot 3 \cdot 4 \cdot 5 \cdot 6} + \cdots,$$

$$\frac{\sin x}{x} = 1 - \frac{x^2}{2 \cdot 3} + \frac{x^4}{2 \cdot 3 \cdot 4 \cdot 5} - \frac{x^6}{2 \cdot 3 \cdot 4 \cdot 5 \cdot 6 \cdot 7} + \cdots,$$

$$\frac{2J_1(x)}{x} = 1 - \frac{x^2}{2 \cdot 4} + \frac{x^4}{2 \cdot 4 \cdot 4 \cdot 6} - \frac{x^6}{2 \cdot 4 \cdot 4 \cdot 6 \cdot 6 \cdot 8} + \cdots,$$

one can see pronounced similarities. Like the sine and cosine functions, the Bessel functions have an infinite sequence of zeros (see Figure 7.1).

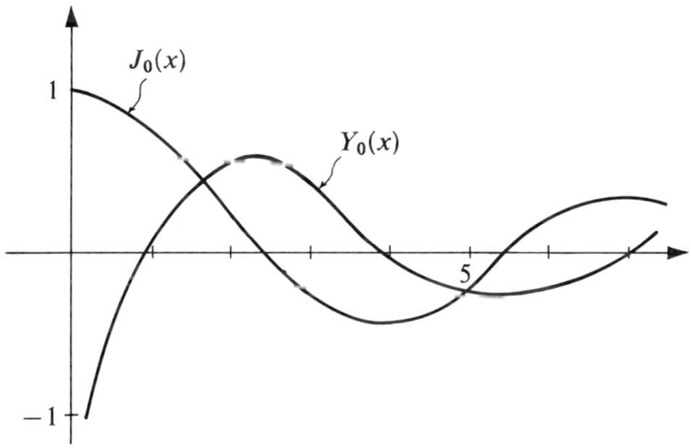

Figure 7.1

Now, let us review in outline form the method we have devised to solve this type of equation:

1. Recognizing a regular singular point, we make the substitution $y = \Sigma_{n=0}^{\infty} c_n x^{n+r}$ and perform the usual simplifications, whereupon we find that $r_1 = r_2$.

2. Instead of substituting the indicial equation root for r, we leave r in as a parameter, set up the recursion formula, and find $c_n = c_n(r)$ as a function of r and c_0 (which we normally choose to be 1, for a particular solution).

3. We next compute $c_n'(r)$, the derivative of c_n with respect to r, a process which is usually made simpler by using logarithmic differentiation.

4. Once we have $c_n(r)$ and $c_n'(r)$, we can write down the two solutions of the differential equation immediately:

$$y_1(x) = x^{r_1} + \sum_{n=0}^{\infty} c_n(r_1)x^{n+r_1}; \quad y_2(x) = y_1(x)\ln x + \sum_{n=1}^{\infty} c_n'(r_1)x^{n+r_1}.$$

Note that only in this final step do we insert the root r_1 in place of the parameter r.

One further example will serve to illustrate the normal solution process once the four preceding rules have been drawn up.

Example 2

Consider the equation

$$x^2y'' + (7x - 3x^2)y' + 9y = 0.$$

Substituting $y = \Sigma_{n=0}^{\infty} c_n x^{n+r}$, we obtain

$$\sum_{n=0}^{\infty} [(n + r)(n + r - 1) + 7(n + r) + 9]c_n x^{n+r} - \sum_{n=0}^{\infty} 3(n + r)c_n x^{n+r+1} = 0$$

or

$$\sum_{n=0}^{\infty} (n + r + 3)^2 c_n x^{n+r} - \sum_{n=1}^{\infty} 3(n + r - 1)c_{n-1}x^{n+r} = 0.$$

Hence

$$(r + 3)^2 c_0 x^r + \sum_{n=1}^{\infty} [(n + r + 3)^2 c_n - 3(n + r - 1)c_{n-1}]x^{n+r} = 0.$$

Since both roots of the indicial equation are $r_1 = r_2 = -3$, we leave r in as a parameter and use the recursion formula

$$c_n = \frac{3(n + r - 1)}{(n + r + 3)^2}c_{n-1} \qquad (n = 1, 2, 3, \ldots)$$

to find c_n as a function of r:

$$c_0 = 1, \quad c_1 = \frac{3r}{(r + 4)^2}, \quad c_2 = \frac{3(r + 1)}{(r + 5)^2}c_1, \quad c_3 = \frac{3(r + 2)}{(r + 6)^2}c_2,$$

$$c_4 = \frac{3(r + 3)}{(r + 7)^2}c_3, \ldots$$

leading to the formula

$$c_n(r) = \frac{3^n r(r+1)(r+2)(r+3)\cdots(r+n-1)}{(r+4)^2(r+5)^2(r+6)^2(r+7)^2\cdots(r+n+3)^2}.$$

Note that when we substitute $r = -3$, all the coefficients from c_4 on will vanish because of the factor $(r+3)$ in the numerator. However, before we do this we shall have to calculate the derivatives of the functions $c_n(r)$. The formula for $c_n(r)$ contains the factor $r+3$ only for $n \geqq 4$; thus for $n \geqq 4$,

$$c_n'(r) = \frac{3^n r(r+1)(r+2)(r+3)\cdots(r+n-1)}{(r+4)^2(r+5)^2\cdots(r+n+3)^2}\left(\frac{1}{r} + \frac{1}{r+1} + \frac{1}{r+2}\right.$$
$$\left. + \frac{1}{r+3} + \cdots + \frac{1}{r+n-1} - \frac{2}{r+4} - \frac{2}{r+5} - \cdots - \frac{2}{r+n-3}\right).$$

When -3 is inserted in place of r, all terms of the above product will vanish except for the one obtained by multiplying $c_n(r)$ by $1/(r+3)$, since the factors of $(r+3)$ in numerator and denominator will have canceled. Thus for $n \geqq 4$,

$$c_n'(-3) = \left.\frac{3^n r(r+1)(r+2)(r+4)\cdots(r+n-1)}{(r+4)^2(r+5)^2\cdots(r+n+3)^2}\right|_{r=-3} = \frac{3^n(-6)(n-4!)}{(n!)^2}.$$

For $n < 4$, we shall have to write out and differentiate $c_1(r)$, $c_2(r)$ and $c_3(r)$ separately:

$$c_1(r) = \frac{3r}{(r+4)},$$

so

$$c_1'(r) = \frac{3r}{(r+4)^2}\left(\frac{1}{r} - \frac{2}{r+4}\right);$$

$$c_2(r) = \frac{3r(r+1)}{(r+4)^2(r+5)^2},$$

so

$$c_2'(r) = \frac{3r(r+1)}{(r+4)^2(r+5)^2}\left(\frac{1}{r} + \frac{1}{r+1} - \frac{2}{r+4} - \frac{2}{r+5}\right);$$

$$c_3(r) = \frac{3r(r+1)(r+2)}{(r+4)^2(r+5)^2(r+6)^2},$$

so

$$c_3'(r) = \frac{3r(r+1)(r+2)}{(r+4)^2(r+5)^2(r+6)^2}\left(\frac{1}{r} + \frac{1}{r+1} + \frac{1}{r+2} - \frac{2}{r+4} - \frac{2}{r+5} - \frac{2}{r+6}\right).$$

Evaluating these coefficients at $r = -3$, we obtain

$$c_1(-3) = -9;$$

$$c_1'(-3) = -9(-\tfrac{1}{3} - 2) = 3 + 18 = 21;$$

$$c_2(-3) = -9 \cdot \frac{3(-2)}{4} = \frac{27}{2};$$

$$c_2' = \tfrac{27}{2}(-\tfrac{1}{3} - \tfrac{1}{2} - 2 - 1) = -\tfrac{27}{2}(\tfrac{23}{6}) = -\tfrac{207}{4};$$

$$c_3(-3) = \tfrac{27}{2}(-\tfrac{1}{9}) = -\tfrac{3}{2};$$

$$c_3' = -\tfrac{3}{2} - \tfrac{1}{3} - \tfrac{1}{2} - 1 - 2 - 2 - \tfrac{2}{3} = -\tfrac{3}{2}(-\tfrac{11}{2}) = \tfrac{33}{4}.$$

Hence, since $c_4 = c_5 = c_6 = \cdots = 0$, as noted above, we have

$$y_1(x) = x^{-3} - 9x^{-2} + \tfrac{27}{2}x^{-1} - \tfrac{3}{2},$$

and

$$y_2(x) = y_1(x)\ln x + 21x^{-2} - \tfrac{207}{4}x^{-1} + \tfrac{33}{4}x + \sum_{n=4}^{\infty} \frac{(-6)3^n(n-4)!}{(n!)^2}x^{n-3}.$$

This example clearly shows that in some cases, all but a finite number of the c_n may vanish; however, this does not at all imply that all of the c_n' must vanish. The reason for this is that the factor in $c_n(r)$ which causes it to vanish will cancel a similar factor in a denominator after differentiation and leave an expression for $c_n'(r)$ which does not vanish when we put $r = r_1$. In such a case one should proceed as we did above: Differentiate the general expression for $c_n(r)$ to obtain the one for $c_n'(r)$; then write out and differentiate separately—as special cases—all the nonvanishing c_n. After this is done, all coefficients can be evaluated and the finite series for y_1 and the infinite series for y_2 can be written out. There are a number of problems of this sort at the end of the section; 1, 3, and 5 are three such.

Problems

In 1–16, find two linearly independent solutions around $x = 0$.

1. $x^2y'' + x(x + 1)y' = 0.$
2. $2x(2x + 1)y'' + 2y' + y = 0.$
3. $xy'' + y' = 0.$
4. $x^2y'' + 3x(1 - x)y' + y = 0.$
5. $x(3x + 2)y'' + 2y' - 6y = 0.$
6. $2xy'' + 2y' - 3y = 0.$
7. $x^2y'' + x(x + 5)y' + 4y = 0.$
8. $x(2x + 3)y'' + 3y' - 4y = 0.$
9. $xy'' + y' - xy = 0$ (Bessel's modified equation of order zero).
10. $x^2y'' - 3x(1 - 2x)y' + 4y = 0.$
11. $2x(2x + 3)y'' + 6y' + y = 0.$
12. $xy'' + y' + 2y = 0.$
13. $x^2y'' + 5x(2x - 1)y' + 9y = 0.$
14. $xy + y' - 4xy = 0.$
15. $2xy'' + 2y' - xy = 0.$
16. $4x^2y'' + x(16 - x)y' + 9y = 0.$

In 17–20, find two linearly independent solutions around the indicated point $x = x_0$.

17. $4(x - 1)^2y'' - 8x(x - 1)y' + y = 0, x_0 = 1.$
18. $(x - 2)(x - 3)y'' - y' = 0, x_0 = 2.$
19. $(x + 3)y'' + y' + 2y + 0, x_0 = -3.$
20. $4(x + 2)^2y'' + (x^2 - 4)y' + 4y = 0, x_0 = -2.$

7.2 Solutions of Homogeneous Linear Equations near a Regular Singular Point (part 3)

In this section and the following one we shall discuss what happens when the roots of the indicial equation differ by an integer. It will turn out that there are two possibilities: Either two arbitrary constants are implied by the recursion formula, in which case we obtain two series solutions; or else only one arbitrary constant is present, in which case we use the method of Frobenius to obtain a second solution with a logarithmic term. We shall consider the first case in this section.

In order to see what might go wrong when the roots of the indicial equation differ by an integer, let us suppose that $r_2 - r_1 = k$ (a positive integer, since we have agreed to number the roots so that $r_1 \leq r_2$).

The indicial equation will have come from the $n = 0$ term, which usually looks as follows:

$$(r - r_1)(r - r_2)c_0x^r = 0. \tag{7.6}$$

The recursion formula will usually be of the form

$$(n + r - r_1)(n + r - r_2)c_n - f(n; r)c_{n-1} = 0. \tag{7.7}$$

where $f(n; r)$ is some function of n and r. If we solve this for c_n, we obtain

$$c_n = \frac{f(n; r)}{(n + r - r_1)(n + r - r_2)} \tag{7.8}$$

or, putting $r = r_1$, we have

$$c_n = \frac{f(n; r_1)}{n(n + r_1 - r_2)} c_{n-1} = \frac{f(n; r_1)}{n(n - k)} c_{n-1}$$

since $r_1 - r_2 = -k$. If we attempt to calculate c_k from this relationship, we will be dividing by zero; so, in other words, c_k is indeterminate. How can we correct this problem?

We reason as follows: Equation (7.6) tells us c_0 is an arbitrary constant. Furthermore, in passing from (7.7) to (7.8) we created a major difficulty by dividing by zero, when $n = k$. But equation (7.8) creates no difficulty when $n \neq k$. When $n = k$, instead of dividing, we use equation (7.7), which becomes

$$0 \cdot c_k - f(k; r_1)c_{k-1} = 0.$$

If either $f(k; r_1)$ or c_{k-1} is zero, then this says that $0 \cdot c_k = 0$; in other words, c_k is the second arbitrary constant (c_0 being the other). We can then use (7.8) safely, to find $c_{k+1}, c_{k+2}, \ldots$ in terms of c_k, and to compute $c_1, c_2, \ldots, c_{k-1}$ in terms of c_0.

Hence our procedure should be roughly as follows: after determining r_1 and r_2, and that $r_2 - r_1 = k$ is an integer, we calculate the recursion formula (7.7) with $r = r_1$ and evaluate it for $n = 1, 2, \ldots, k$. If this formula implies that c_k is an arbitrary constant, then we can proceed to use (7.8) to evaluate the rest of the coefficients and complete the solution of the problem. To illustrate this, let us now work some examples.

Example 1

Let us solve

$$x^2y'' + (2x - x^2)y' - (x + 2)y = 0.$$

After making the usual substitution

$$y = \sum_{n=0}^{\infty} c_n x^{n+r},$$

we obtain

$$\sum_{n=0}^{\infty} [(n + r)(n + r - 1) + 2(n + r) - 2]c_n x^{n+r} - \sum_{n=0}^{\infty} [(n + r) + 1]c_n x^{n+r+1} = 0.$$

After simplification and changing the summation index on the second series, we have

$$\sum_{n=0}^{\infty} (n + r - 1)(n + r + 2)c_n x^{n+r} - \sum_{n=1}^{\infty} (n + r)c_{n-1} x^{n+r} = 0.$$

The $n = 0$ term gives the indicial equation

$$(r - 1)(r + 2)c_0 = 0$$

and the $n \geqq 1$ terms give the recursion formula

$$(n + r - 1)(n + r + 2)c_n - (n + r)c_{n-1} = 0.$$

We see that the roots of the indicial equation are $r_1 = -2$ and $r_2 = 1$, with difference $k = r_2 - r_1 = 3$. Thus the crucial coefficient, the one which we wish to be arbitrary in addition to c_0, is c_3. Placing $r_1 = -2$ into the recursion formula, we have

$$(n - 3)nc_n - (n - 2)c_{n-1} = 0 \tag{7.9}$$

and, taking c_0 arbitrary, we have

$$n = 1: \quad (-2) \cdot 1 \cdot c_1 - (-1)c_0 = 0 \Rightarrow c_1 = \tfrac{1}{2}c_0;$$

$$n = 2: \quad (-1) \cdot 2 \cdot c_2 - 0 \cdot c_1 = 0 \Rightarrow c_2 = 0;$$

$$n = 3: \quad 0 \cdot c_3 - c_2 = 0 \cdot c_3 = 0,$$

and hence c_3 is indeed an arbitrary constant. Thus we can solve (7.9) for c_n if $n > 3$, obtaining

$$c_n = \frac{(n-2)}{n(n-3)} c_{n-1};$$

therefore with c_3 arbitrary we have

$$c_4 = \frac{2}{4 \cdot 1} c_3, c_5 = \frac{3}{5 \cdot 2} c_4, c_6 = \frac{4}{6 \cdot 3} c_5, \ldots, c_n = \frac{(n-2)}{n(n-3)} c_{n-1}, \ldots .$$

Multiplying these all together we have

$$c_n = \frac{(n-2)!}{4 \cdot 5 \cdot 6 \cdots n(n-3)!} c_3 = \frac{6(n-2)}{n!} c_3.$$

Hence, since

$$y = \sum_{n=0}^{\infty} c_n x^{n+r_1} = (c_0 x^{r_1} + c_1 x^{r_1+1} + c_2 x^{r_1+2}) + \sum_{n=3}^{\infty} c_n x^{r_1+n},$$

we have

$$y = c_0(x^{-2} + \tfrac{1}{2}x^{-1}) + c_3 \sum_{n=3}^{\infty} \frac{6(n-2)}{n!} x^{n-2}.$$

Note that we used $r_1 = -2$, the smaller value, to find both solutions; however, the second solution actually starts with a term in $x^1 = x^{r_2}$, so that when we use the smaller root of the indicial equation, the solution corresponding to the larger root of the indicial equation comes out in the wash. Therefore whenever the roots differ by an integer k and c_0 and c_k both turn out to be arbitrary, the complete solution is always obtained by using the smaller root $r = r_1$.

Example 2
Let us again consider Bessel's equation,

$$x^2 y'' + xy' + (x^2 - v^2)y = 0,$$

where this time the index v is half an odd integer, that is, $v = (2m + 1)/2$

for $m = 0, 1, 2, \ldots$. Referring back to example 2 of Section 5.3, we know that the substitution

$$y = \sum_{n=0}^{\infty} c_n x^{n+r}$$

leads to

$$(r + v)(r - v)c_0 x^r + (r + v + 1)(r - v + 1)c_1 x^{r+1}$$
$$+ \sum_{n=2}^{\infty} [(r + n - v)(r + n + v)c_n + c_{n-2}]x^{n+r} = 0.$$

Thus c_0 is arbitrary and the roots of the indicial equation are

$$r_1 = -v = -(2m + 1)/2$$

and

$$r_2 = v = (2m + 1)/2.$$

Thus $r_2 - r_1 = 2m + 1$, an (odd) integer. The second term above for $r = r_1 = -v$ gives us the equation

$$(-2v + 1)c_1 = -2mc_1 = 0.$$

If $m = 0$, in which case $r_2 - r_1 = 1$, this tells us that c_1 is arbitrary. If $m \neq 0$, then this tells us that $c_1 = 0$. The recursion formula, for $r = r_1 = -v$, is

$$(n - 2v)nc_n + c_{n-2} = (n - 2m - 1)nc_n + c_{n-2} = 0. \tag{7.10}$$

Thus the odd coefficients are all determined by c_1; for $n = 2k + 1$ the recursion formula is

$$2(k - m)(2k + 1)c_{2k+1} + c_{2k-1} = 0,$$

implying that for $k < m$, we have $c_{2k+1} = 0$, and for $k = m$,

$$0 \cdot c_{2m+1} + c_{2m-1} = 0 \cdot c_{2m+1} = 0,$$

and hence c_{2m+1} is arbitrary. Thus, we will be able to find two series-type solutions from the recursion formula (7.10), one with even coefficients based on c_0 and the other odd coefficients based on c_{2m+1} (Table 7.1). So, for the even coefficients, we have

$$c_2 = \frac{(-1)^k c_0}{(2-2v)(4-2v)\cdots(2k-2v)2\cdot 4\cdots 2k}$$
$$= \frac{(-1)^k c_0}{2^{2k}(1-v)(2-v)\cdots(k-v)k!}$$

and therefore

$$y_1(x) = c_0 x^{-v}\left[1 + \sum_{k=1}^{\infty} \frac{(-1)^k x^{2k}}{2^{2k}\cdot(1-v)(2-v)\cdots(k-v)k!}\right].$$

If, as before, we choose $c_0 = 1/2^{-v}\Gamma(1-v)$, we obtain

$$y_1(x) = \left(\frac{x}{2}\right)^{-v}\left[\frac{1}{\Gamma(1-v)} + \sum_{k=1}^{\infty} \frac{(-1)^k}{\Gamma(k-v+1)k!}\left(\frac{x}{2}\right)^{2k}\right] = J_{-v}(x).$$

Thus $J_{-v}(x)$ is one solution of the Bessel equation with $v = (2m+1)/2$. What is the other solution? We have

$$c_{2m+2+1}$$
$$= \frac{(-1)^k c_{2m+1}}{(2m+3-2v)\cdots(2m+2k+1-2v)(2m+3)(2m+5)\cdots(2m+2k-1)},$$

Table 7.1

c_0 arbitrary	c_{2m+1} arbitrary
$c_2 = -\dfrac{c_0}{(2-2v)2}$	$c_{2m+3} = -\dfrac{c_{2m+1}}{(2m+3-2v)(2m+3)}$
$c_4 = -\dfrac{c_2}{(4-2v)4}$	$c_{2m+5} = -\dfrac{c_{2m+3}}{(2m+5-2v)(2m+5)}$
$\vdots$	$\vdots$
$c_{2k} = -\dfrac{c_{2k-2}}{(2k-2v)2k}$	$c_{2m+2k+1} = -\dfrac{c_{2m+2k-1}}{(2m+2k+1-2v)(2m+2k-1)}$

so that

$$y_2(x) = c_{2m+1}x^{2m+1-v}$$
$$+ \sum_{k=1}^{\infty} \frac{(-1)^k x^{2m+2k+1-v}}{(2m+3-2v)\cdots(2m+2k+1-v)(2m+3)(2m+5)\cdots(2m+2k-1)}$$

Noting that $2m + 1 = 2v$, we can rewrite this as

$$y_2(x) = c_{2v}\left[x^v + \sum_{k=1}^{\infty} \frac{(-1)^k x^{2k+v}}{2\cdot 4\cdots 2k(2v+2)(2v+4)\cdots(2v+2k)}\right],$$

because $2k + 1 = 2m + 2(k - m) = 2v + 2(k - m)$; and this reduces to

$$y_2(x) = c_{2v}\left[x^v + \sum_{k=1}^{\infty} \frac{(-1)^k x^{2k+v}}{2^{2k}k!(v+1)(v+2)\cdots(v+k)}\right],$$

and, by choosing $c_{2v} = 1/2^v\Gamma(v + 1)$, we have

$$y_2(x) - \left(\frac{x}{2}\right)^v \left[\frac{1}{\Gamma(v+1)} + \sum_{k=1}^{\infty} \frac{(-1)^k}{2^{2k}k!\Gamma(k+v+1)}\left(\frac{x}{2}\right)^{2k}\right] = J_v(x).$$

Hence $J_v(x)$ is the other solution. Combining this with the result of example 2, Section 5.3, we obtain the fact that *$J_v(x)$ and $J_{-v}(x)$ are two linearly independent solutions of Bessel's equation of index v, as long as v is not an integer.*

Problems

In problems 1–16, find series solutions about $x = 0$.

1. $2y'' + 3xy' - y = 0.$
2. $x(x + 2)y'' - 4y' - 2y = 0.$
3. $4x^2y'' + 4xy' + (4x^2 - 9)y = 0.$ (Bessel's equation of order $\frac{3}{2}$.)
4. $xy'' + 2(x + 1)y' + 2y = 0.$
5. $x^2y'' + (x^2 + 2x)y' - 2y = 0.$
6. $x^2y'' + 2x^2y' - 2y = 0.$
7. $xy'' + (x^3 - 2)y' = 0.$
8. $y'' - 4xy' - 6y = 0.$
9. $(2x + 3)xy'' + 4(2x - 3)y' - 20y = 0.$
10. $4x^2y'' + 12xy' + (2x^2 + 3) = 0.$
11. $2xy'' + (3x - 2)y' - 3y = 0.$
12. $x^2y'' - x(1 + 3x)y' - 3y = 0.$

13. $2x^2y'' + 3x^2y' - 16 = 0.$
14. $2xy'' + (x^3 - 2)y' + 2x^2y = 0.$
15. $x(x + 1)y'' + (x + 8)y' - y = 0.$
16. $xy'' + (x + 4)y' + y = 0.$

In problems 17–20, find series solutions about the indicated value of x_0.

17. $3y'' + 2(x - 3)y' + 5y = 0,\ x_0 = 3.$
18. $(x + 1)^2y'' + 2(x + 1)y' + (x^2 + 2x - 1)y = 0, x_0 = -1.$
19. $(x + 4)^2y'' + (x + 4)(2x + 15)y' + (6x + 29)y + 0, x_0 = 4.$
20. $(x - 2)y'' + 2(x^3 - 6x^2 + 12x - 6)y' + 10(x - 2)^2y = 0, x_0 = 2.$

7.3 Solutions of Homogeneous Linear Equations near a Regular Singular Point (part 4)

We shall now consider what happens in the case when the roots of the indicial equation differ by an integer, but the recursion formula yields up only one arbitrary constant. When this happens the technique which we use is strikingly similar to that used when the roots of the indicial equation are equal, which we discussed in Section 7.1. We shall describe the general procedure by means of a detailed example, and then work a couple of further examples utilizing various short cuts.

Example 1

Let us consider

$$x^2y'' + xy' + (x^2 - 4)y = 0.$$

This is Bessel's equation with index 2, and will illustrate what happens with Bessel's equation for arbitrary integral index ν. First, we make the usual substitution

$$y = \sum_{n=0}^{\infty} c_n x^{n+r}$$

and observe that this leads to

$$(r + 2)(r - 2)c_0x^r + (r + 3)(r - 1)c_1x^{r+1} + \sum_{n=2}^{\infty} [(n + r + 2)(n + r - 2)c_n + c_{n-2}]x^{n+r} = 0. \tag{7.11}$$

The first term gives the equation $(r + 2)(r - 2)c_0 = 0$ with indicial roots $r_1 = -2$ and $r_2 = 2$, and their difference is $r_2 - r_1 = 4$. Hence, as we did in the last section, we must use the recursion formula

$$(n + r + 2)(n + r - 2)c_n + c_{n-2} = 0 \qquad (n = 2, 3, 4, \ldots)$$

with $r = r_1 = -2$ to determine whether c_4 is an arbitrary constant. We have $n(n - 4)c_n = -c_{n-2}$, so that $2(-2)c_2 = -c_0$ and $4(0)c_4 = -c_2$. The latter implies that $c_2 = 0$, so the first equation implies that $c_0 = 0$; but c_0 is supposed to be an arbitrary constant. Thus our attempt to prove that c_4 is arbitrary leads only to a contradiction of the fact that c_0 is arbitrary. Our procedure then, must be similar to that of Section 7.1: find a formula for c_n as a function of r, set up

$$y(x; r) = \sum_{n=0}^{\infty} c_n(r)x^{n+r},$$

and then see if we can, by some device similar to that used in Section 7.1, arrive at two different solutions of the equation.

First, we note that the second term of (7.11) implies that $(r + 3)(r - 1)c_1 = 0$, so that for either value of r, c_1 vanishes. The recursion formula for odd values $n = 2k + 1$ of the subscript is

$$c_{2k+1} = -\frac{c_{2k-1}}{(2k + r + 3)(2k + r - 1)},$$

so that for either value of r, all the coefficients with odd subscripts are zero. Thus we need only calculate a formula for the coefficients with even subscripts:

$$c_2 = -\frac{c_0}{(r + 4)r},$$

$$c_4 = -\frac{c_2}{(r + 6)(r + 2)},$$

$$c_6 = -\frac{c_4}{(r + 8)(r + 4)},$$

$$\vdots$$

$$c_{2k} = -\frac{c_{2k-2}}{(r + 2k + 2)(r + 2k - 2)},$$

and therefore

$$c_{2k}(r) = \frac{(-1)^k c_0}{(r+4)(r+6)\cdots(r+2k+2)\cdot r(r+2)(r+4)\cdots(r+2k-2)}.$$

Writing

$$y(x;r) = c_0 x^r + \sum_{k=1}^{\infty} c_2(r) x^{2k+r},$$

we see that (7.11) implies that

$$[x^2D^2 + xD + (x^2 - 4)]y(x;r) = (r+2)(r-2)c_0 x^r. \tag{7.12}$$

All the other terms vanish because $c_{2k}(r)$ satisfies the recursion formula derived from (7.11).

In Section 7.1, at this point we were able to show that $y(x;r_1)$ was a solution because the right side vanished at $r = r_1$; and, because $(r - r_1)^2$ occurred as a factor on the right side, we were also able to show that $\partial y(x;r)/\partial r$ was a solution. We cannot do that in this case, however, for two reasons:

1. $y(x;-2)$ is not defined because $c_{2k}(r)$ is not defined at $r = -2$ for $k \geqq 2$, since the factor $(r+2)$ occurs in its denominator; and
2. because $(r+2)$ occurs only to the first power on the right side of (7.12), its partial derivative with respect to r will not vanish at $r = -2$.

However, the following rather simple observation will enable us to solve the problem: We could use a factor of $(r+2)$ in the numerator of the expression for $c_{2k}(r)$ and on the right side of (7.12). The problem would be solved if this factor were present. Since c_0 is present in both of the crucial spots, let us try writing $c_0 = d_0(r+2)$, where d_0 is another arbitrary constant, to see if this will clear up our difficulty. The expression for $c_{2k}(r)$ then becomes

$$c_{2k}(r) = \frac{(-1)^k d_0}{(r+4)(r+6)\cdots(r+2k+2)\cdot r(r+4)(r+6)\cdots(r+2k-2)} \qquad (k \geqq 2),$$

with $c_0 = d_0(r+2)$ and $c_2 = -d_0(r+2)/r(r+4)$. At the value $r = -2$ we have $c_0 = c_2 = 0$ and

$$c_{2k}(-2) = \frac{(-1)^k d_0}{2\cdot 4\cdots 2k(-2)\cdot 2\cdot 4\cdots(2k-4)} = \frac{(-1)^{k+1} d_0}{2^{2k-1} k!(k-2)!}$$

so that $y(x; -2)$ is defined and gives us the solution

$$y_1(x) = y(x; -2) = \sum_{k=2}^{\infty} \frac{(-1)^{k+1}x^{2k-2}}{2^{2k-1}k!(k-1)} = \sum_{k=0}^{\infty} \frac{(-1)^{k+1}x^{2k+2}}{2^{2k+3}(k+2)!k!},$$

where we have set $d_0 = 1$ for a particular solution. Retaining $c_0 = r + 2$, with $d_0 = 1$, equation (7.12) now becomes

$$[x^2D^2 + xD + (x^2 - 4)]y(x; r) = (r+2)^2(r-2)x^r.$$

Now that we have a factor of $(r + 2)^2$ on the right side, we can take the partial derivative of both sides with respect to r and obtain

$$[x^2D^2 + xD + (x^2 - 4)]\frac{\partial y(x; r)}{\partial r} = 2(r+2)(r-2)x^r + (r+2)^2x^r + (r+2)^2(r-2)x^r \ln x,$$

and since the common factor $(r + 2)$ occurs throughout the right side, we see that $\partial y(x; r)/\partial r|_{r=-2}$ is a second solution of the equation.

Now we already know that

$$\frac{\partial y}{\partial r} = y(x; r) \ln x + c_0'(r)x^{-2} + c_2'(r) + \sum_{k=2}^{\infty} c_{2k}'(r)x^{2k-2}.$$

Evaluating this at $r = -2$ will give us the solution. Let us now compute the derivatives of the coefficients:

$$c_0'(r) = 1; \quad c_2'(r) = -\frac{r+2}{r(r+4)}\left(\frac{1}{r+2} - \frac{1}{r} - \frac{1}{r+4}\right);$$

$$c_{2k}'(r) = \frac{(-1)^k}{(r+4)(r+6)\cdots(r+2k+2)r(r+4)(r+6)\cdots(r+2k-2)}\left(\frac{-1}{r+4} - \frac{1}{r+6} - \cdots - \frac{1}{r+2k+2} - \frac{1}{r} - \cdots - \frac{1}{r+2k-2}\right).$$

We can rewrite $c_2'(r)$ as

$$c_2'(r) = -\frac{1}{r(r+4)}\left[1 - (r+2)\left(-\frac{1}{r} - \frac{1}{r+4}\right)\right],$$

and hence $c_2'(-2) = -1/(-2)(2) = \frac{1}{4}$. And, finally,

$$c_{2k}'(-2) = \frac{(-1)^k}{2\cdot 4\cdots 2k(-2)2\cdot 2\cdots(2k-4)}\left(-\frac{1}{2}-\frac{1}{4}\cdots -\frac{1}{2k}+\frac{1}{2}-\frac{1}{2}\cdots-\frac{1}{2k-4}\right)$$

$$= \frac{(-1)^k}{2^{2k}k!(k-2)!}(-H_k + 1 - H_{k-2});$$

therefore,

$$y_2(x) = \left.\frac{\partial y}{\partial r}\right|_{r=-2} = y_1(x)\ln x + x^{-2} + \frac{1}{4} + \sum_{k=2}^{\infty}\frac{(-1)^k(1-H_k-H_{k-2})}{2^{2k}k!(k-2)!}x^{2k-2}$$

We have thus determined two linearly independent solutions of Bessel's equation of index 2. Note that

$$-2y_1(x) = \sum_{k=0}^{\infty}\frac{(-1)^k}{k!(k+2)!}\left(\frac{x}{2}\right)^{2k+2} = J_2(x)$$

and $-2y_2(x) = Y_2(x)$, the so-called *Bessel functions of first and second kind*, are two other linearly independent solutions of the equation. Having now studied all possible cases of Bessel's equation of index

$$x^2y'' + xy' + (x^2 - v^2)y = 0,$$

we can conclude the following:

1. If v is not an integer, then $J_v(x)$ and $J_{-v}(x)$ are two linearly independent solutions of Bessel's equation.
2. If v is an integer, then $J_v(x)$ and $Y_v(x)$ are two linearly independent solutions of Bessel's equation.

Having noted these facts about Bessel's equation, let us return to note that the main points of the procedure for working out the second solution in the case where $r_2 - r_1$ is an integer and two arbitrary constants are not present. Naturally, once we have determined that $k = r_2 - r_1$ is an integer, there is no way of knowing which type of solution might be present without further work. We must first determine whether c_k is an arbitrary constant

(c_0 always is, and we must have two). If c_k is indeed arbitrary, then we proceed as in the previous section. If c_k is not arbitrary, then we proceed as follows.

1. Calculate $c_n = c_n(r)$ from the recursion formula, put $c_0(r) = r - r_1$, and write

$$y(x;r) = c_0(r)x^r + \sum_{n=1}^{\infty} c_n(r)x^{n+r}$$

as in Section 7.1.

2. One solution of the equation is $y_1(x) = y(x;r_1)$, and the other is $y_2(x) = \partial y(x;r)/\partial r|_{r=r_1}$ exactly as in Section 7.1. Note that the device that enables us to apply the same technique as in the case when $r_1 = r_2$ is the substitution $c_0 = r - r_1$. Note also that it is generally necessary to work out the derivatives $c_0'(r), c_1'(r), \ldots, c_{k-1}'(r)$ separately from the rest, which are all encompassed by one formula, because it is not until the c_k coefficient that the factors of $r - r_1$ in numerator and denominator begin to cancel.

Let us work one further example to make the procedure a bit more clear.

Example 2

Now let us consider the equation

$$x^2y'' - x(1 + 2x)y' - 3(1 + 2x)y = 0.$$

The usual substitution

$$y = \sum_{n=0}^{\infty} c_n x^{n+r}$$

leads to

$$\sum_{n=0}^{\infty} (n + r + 1)(n + r - 3)c_n x^{n+r} - \sum_{n=0}^{\infty} 2(n + r + 3)c_n x^{n+r+1} = 0$$

or

$$(r + 1)(r - 3)c_0 x^r + \sum_{n=0}^{\infty} [(n + r + 1)(n + r - 3)c_n - 2(n + r + 2)c_{n-1}]x^{n+r} = 0.$$

Hence the two solutions of the indicial equation are $r_1 = -1$, $r_2 = 3$ and they differ by the integer 4. Examining the recursion formula for $n = 1, 2, 3,$ and 4, we see that

$$\begin{aligned}
n = 1:&\quad 1(-3)c_1 - 2(2)c_0 = 0,\\
n = 2:&\quad 1(-2)c_2 - 2(3)c_1 = 0,\\
n = 3:&\quad 3(-1)c_3 - 2(4)c_2 = 0,\\
n = 4:&\quad 4(0)c_4 - 2(5)c_3 = 0,
\end{aligned}$$

and the latter can certainly not be satisfied if c_0 is an arbitrary nonzero constant. Hence we use the device of the previous example: put $c_0 = (r - r_1) = (r + 1)$, and find a formula for c_n as a function of r from the recursion formula:

$$\begin{aligned}
c_0 &= r + 1,\\
c_1 &= \frac{2(r + 3)}{(r + 2)(r - 2)} c_0,\\
c_2 &= \frac{2(r + 4)}{(r + 3)(r - 1)} c_1,\\
c_3 &= \frac{2(r + 5)}{(r + 4)(r)} c_2,\\
c_4 &= \frac{2(r + 6)}{(r + 5)(r + 1)} c_3,\\
&\;\;\vdots\\
c_n &= \frac{2(r + n + 2)}{(r + n + 1)(r + n - 3)} c_{n-1}, \qquad n \geqq 4,
\end{aligned}$$

and therefore

$$\begin{aligned}
c_n(r) &= \frac{2^n(r+3)(r+4)\cdots(r+n+1)(r+n+2)}{(r+2)(r+3)\cdots(r+n+1)(r-2)(r-1)r(r+1)(r+2)\cdots(r+n-3)}(r+1)\\
&= \frac{2^n(r+n+2)}{(r+3)(r-2)(r-1)r(r+1)(r+2)\cdots(r+n-3)}(r+1).
\end{aligned}$$

The common factor of $(r + 1)$ in numerator and denominator cancels only for $n \geqq 4$. Thus we have the following relationships for the coefficients:

$$c_0 = r + 1, \quad c_1 = \frac{2(r+3)}{(r+2)(r-2)}(r+1),$$

$$c_2 = \frac{2^2(r+4)}{(r+2)(r-2)(r-1)}(r+1),$$

$$c_3 = \frac{2^3(r+5)}{(r+2)(r-2)(r-1)r}(r+1),$$

$$c_n = \frac{2^n(r+n+2)}{(r+2)(r-2)(r-1)r(r+2)\cdots(r+n-3)} \qquad \text{for } n \geqq 4.$$

The first four coefficients vanish for $r = -1$; and for $n \geqq 4$ we have

$$c_n = \frac{2^n(n+1)}{(-3)(-2)(-1)(n-4)!} = -\frac{2^n(n+1)}{6(n-4)!}.$$

Therefore the first solution is

$$y_1(x) = -\sum_{n=4}^{\infty} \frac{2^n(n+1)x^{n-1}}{6(n-4)!} = \sum_{n=0}^{\infty} \frac{2^{n+4}(n+5)x^{n+3}}{6n!}.$$

To find the second solution we need to find the values of the $c_n'(r)$ at $r = -1$. The first four of these must be computed separately:

$$c_0' = 1, \quad c_1'(-1) = \frac{2 \cdot 2}{1(-3)} = -\frac{4}{3},$$

$$c_2'(-1) = \frac{2^2 \cdot 3}{1(-3)(-2)} = 2, \quad c_3'(-1) = \frac{2^3 \cdot 4}{1(-3)(-2)(-1)} = -\frac{16}{3}.$$

[Note that these coefficients are evaluated simply by putting $r = -1$ into the factors which multiply $(r + 1)$.] For the general coefficient,

$$c_n'(r) = c_n(r)\left(\frac{1}{r+n+2} - \frac{1}{r+2} - \frac{1}{r-2} - \frac{1}{r-1} - \frac{1}{r} - \frac{1}{r+2} - \cdots - \frac{1}{r+n-3}\right),$$

so that

$$c_n'(-1) = \frac{-2^n(n+1)}{6(n-4)!}\left(\frac{1}{n+1} - 1 + \frac{1}{3} + \frac{1}{2} + 1 - 1 - \frac{1}{2} - \cdots - \frac{1}{n-4}\right)$$
$$= \frac{2^n(n+1)}{6(n-4)!}\left(\frac{1}{n+1} + \frac{5}{6} - H_{n-4}\right) = \frac{2^n[6(n+1)H_{n-4} - (5n+11)]}{36(n-4)!},$$

and therefore

$$y_2(x) = y_1(x)\ln x + x^{-1}\left\{1 - \frac{4}{3}x + 2x^2 - \frac{16}{3}x^3 + \sum_{n=4}^{\infty} \frac{2^n[6(n+1)H_{n-4} - (5n+11)]x^n}{36(n-4)!}\right\}.$$

Problems

In 1–16, find two linearly independent solutions valid near $x = 0$.

1. $2x(x+1)y'' + (11x+4)y' - 2y = 0.$
2. $4x^2y'' + 4xy' + (3x-1)y = 0.$
3. $x^2y'' + xy' + (x^2-1)y = 0.$ (Bessel's equation of order 1.)
4. $xy'' + (3x+4)y' + 6y = 0.$
5. $x^2y'' + x(2x-3)y' + 3y = 0.$
6. $xy'' + (x^3+4)y' + 7x^2y = 0.$
7. $3x(x+1)y'' + (10x-3)y' - 6y = 0.$
8. $9x^2y'' + 12xy' - (x+2)y = 0.$
9. $x^2y'' + 5xy' + (3-2x^2)y = 0.$
10. $2xy'' + 2(x-2)y' + 3y = 0.$
11. $4x^2y'' + x(x+4)y' - y = 0.$
12. $xy'' + 2(x^3-1)y' + x^2y = 0.$
13. $(x+1)xy'' - 2(x+1)y' - 10y = 0.$
14. $x^2y'' - 3xy' + (2x-5)y = 0.$
15. $4x^2y'' + 3(x-1)y = 0.$
16. $3xy'' + (x-3)y' + 2y = 0.$

In 17–20, find two linearly independent solutions valid near the x_0 indicated.

17. $(x+1)^2y'' - 2(x+1)(x+4)y' + 10y = 0, x_0 = -1.$
18. $2(x-3)y'' - (3x^3 - 27x^2 + 81x - 77)y' - 2(x-3)^2y = 0, x_0 = 3.$
19. $(x^2-1)y'' + (11x+1)y' + 25y = 0, x_0 = -1.$
20. $(x+2)^2y'' + 10(x+2)y' + 6(x+5)y = 0, x_0 = -2.$

7.4 Applications to Vibrations of a Circular Membrane and Heat Conduction

Let us consider the problem of describing the vibrations of a frictionless circular membrane (such as a drumhead) of radius a. For simplicity we assume that it is centered at the origin in three-dimensional space and that its circumference is pinned down in the (x, y) plane. If it is initially at rest, with a given initial displacement $f(r, \theta)$, then the initial-boundary value problem for its vertical displacement $z(r, \theta, t)$ is given by

$$\frac{\partial^2 z}{\partial t^2} = k^2\left(\frac{\partial^2 z}{\partial r^2} + \frac{1}{r}\frac{\partial z}{\partial r} + \frac{1}{r^2}\frac{\partial^2 z}{\partial \theta^2}\right),$$

$$z(a, \theta, t) = 0,$$

$$z(r, \theta, 0) = f(r, \theta), \frac{\partial z(r, \theta, 0)}{\partial t} = 0.$$

Here k^2 is a constant which embodies the physical properties of the membrane (tension, density, etc.). The partial differential equation is the equation of motion; the first condition is that $z = 0$ on the circumference (the membrane is pinned down); and the last two conditions are the initial conditions.

One of the first steps in completely solving this equation is the separation of variables which we have discussed elsewhere in this work. That is, we seek functions of the form

$$z(r, \theta, t) = R(r)\Theta(\theta)T(t),$$

which are particular solutions of the partial differential equation; in this case, we must also insist that each of these particular solutions satisfy the boundary condition and the second initial condition. Thus we must have

$$R(a)\Theta(\theta)T(t) = 0$$

for all θ and t, and

$$R(r)\Theta(\theta)T'(0) = 0$$

for all r and θ. If we are to have nontrivial solutions, then none of the functions R, Θ, T can be identically zero. Hence we conclude that we must have

$$R(a) = 0 \quad \text{and} \quad T'(0) = 0.$$

Now, substituting the expression $z = R\Theta T$ into the partial differential equation and then dividing through by $k^2 R\Theta T$, we obtain

$$\frac{T''}{k^2T} = \frac{R''}{R} + \frac{1}{r}\frac{R'}{R} + \frac{1}{r^2}\frac{\Theta''}{\Theta}.$$

Since the left side is a function of t alone and the right is a function of the other two independent variables, both sides must be equal to the same constant, say $-\lambda^2$. Thus

$$\frac{T''}{k^2T} = -\lambda^2$$

and hence

$$T'' + k^2\lambda^2 T = 0, \quad T'(0) = 0$$

is the initial value problem to be satisfied by T. The reason that we chose to exhibit the constant parameter as $-\lambda^2$ was that this choice of sign (and *not* the positive sign) leads to a periodic solution for T, and physical considerations tell us that lacking friction, the drumhead should vibrate periodically. From our previous work with equations having constant coefficients, we know that the solution of this equation and initial condition is

$$T(t) = A\cos\lambda kt,$$

where A is an arbitrary constant. Note that the value of the parameter λ is still unknown.

Next, recalling that

$$\frac{R''}{R} + \frac{1}{r}\frac{R'}{R} + \frac{1}{r^2}\frac{\Theta''}{\Theta} = -\lambda^2.$$

we have, after a bit of rearranging,

$$r^2\frac{R''}{R} + r\frac{R'}{R} + \lambda^2 r^2 = -\frac{\Theta''}{\Theta}.$$

The left side is a function of r alone and the right is a function of Θ alone; hence both sides must equal the same constant, say μ^2. In particular

$$-\Theta''/\Theta = \mu^2,$$

so

$$\Theta'' + \mu^2\Theta = 0.$$

Once again we chose the sign of the constant so that a periodic solution would result, since clearly the displacement $z(r, \theta, t)$ must be a continuous function—a discontinuity, which would be implied if there were any θ for which $z(r, \theta, t) \neq z(r, \theta + 2\pi, t)$, would represent a break in the membrane. In fact, because Θ must be periodic with period 2π, we recognize that μ must be an integer, say n, and hence

$$\Theta'' + n^2\Theta = 0$$

has the general solution

$$\Theta_n(\theta) = B_n \cos n\theta + C_n \sin n\theta \qquad (n = 0, 1, 2, \ldots)$$

corresponding to the eigenvalue $\mu = n$.

Finally, we consider the equation for R:

$$r^2 \frac{R''}{R} + r \frac{R'}{R} + \lambda^2 r^2 = n^2,$$

which becomes

$$r^2 R'' + rR' + (\lambda^2 r^2 - n^2)R = 0.$$

This equation is very nearly Bessel's equation—it would be, in fact, if it were not for λr instead of r in the parentheses. This prompts us to make the change of variable $r = u/\lambda$, so that

$$R' = \frac{dR}{dr} = \frac{dR}{du}\frac{dy}{dr} = \lambda \frac{dR}{du}$$

and

$$R'' = \lambda \frac{d}{du}\left(\lambda \frac{dR}{du}\right) = \lambda^2 \frac{d^2R}{du^2}.$$

Hence the differential equation for R as a function of u is

$$u^2 \frac{d^2R}{du^2} + u \frac{dR}{du} + (u^2 - n^2)R = 0,$$

which is certainly Bessel's equation of index n, having as its general solution

$$R = DJ_n(u) + EY_n(u)$$

or

$$R = DJ_n(\lambda r) + EY_n(\lambda r).$$

Now the Bessel function of second kind, $Y_n(\lambda r)$, has a logarithmic singularity at $r = 0$, so, because we require continuous solutions, we must have $E = 0$, and we conclude that

$$R(r) = DJ_n(\lambda r).$$

The boundary condition $R(a) = 0$ now requires that

$$J_n(\lambda a) = 0.$$

Like trigonometric functions, Bessel functions have an infinite sequence of zeros.* Calling this sequence $\sigma_{nm}(m = 1, 2, 3, \ldots)$, where n represents the order of the Bessel function and where m numbers these zeros in an increasing sequence, we have

$$\lambda_{nm} = \sigma_{nm}/a$$

as the eigenvalues for λ. Hence

$$R_{nm}(r) = D_{nm}J_n(\lambda_{nm}r).$$

Combining all of these solutions together, we have

$$z_{nm}(r, \theta, t) = R_{nm}\theta_n T = J_n(\lambda_{nm}r)\cos k\lambda_{nm}t(\alpha_{nm}\cos n\theta + \beta_{nm}\sin n\theta),$$
$$n = 0, 1, 2, \ldots; m = 1, 2, 3, \ldots .$$

Here we have written

$$\alpha_{nm} = AB_nD_{nm}, \quad \beta_{nm} = AC_nD_{nm}.$$

In more advanced courses, we learn how to superpose such particular solutions in order to satisfy the other initial condition.

Problems

1. The temperature $u(r, t)$ in a homogeneous, isotropic circular plate of radius a with top and bottom insulated and with the circumference kept at

* For proof of this, see R. V. Churchill, *Fourier Series and Boundary Value Problems* (McGraw-Hill Book Company, New York, 1941), pp. 153–157.

zero degrees and initial temperature $f(r)$ is given by the solution of the problem

$$\frac{\partial u}{\partial t} = k\left(\frac{\partial^2 u}{\partial r^2} + \frac{1}{r}\frac{\partial u}{\partial r}\right),$$

$$u(a, t) = 0,$$

$$u(r, 0) = f(r).$$

Find the particular solutions of this problem and the boundary conditions. *Hint*: Physically, what would you expect to happen as $t \to \infty$?

2. The steady-state temperature $u = u(r, \theta, z)$ of a homogeneous, isotropic cylinder of radius a and height d, with the top and bottom insulated and the sides kept at a prescribed temperature $f(\theta, z)$ is given by the solution of the problem

$$\frac{\partial^2 u}{\partial r^2} + \frac{1}{r}\frac{\partial u}{\partial r} + \frac{1}{r^2}\frac{\partial^2 u}{\partial \theta^2} + \frac{\partial^2 u}{\partial z^2} = 0,$$

$$\frac{\partial u}{\partial z}(r, \theta, 0) = \frac{\partial u}{\partial z}(r, \theta, d) = 0,$$

$$u(a, \theta, z) = f(\theta, z).$$

Find particular solutions of this problem and the homogeneous conditions. *Note*: The equation for R is not quite Bessel's equation; it can be converted into Bessel's equation by the substitution $ir = u$, where $i^2 = -1$.

7.5 Series Valid for Large Values of the Variable

On some occasions one might desire solutions of a given differential equation which are valid for arbitrarily large values of x, or, as we sometimes say, "for x near the point at infinity." An example of this would be studying the temperature of a medium outside (rather than inside) a given surface. By the simple mathematical device of noting that $1/x$ is small when x is large, we see that if we make the change of independent variables from x to $t = 1/x$, then we can study the new equation near the origin $t = 0$ and by doing so we shall have determined the behavior of the original equation for large values of x.

Example 1

For example, let us study the equation

$$x^4 y'' + 2x(x^2 - 1)y' - 2y = 0 \tag{7.13}$$

for large values of x. Under the substitution $t = 1/x$, we have

$$\frac{dy}{dx} = \frac{dy}{dt}\frac{dt}{dx} = \frac{-1}{x^2}\frac{dy}{dt} = -t^2\frac{dy}{dt}$$

and

$$\frac{d^2y}{dx^2} = -t^2\frac{d}{dt}\left(-t^2\frac{dy}{dt}\right) = t^2\left(t^2\frac{d^2y}{dt^2} + 2t\frac{dy}{dt}\right) = t^4\frac{d^2y}{dt^2} + 2t^3\frac{dy}{dt}.$$

Hence the above equation becomes

$$\frac{d^2y}{dt^2} + 2t\frac{dy}{dt} - 2y = 0. \tag{7.14}$$

An ordinary point for this equation is $t = 0$; hence we say that *the original equation (7.13) has an ordinary point at infinity.* [Note, by the way, that $x = 0$ is an irregular singular point of (7.13).] Since $t = 0$ is an ordinary point of (7.14), we can solve it in the usual way, by substituting

$$y(t) = \sum_{n=0}^{\infty} c_n t^n,$$

and after working out the details, we find that

$$y(t) = c_0\left[1 + \sum_{k=1}^{\infty}\frac{(-1)^{k+1}}{k!(2k-1)}t^{2k}\right] + c_1 t,$$

and therefore in terms of the variable x, we have

$$y(x) = c_0\left[1 + \sum_{k=1}^{\infty}\frac{(-1)^{k+1}}{k!(2k-1)}x^{-2k}\right] + c_1 x^{-1}.$$

This expression tells us how $y(x)$ behaves for large values of x. Note, however, that both the infinite series and the x^{-1} term are defined and converge for all $x \neq 0$. Thus the equation (7.13) with an irregular singular point at $x = 0$ has two linearly independent solutions, neither of which is defined at the origin because each contains only negative powers of x. It is sometimes possible, as it is here, to consider the case of an equation with an irregular singular point at $x = 0$ by solving it near the point at infinity.

Example 2

Now let us consider

$$2x^3y'' + x(x + 3)y' + 3(1 - x)y = 0.$$

(Once again, note that $x = 0$ is an irregular singular point for this equation.) Upon substituting $t = 1/x$, we obtain

$$2t^2\frac{d^2y}{dt^2} + 3t(1 - t)\frac{dy}{dt} + 3(t - 1)y = 0,$$

which has a regular singular point at $t = 0$. Hence we attempt the usual

$$y = \sum_{n=0}^{\infty} c_n t^{n+r};$$

we find that the indicial roots are $r_1 = -\frac{3}{2}$, $r_2 = 1$, and the usual computation leads to a general solution of the form

$$y(t) = a_0 t^{-3/2}\left[1 + \sum_{n=1}^{\infty}\left(\frac{3}{2}\right)^n \frac{(-5)}{n!(2n - 5)} t^n\right] + b_0 t$$

so that

$$y(x) = a_0 x^{3/2}\left[1 + \sum_{n=1}^{\infty}\left(\frac{3}{2}\right)^n \frac{(-5)}{n!(2n - 5)} x^{-n}\right] + b_0 x^{-1}.$$

Again, as in example 1, the solution is defined for all values of x except for $x = 0$, and $x < 0$ in the first portion. The above examples suffice to show that an equation with an irregular singular point at the origin can sometimes be solved for large values of the independent variable, provided that the point at infinity is not also an irregular singular point.

Problems

Solve each of the following equations near the point at infinity.

1. $x^2(x^2 + 2)y'' + 2x(3 + 2x^2)y' + 6y = 0.$
2. $2x^2(x - 5)y'' + x(x - 25)y' - (3x - 10)y = 0.$
3. $x^4y'' + x^2(2 + 3x)y' + 2y = 0.$
4. $x^2(x + 2)y'' + 3x(2 - x)y' + (4x + 2)y = 0.$
5. $x^4y'' + 2x^3y' + y = 0.$
6. $x^2(x + 1)y'' + 2x(5x + 6)y' + 2(9x + 14)y = 0.$
7. $x^4y'' + 2x^2(x + 2)y' + 13y = 0.$
8. $2x^3y'' + 5x^2y' + (5 + x)y = 0.$
9. $x^2(x^3 - 2)y'' + 2x(x^3 + 7)y' - 48y = 0.$
10. $9x^3y'' + x(12x - 1)y' + 2(1 - 2x)y = 0.$

7.6 Series Solutions when the Recursion Formula Contains More than One Term

The previous sections of this chapter and Chapter 5 should not have lulled the reader into believing that series solutions always come out nicely. In fact, in general, the calculation of series solutions can turn out to be quite a mess, as we shall see. Why, then, have we spent so much time on problems which were so carefully contrived to come out nicely? The answer is that we have to know how to solve nice problems before we can do a decent job on the messy ones. Nice problems build up confidence in our abilities and faith in the ultimate justice of mathematics. More importantly, they also teach us what forms of solutions to expect in more general situations. Thus, in the problems which we shall consider in this section, we shall discover recursion formulas that do not yield simple equations for a_n because more than one of the earlier coefficients occurs on the right side. However, we shall still know what form to expect for the solution—ordinary points and singular points have the same definitions, and in the case of a singular point, the roots of the indicial equation in conjunction with the behavior of the recursion formula will still tell us whether or not we can expect two "series-type" solutions or one "series-type" and one logarithmic solution. Let us work some examples to see how straightforward the process can be.

Example 1

Solve the equation

$$(x^2 + x - 1)y'' + 3xy' - 3y = 0$$

in the neighborhood of $x = 0$. Now $x = 0$ is an ordinary point; therefore we can expect to find a power series solution of the form

$$y = \sum_{n=0}^{\infty} c_n x^n,$$

where two of the c_n's will be arbitrary. Substituting the series into the equation, we find that

$$\sum_{n=0}^{\infty} n(n-1)c_n x^n + \sum_{n=0}^{\infty} n(n-1)c_n x^{n-1} - \sum_{n=0}^{\infty} n(n-1)c_n x^{n-2}$$
$$+ \sum_{n=0}^{\infty} 3nc_n x^n - \sum_{n=0}^{\infty} 3c_n x^n = 0$$

or

$$\sum_{n=0}^{\infty} n(n-1)c_n x^{n-2} - \sum_{n=0}^{\infty} n(n-1)c_n x^{n-1} - \sum_{n=0}^{\infty} (n+3)(n-1)c_n x^n = 0.$$

Now, we change the summation index from n to $n - 1$ in the second series and $n - 2$ in the third, obtaining

$$\sum_{n=0}^{\infty} n(n-1)c_n x^{n-2} - \sum_{n=1}^{\infty} (n-1)(n-2)c_{n-1}x^{n-2}$$

$$- \sum_{n=2}^{\infty} (n+1)(n-3)c_{n-2}x^{n-2} = 0,$$

which becomes, after removing the $n = 0$ and $n = 1$ terms, and combining the rest under one summation,

$$0 \cdot c_0 x^{-2} + 0 \cdot c_1 x^{-1} - 0 \cdot c_0 x^{-1} + \sum_{n=2}^{\infty} [n(n-1)c_n - (n-1)(n-2)c_{n-1}$$

$$- (n+1)(n-3)c_{n-2}]x^{n-2} = 0.$$

Equating the coefficients of like powers of x to zero, we obtain that c_0 and c_1 are arbitrary, and that

$$n(n-1)c_n - (n-1)(n-2)c_{n-1} - (n+1)(n-3)c_{n-2} = 0$$

for $n \geqq 2$. Solving this for c_n, we have

$$c_n = \frac{(n-2)c_{n-1}}{n} + \frac{(n+1)(n-3)c_{n-2}}{n(n-1)}, \qquad n \geqq 2. \tag{7.15}$$

By substituting $n = 2, 3, \ldots$ we can obtain individual formulas for the c_n's in terms of c_0 and c_1 occur on the right side. Thus, for example,

$$c_2 = \frac{3(-1)c_0}{2 \cdot 1} = -\frac{3}{2}c_0,$$

$$c_3 = \frac{1}{3}c_1,$$

$$c_4 = \frac{2}{4}c_3 + \frac{5 \cdot 1}{4 \cdot 3}c_2 = \frac{1}{6}c_1 - \frac{5}{8}c_0, \text{ etc.}$$

The best we can do in a case like this is to write that

$$y = \sum_{n=0}^{\infty} c_n x^n,$$

where c_0 and c_1 are arbitrary, and where c_n is given by (7.15) for $n \geqq 2$. Equation (7.15) is actually an algorithm which can be repeatedly applied, for example on an electronic digital computer. Normally, in a problem which arises from applications, c_0 and c_1 will be given, and as many c_n's can be calculated from (7.15) as are necessary for the finite sum to be a reasonable approximation to y at the value of x under consideration. We have seen before how a series solution can be considered an approximation to the actual solution, in that the infinite process of summation can never really be completed. However, in a case like the present one, not even the process of calculating the coefficients can be complete; thus it is even more apparent how much the series method of solution is truly an approximate method.

Example 2

Solve the equation

$$2x^2y'' + x(1 + x)y' - (1 - 4x^2)y = 0$$

near $x = 0$. Here, we have a regular singular point at $x = 0$, so we recognize that the solution will be of the form

$$y = \sum_{n=0}^{\infty} c_n x^{n+r-1}.$$

Substituting this into the equation, we obtain

$$\sum_{n=0}^{\infty} 2(n + r)(n + r - 1)c_n x^{n+r} + \sum_{n=0}^{\infty} (n + r)c_n x^{n+r} + \sum_{n=0}^{\infty} (n + r)c_n x^{n+r+1}$$
$$- \sum_{n=0}^{\infty} c_n x^{n+r} + \sum_{n=0}^{\infty} 4c_n x^{n+r+3} = 0,$$

which reduces to

$$\sum_{n=0}^{\infty} [2(n + r) + 1](n + r - 1)c_n x^{n+r} + \sum_{n=1}^{\infty} (n + r + 1)c_{n-1} x^{n+r}$$
$$+ \sum_{n=3}^{\infty} 4c_{n-3} x^{n+r} = 0$$

or

$$(2r + 1)(r - 1)c_0 x^r + r(2r + 3)c_1 + c_0 x^{r+1} + (r + 1)(2r + 5)c_2 + c_1 x^{r+2}$$
$$+ \sum_{n=3}^{\infty} [2(n + r) + 1(n + r - 1)c_n + (n + r - 1)c_{n-1} + 4c_{n-3}]x^{n+r} = 0.$$

The first term gives us the two roots of the indicial equation $r = -\frac{1}{2}, r = 1$, and that c_0 is an arbitrary constant. The next two terms give

$$c_1 = -\frac{c_0}{(2r + 3)} \quad \text{and} \quad c_2 = -\frac{c_1}{(2r + 5)}.$$

For $n \geqq 3$ we then have the recursion formula

$$c_n = -\frac{1}{2(n + r) + 1}\left[c_{n-1} + \frac{4c_{n-3}}{(n + r - 1)}\right].$$

Once again, we cannot obtain a general formula for c_n in terms of n and c_0 because this recursion formula contains both c_{n-1} and c_{n-3} on the right side. However, as in the previous example, we can calculate as many terms of the series as desired. The general solution is

$$y = x^{-\frac{1}{2}} \sum_{n=0}^{\infty} c_n x^n + x \sum_{n=0}^{\infty} d_n x^n,$$

where c_0 and d_0 are arbitrary, and

$$c_1 = -c_0/2,$$

$$c_2 = -c_1/4,$$

$$\vdots$$

$$c_n = -\frac{1}{2}n\left(c_{n-1} + \frac{8c_{n-3}}{2n - 3}\right) \qquad (n \geqq 3),$$

$$d_1 = -d_0/5,$$

$$d_2 = -d_1/7,$$

$$\vdots$$

$$d_n = -\frac{1}{2n + 3}\left(d_{n-1} + \frac{4d_{n-3}}{n}\right) \qquad (n \geqq 3).$$

Example 3
Consider

$$x^2y'' - (5x + x^2 + 3x^3)y' + (3x^2 - 2x + 9)y = 0.$$

Once again $x = 0$ is a regular singular point, so we attempt

$$y = \sum_{n=0}^{\infty} c_n x^{n+r}$$

and obtain

$$\sum_{n=0}^{\infty} (n + r)(n + r - 1)c_n x^{n+r} - \sum_{n=0}^{\infty} 5(n + r)c_n x^{n+r} - \sum_{n=0}^{\infty} (n + r)c_n x^{n+r+1}$$

$$- \sum_{n=0}^{\infty} 3(n + r)6_n x^{n+r+2} + \sum_{n=0}^{\infty} 3c_n x^{n+r+2} - \sum_{n=0}^{\infty} 2c_n x^{n+r+1}$$

$$+ \sum_{n=0}^{\infty} 9c_n x^{n+r} = 0.$$

After some simplification we obtain

$$\sum_{n=0}^{\infty} (n + r - 3)^2 c_n x^{n+r} - \sum_{n=0}^{\infty} (n + r + 2)c_n x^{n+r+1}$$

$$- \sum_{n=0}^{\infty} 3(n + r - 1)c_n x^{n+r+2} = 0$$

or

$$(r - 3)^2 c_0 x^r + [(r - 2)^2 c_1 - (r + 2)c_0]x^{r+1} + \sum_{n=0}^{\infty} (n + r - 3)^2 c_n$$

$$- (n + r + 1)c_{n-1} - 3(n + r - 3)c_{n-2}]x^{n+r} = 0,$$

so that the roots of the indicial equation are $r_1 = r_2 = 3$, and we have

$$c_1 = \frac{(r + 2)c_0}{(r - 2)^2} \quad \text{and} \quad c_n = \frac{(n + r + 1)c_{n-1}}{(n + r - 3)^2} + \frac{3c_{n-2}}{(n + r - 3)}, \qquad n \geqq 2.$$

Since the roots of the indicial equation are equal, we can expect to obtain only one solution from these relations by putting $r = 3$. Indeed, if we do so, and put $c_0 = 1$ for a particular solution, we have

$$y_1(x) = x^3\left(1 + \sum_{n=1}^{\infty} c_n x^n\right), \qquad \text{where } c_1 = 5$$

and

$$c_n = \frac{(n+4)c_{n-1}}{n^2} + \frac{3c_{n-2}}{n}, \qquad n \geqq 2.$$

We know from our previous studies that the differential equation has a second solution of the form

$$y_2(x) = y_1(x) \ln x + \sum_{n=1}^{\infty} d_n x^{n+3}.$$

Since we cannot determine a formula for $c_n(r)$ because of the complicated recursion formula given above, it is best to substitute $y_2(x)$ directly into the differential equation in order to obtain recursion formulas for the d_n's in terms of the known c_n's. First, note that

$$\begin{aligned}
&x^2(y_1 \ln x)'' - (5x + x^2 + 3x^3)(y_1 \ln x)' + (3x^2 - 2x + 9)y_1 \ln x \\
&= [x^2 y_1'' - (5x + x^2 + 3x^3)y_1' + (3x^2 - 2x + 9)y_1] \ln x \\
&+ x^2\left(\frac{2y_1}{x} - \frac{y_1}{x^2}\right) - (5x + x^2 + 3x^3)\left(\frac{y_1}{x}\right);
\end{aligned}$$

the term in square brackets vanishes, because y_1 is a solution of the equation; the rest is equal to

$$\begin{aligned}
2xy_1' - y_1(6 + x + 3x^2) &= \sum_{n=0}^{\infty} 2(n+3)c_n x^{n+3} - \sum_{n=0}^{\infty} 6c_n x^{n+3} - \sum_{n=0}^{\infty} c_n x^{n+4} \\
- \sum_{n=0}^{\infty} 3c_n x^{n+5} &= (2c_1 - 1)x^4 + \sum_{n=2}^{\infty} (2nc_n - c_{n-1} - 3c_{n-2})x^{n+3}.
\end{aligned}$$

Hence, upon substituting $y_2(x)$ into the differential equation, we obtain, after simplifying,

$$\begin{aligned}
&d_1 x^4 + \sum_{n=0}^{\infty} [n^2 d_n - (n+r)d_{n-1} - 3nd_{n-2}]x^{n+3} + (2c_1 - 1)x^4 \\
&+ \sum_{n=2}^{\infty} (2nc_n - c_{n-1} - 3c_{n-2})x^{n+3} = 0.
\end{aligned}$$

Equating the coefficients of like powers of x to zero, we obtain recursion formulas for the d_n's in terms of the c_n's as well as of d_n's with lower subscript, namely

$$d_1 = 1 - 2c_1,$$
$$d_n = 1/n^2(n + 4)d_{n-1} - 3nd_{n-2} - 2nc_n + c_{n-1} + 3c_{n-2}, \qquad n \geqq 2.$$

Thus, once we have calculated $c_1, c_2, \ldots, c_n$, we can then calculate $d_1, d_2, \ldots, d_n$. In this way we have succeeded in approximating both of the linearly independent solutions of our differential equation.

Let us emphasize again that when a logarithmic solution occurs, we simply write down an expression for it and substitute directly into the differential equation, and obtain its coefficients in terms of those of the nonlogarithmic solution. Thus, when $r_1 = r_2$ as in the above case, we first try

$$y_1(x) = \sum_{n=0}^{\infty} c_n x^{n+r_1},$$

then

$$y_2(x) = y_1 \ln x + \sum_{n=1}^{\infty} d_n x^{n+r_1},$$

taking $c_0 = 1$ for a particular solution. When $r_2 - r_1 = k$ is an integer and c_k is not an arbitrary constant, once again a logarithmic solution must occur; in this case, the two solutions to use are

$$y_1(x) = \sum_{n=0}^{\infty} c_n x^{n+r_2}$$

and

$$y_2(x) = y_1 \ln x + \sum_{n=0}^{\infty} d_n x^{n+r_1},$$

again taking $c_0 = 1$ for a particular solution. Recall from our study of this type of problem that the lowest power of x in y_1 is always x^{r_2}, and in the series portion of y_2 it is x^{r_1}. The following example is of this type:

Example 4
Consider

$$2x^2y'' + (x^2 - 4x)y' + (x - 3x^2)y = 0.$$

Upon making the usual substitution

$$y = \sum_{n=0}^{\infty} c_n x^{n+r}$$

and simplifying, we find that

$$2r(r-3)c_0x^r + [2(r+1)(r-2)c_1 + rc_0]x^{r+1}$$

$$+ \sum_{n=2}^{\infty} [2(n+r)(n+r-2)c_n + (n+r)c_{n-1} - 3c_{n-2}]x^{n+r} = 0.$$

The first term yields $r_1 = 0$, $r_2 = 3$, so that $r_2 - r_1 = k = 3$. c_0 is an arbitrary constant; from the second term we see that $c_1 = 0$ regardless of the value of r. Is c_3 arbitrary? We have

$$n = 2: \quad 2(2)(-1)c_2 - 3c_0 = 0,$$
$$n = 3: \quad 2(3)(0)c_3 + 3c_2 = 0,$$

which is nonsense, since c_0 is arbitrary; hence c_3 is not arbitrary. Therefore our first solution (taking $c_0 = 1$) is

$$y_1(x) = x^3\left(1 + \sum_{n=1}^{\infty} c_n x^n\right),$$

where $c_1 = 0$ and

$$c_n = -\frac{(n+3)c_{n-1} - 3c_{n-2}}{2n(n+3)}, \qquad \text{for } n \geqq 2.$$

The second solution will be of the form

$$y_2(x) = y_1(x) \ln x + \sum_{n=0}^{\infty} d_n x^n.$$

By proceeding as in the previous example,

$$2x^2(y_1 \ln x)' - (x^2 - 4x)(y_1 \ln x)' + (x - 3x^2)y_1 \ln x$$

$$= 2x^2\left(\frac{2y_1'}{x} - \frac{y_1}{x^2}\right) + (x^2 - 4x)\frac{y_1}{x} = 4xy_1' - 2y_1 + xy_1 - 4y_1$$

$$= 4xy_1' + (x - 6)y_1 = \sum_{n=0}^{\infty} 4(n + 3)c_n x^{n+3} = \sum_{n=0}^{\infty} c_n x^{n+4} - \sum_{n=0}^{\infty} 6c_n x^{n+3}$$

$$= 6x^3 + \sum_{n=1}^{\infty} [2(2n + 3)c_n + c_{n-1}]x^{n+3}.$$

Hence placing y_2 into the differential equation yields

$$(-4d_1 + d_0)x + \sum_{n=2}^{\infty} [2n(n - 3)d_n + nd_{n-1} - 3d_{n-2}]x^n + 6x^3$$

$$+ \sum_{n=1}^{\infty} [2n(2n + 3)c_n + c_{n-1}]x^{n+3} = 0.$$

When we go to equate coefficients of like powers of x to zero, we will have for $n = 3$

$$0 \cdot d_3 + 3d_2 - 3d_1 + 6 = 0,$$

which implies that d_3 is arbitrary provided that $3d_2 - 3d_1 + 6 = 0$. The two formulas for $n = 1$ and $n = 2$,

$$-4d_1 + d_0 = 0 \quad \text{and} \quad -4d_2 + 2d_1 - 3d_0 = 0$$

together imply that $d_0 = \frac{16}{7}$, $d_1 = \frac{4}{7}$, $d_2 = -\frac{10}{7}$, d_3 is arbitrary; we may as well choose it to be zero, for ease of calculation. Then for $n > 3$, we have

$$d_n = \frac{1}{2n(n - 3)}[-nd_{n-1} + 3d_{n-2} - 2n(2n + 3)c_n - c_{n-1}],$$

which gives the remaining d_n's in terms of the previous ones and of the c_n's. Thus in principle the problem is completely solved. The reader may be justifiably skeptical at this point; what, he may ask, allowed us to choose $d_3 = 0$? What would have happened if we had chosen some other value, say $d_3 = 1$? Naturally, we would have obtained a different solution $y_2(x)$, although it would have begun with the terms $y_1 \ln x + \frac{16}{7} + \frac{4}{7}x - \frac{10}{7}x^2$, and only the series portions from $n = 3$ on would have differed. In fact, the difference between two such y_2's would simply be a constant multiple of y_1. We leave this to the reader to verify as an exercise.

It is also quite possible that when $k = r_2 - r_1$ is an integer, then c_k turns out to be arbitrary even in this case of recursion formulas containing several of the c_n's on the right side. When this happens, we can find two independent series solutions, just as in the simple case. At the risk of seeming cavalier, however, we shall not work an example of this type.

Problems

Find recursion formulas for coefficients of the solutions of problems 1–20, near the point $x = 0$.

1. $(x + 1)y'' + y' - xy = 0$.
2. $x^2y'' - 3xy' + (x - 2)^2y = 0$.
3. $x^2y'' + (x - 3x^2 + x^3)y' - 2y = 0$.
4. $x^2y'' + 2x(1 + x^2)y' - 2(1 + 3x^2)y = 0$.
5. $y'' + (2 - x)y' - y = 0$.
6. $2y'' - x^2y' - 3y = 0$.
7. $x^2y'' + x(1 + x^3)y' + 3x^3y = 0$.
8. $4x^2y'' - 7xy' + (6 + 2x - 5x^3)y = 0$.
9. $x^2y'' + (4x - x^2)y' + 5xy = 0$.
10. $x^2y'' - x^2(3 - 2x^2)y'' - 6(1 - x)y = 0$.
11. $(x^2 + 1)y'' + 2x^2y' - y = 0$.
12. $x^2y'' + 11xy' + (x^2 - x + 25)y = 0$.
13. $3x^2y'' + 5x(2 - x)y' - (15x + 2)y = 0$.
14. $x^2y'' - 4xy' + (6 + x - 3x^2)y = 0$.
15. $x^2y'' + x(1 + x^2 - x^3)y' - (1 - x^2 + 2x^3)y = 0$.
16. $(2 - x)y'' + 3xy' - 5y = 0$.

17. $x^2y'' - (x + x^3 - x^4)y' + (1 - 2x^2)y = 0.$

18. $2x^2y'' + xy' + (x^2 + x - 1)y = 0.$

19. $x^2y'' - 2xy' - (4 - 3x + 2x^3)y = 0.$

20. $xy'' + (4 - x - x^2 + x^3)y' - (1 + 2x^2 - 3x^3)y = 0.$

21. Prove the contention of example 4: that two solutions

$$y_2(x) = y_1(x) \log x + \sum_{n=0}^{\infty} d_n x^{n+r_1}$$

calculated with different values of d_3 will differ by a multiple of $y_1(x)$.

We have not considered equations with irregular singular points at all in this chapter. In problems 22–25, attempt a solution of the form

$$y(x) = \sum_{n=0}^{\infty} c_n x^{n+r}$$

and carry out the details to see what happens:

22. $x^3y'' + 2x^2y' + 3y = 0.$

23. $x^2y'' + (5x + 1)y' - 5y = 0.$

24. $x^2y'' - 7x^3y' - 2y = 0.$

25. $x^3y'' + 3xy' - 3y = 0.$

26. How would you solve an inhomogeneous linear equation, for example, $xy'' + y' + xy = x$ and $xy'' + y' + xy = x^2$?

Appendix

This appendix contains the statements, and a few brief proofs, of some of the principles referred to in the text regarding integrals and infinite series, and a brief introduction to complex numbers. For more details the reader should refer to a standard advanced calculus, intermediate analysis, or complex variables text (see the Bibliography).

A.1 Integral Inequalities and Differentiation under the Integral Sign

Before we prove our main results, we shall need a couple of basic inequalities:

Theorem A.1 (The Triangle Inequality)
For all real numbers x and y,

$$|x + y| \leqq |x| + |y|.$$

Proof
If x and y are the same sign, either both positive or both negative, then equality obviously holds; e.g., if both are positive, then

$$|x + y| = x + y \quad \text{and} \quad |x| + |y| = x + y.$$

If x is positive and y is negative, say $y = -u$, then there are two possibilities:

Case 1
If $x > u$, then

$$|x + y| = |x - u| = x - u$$

and

$$|x| + |y| = x + u;$$

and clearly

$$x - u \leqq x + u$$

with equality only if $u = 0$.

Case 2
If $x < u$, then

$$|x + y| = |x - u| = u - x,$$

and again, as in case 1,

$$|x| + |y| = x + u;$$

and again we easily see that

$$u - x \leqq x + u$$

with equality only if $x = 0$.

Corollary (The Generalized Triangle Inequality)
For any n real numbers $x_1, x_2, \ldots, x_n$ we have

$$|x_1 + x_2 + \cdots + x_n| \leqq |x_1| + |x_2| + \cdots + |x_n|.$$

Proof
By induction.

Theorem A.2
For all $x \geqq 0$, $1 - e^{-x} \leqq x$.

Proof
Certainly for $x = 0$ equality holds. For $x > 0$, the slope of x is 1, while the slope of $1 - e^{-x}$ is e^{-x}, which is less than 1. Since the two functions start out equal and x climbs more rapidly than $1 - e^{-x}$, the inequality follows (Figure A.1).

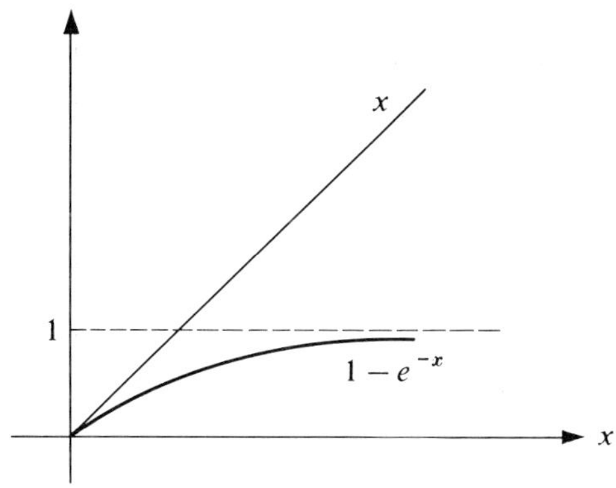

Figure A.1

Theorem A.3

If f is an integrable function on the interval $[a, b]$ then

$$\left|\int_a^b f(x)\,dx\right| \leqq \int_a^b |f(x)|\,dx.$$

Proof

Based on the definition of the integral and on the generalized triangle inequality. In outline, we recall that

$$\int_a^b f(x)\,dx$$

is a limit of sums of the form

$$\sum_{n=0}^{k} f(\tilde{x}_n)(x_{n+1} - x_n),$$

where each $[x_n, x_{n+1}]$ is an interval containing $\tilde{x}_n$, and the limit is taken as the number of intervals becomes infinite while the maximum length goes to zero. By the generalized triangle inequality,

$$\left|\sum f(\tilde{x}_n)(x_{n+1} - x_n)\right| \leqq \sum |f(\tilde{x}_n)|(x_{n+1} - x_n).$$

The limit of the left side is $|\int f\,dx|$ and the limit of the right is $\int |f|\,dx$. Hence the same inequality holds in the limit as well.

Theorem A.4 (The Mean Value Theorem)
If f is differentiable on $[a, b]$ then there exists a number θ between 0 and 1 such that

$$f(x + \eta) - f(x) = \eta f'(x + \theta\eta)$$

for all choices of x and $x + \eta$ lying in the interval $[a, b]$.

The proof of this theorem can be found in any advanced calculus text.

Theorem A.5
If $f(t)$ is of exponential order, that is, if there exist constants A and α such that $|f(t)| \leqq Ae^{\alpha t}$ for all $t \geqq 0$, then

$$D_s \int_0^{\infty} e^{-st} f(t)\, dt = -\int_0^{\infty} te^{-st} f(t)\, dt,$$

for all $s > \alpha$; i.e., Laplace transforms can be differentiated under the integral sign with respect to s.

Proof
By definition of the derivative,

$$D_s \int_0^{\infty} e^{-st} f(t)\, dt = \lim_{\Delta s \to 0} \frac{1}{\Delta s}\left[\int_0^{\infty} e^{-(s+\Delta s)t} f(t)\, dt - \int_0^{\infty} e^{-st} f(t)\, dt\right]$$

$$= \lim_{\Delta s \to 0} \int_0^{\infty} \frac{e^{-(s+\Delta s)t} - e^{-st}}{\Delta s} f(t)\, dt.$$

Our procedure will be to show that the absolute value of the difference between the last integral and

$$-\int_0^{\infty} te^{-st} f(t)\, dt$$

goes to zero as $\Delta s \to 0$. In fact, because of the mean value theorem, there is a θ between 0 and 1 such that

$$\frac{e^{-(s+\Delta s)t} - e^{-st}}{\Delta s} = -te^{-(s+\theta\Delta s)t}.$$

Hence

$$\int_0^\infty \frac{e^{-(s+\Delta s)t} - e^{-st}}{\Delta s} f(t)\,dt = \int_0^\infty -te^{-(s+\theta\Delta s)t} f(t)\,dt,$$

and therefore

$$\left| -\int_0^\infty te^{-st} f(t)\,dt - \int_0^\infty \frac{e^{-(s+\Delta s)t} - e^{-st}}{\Delta s} f(t)\,dt \right|$$

$$= \left| -\int_0^\infty te^{-st} f(t)\,dt + \int_0^\infty te^{-(s+\theta\Delta s)t} f(t)\,dt \right|$$

$$= \left| -\int_0^\infty te^{-st}(1 - e^{-\theta\Delta st}) f(t)\,dt \right|.$$

By Theorem A.3, this is less than or equal to

$$\int_0^\infty te^{-st}(1 - e^{-\theta\Delta st})|f(t)|\,dt \leqq \int_0^\infty te^{-st}(\theta\Delta st)|f(t)|\,dt$$

(from Theorem A.2), and this last integral is less than or equal to

$$\theta\Delta s \int_0^\infty t^2 e^{-st} A e^{\alpha t}\,dt = \Delta s\left[A\theta \int_0^\infty t^2 e^{-(s-\alpha)t}\,dt\right].$$

The quantity in parentheses is bounded for each $s > \alpha$; hence, as $\Delta s \to 0$, the whole quantity approaches zero.

A.2 Brief Summary of Results Concerning Infinite Series

Most of the information in this section is offered without proof; the proofs are standard, and indeed, most of the introductory material can be found in a typical beginning calculus text. Some of the more advanced theorems are available in intermediate analysis texts (see the Bibliography).

(a) Preliminary definitions and results

Definition

A quantity of the form

$$\sum_{n=1}^{\infty} a_n = a_1 + a_2 + a_3 + \cdots + a_n + \ldots$$

is called an *infinite series.* a_1, a_2, etc., are called the *terms* of the series. The sequence

$$s_1 = a_1, s_2 = a_1 + a_2, s_3 = a_1 + a_2 + a_3, \ldots,$$
$$s_n = a_1 + a_2 + a_3 + \cdots + a_n, \ldots$$

is called the *sequence of partial sums* of the infinite series Σa_n.

Definition
An infinite series Σa_n is said to *converge* (*to the sum* s) if and only if the sequence of partial sums has the limit s, i.e.,

$$\lim_{n \to \infty} s_n = s.$$

This means that for every $\varepsilon > 0$, there exists $N(\varepsilon)$ such that

$$|s_n - s| < \varepsilon$$

for every $n \geqq N$.

Definition
An infinite series which does not converge is said to *diverge.*

Theorem
The limit of a sequence $\{s_n\}$ exists if and only if $\{s_n\}$ satisfies the *Cauchy criterion,* i.e., for each $\varepsilon > 0$ there exists $N(\varepsilon)$ such that $|s_n - s_m| < \varepsilon$ for every n and $m \geqq N(\varepsilon)$.

Theorem
If Σa_n converges, then

$$\lim_{n \to \infty} a_n = 0.$$

Definition
A series of the form

$$\sum_{n=k}^{\infty} cr^n \ (c \text{ constant})$$

is said to be a *geometric series with ratio r.*

Theorem
The geometric series in the previous definition (for $c \neq 0$) diverges if $|r| \geqq 1$ and converges if $|r| < 1$ to the sum $cr^k(1 - r)^{-1}$.

(b) Series of positive terms

In this subsection we assume that all series have *positive* terms; that Σc_n represents a convergent series, and that Σd_n represents a divergent series.

The Comparison Test Consider Σa_n. (1) If there exists an integer k and a series Σc_n such that $a_n < c_n$ for all $n \geqq k$, then Σa_n converges. (2) If there exists an integer k and a series Σd_n such that $a_n \geqq d_n$ for all $n \geqq k$, then Σa_n diverges.

The Harmonic Series $\Sigma(1/n)$ diverges.

The Integral Test If $f(x)$ is a positive, decreasing function and $f(n) = a_n$, then $\int_1^\infty f(x)\,dx$ and $\Sigma_{n=1}^\infty a_n$ either both converge or both diverge.

The p-Series $\Sigma(1/n^p)$ converges if $p > 1$, diverges otherwise.

The Ratio Test Consider Σa_n and suppose

$$\lim_{n \to \infty} \frac{a_{n+1}}{a_n} = \rho.$$

Then Σa_n converges if $\rho < 1$, diverges if $\rho > 1$.

The Root Test Considering the same series, suppose that

$$\lim_{n \to \infty} (a_n)^{1/n} = v.$$

Then Σa_n converges if $v < 1$, diverges if $v > 1$.

(c) Series of terms of mixed signs

Definition
A series Σa_n is said to *converge absolutely* if $\Sigma |a_n|$ converges.

Theorem
If Σa_n converges absolutely, then it converges.

The Alternating Series Test If the sequence $\{|a_n|\}$ is monotone decreasing, if

$$\lim_{n\to\infty} a_n = 0,$$

and if a_{n+1} and a_n have opposite signs for all n, then Σa_n converges.

Definition
A series which converges, but not absolutely, is said to be *conditionally convergent*. *Example* $\Sigma(-1)^n/n$, the alternating harmonic series.

(d) Power series

Definition
An infinite series of the form $\Sigma a_n x^n$ is called a *power series*.

The Ratio Test for Power Series Suppose that

$$\lim_{n\to\infty} \left| \frac{a_n}{a_{n+1}} \right| = \rho.$$

Then there are three possibilities regarding the convergence of $\Sigma a_n x^n$:

(1) if $\rho = 0$, then the series converges only when $x = 0$.

(2) if ρ is finite, then the series converges absolutely for all x for which $-\rho < x < \rho$. It may or may not converge conditionally when $x = \pm\rho$.

(3) if $\rho = +\infty$, then the series converges absolutely for all values of x.

For the purposes of this text, we shall only need the above statement. However, it is possible to make a much more general statement about the convergence of power series, because of course it is possible that the above quotient has no limit at all. However, the student of advanced calculus knows that every sequence whatever possesses a "lim inf," which is defined as follows: first, we define a sequence b_n by

$$b_n = \min(|a_1/a_2|, |a_2/a_3|, |a_3/a_4|, \ldots, |a_n/a_{n+1}|).$$

Then we define

$$\liminf_{n\to\infty} |a_n/a_{n+1}| = \lim_{n\to\infty} b_n.$$

The sequence $\{b_n\}$ of course has a limit, since it is a decreasing sequence, bounded from below by 0. If we call this limit ρ, then ρ delimits the convergence of the given power series exactly as in the above statements.

Theorem
$\Sigma n a_n x^{n-1}$ converges for the same values of x as $\Sigma a_n x^n$.

This is easy enough to prove, since the value of ρ must be the same for both series.

Theorem
A power series may be differentiated term by term for any value of x at which it converges absolutely, i.e., $D\Sigma a_n x^n = \Sigma n a_n x^{n-1}$.

A.3 An Introduction to Complex Numbers

One arrives at the complex numbers by adjoining the *imaginary unit* i to the real numbers, and following the usual rules of arithmetic with the added convention that $i^2 = -1$.

Every complex number α can be written in the form $\alpha = a + ib$, where a and b are real; a is called the *real part* of α and b is called the *imaginary part* of α. Using this format, the basic rules of complex arithmetic are as follows:

$$(a + ib) + (c + id) = (a + c) + i(b + d),$$

$$(a + ib)(c + id) = (ac - bd) + i(ad + bc),$$

$$\frac{1}{a + ib} = \frac{a - ib}{a^2 + b^2}.$$

The number $\bar{\alpha} = a - ib$ is called the *complex conjugate* of α. Note the following results:

$$a = \mathrm{Re}(\alpha) = \tfrac{1}{2}(\alpha + \bar{\alpha});$$

$$b = \mathrm{Im}(\alpha) = \frac{1}{2i}(\alpha - \bar{\alpha}).$$

The number $|\alpha| = \sqrt{a^2 + b^2} = \sqrt{\alpha\bar{\alpha}}$ is called the *modulus* or *absolute value* of α and has the same properties as the absolute value of a real number;

for example, for any two complex numbers α and β, their moduli satisfy the triangle inequality:

$$||\alpha| - |\beta|| \leqq |\alpha + \beta| \leqq |\alpha| + |\beta|.$$

Graphically, the complex number $\alpha = a + ib$ is identified with the vector pointing from the origin to the point (a, b) in the Cartesian plane (Figure A.2). The sum of two complex numbers, α and $\beta = c + id$, is seen to be the diagonal of the parallelogram defined by the two vectors from the origin to the points (a, b) and (c, d) (Figure A.3). Thus complex numbers add just like vectors, according to the parallelogram law.

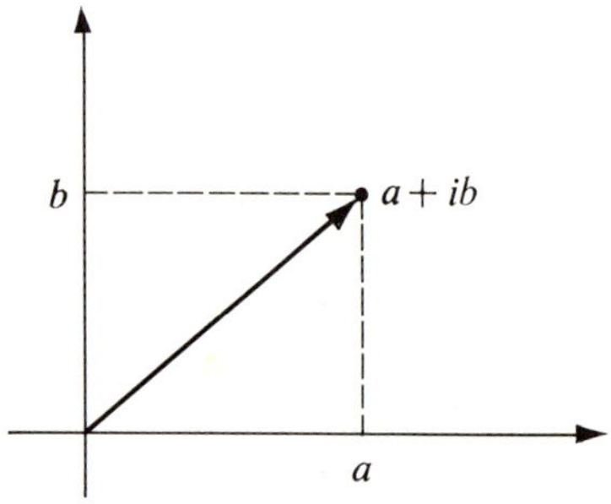

Figure A.2

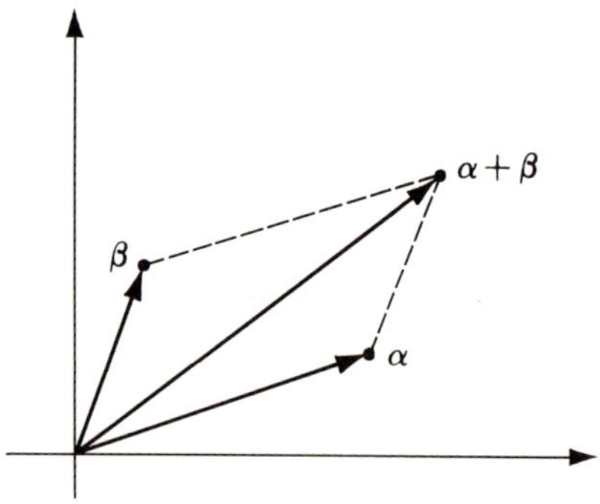

Figure A.3

In polar coordinates, a complex number can be represented as

$$\alpha = r(\cos\theta + i\sin\theta),$$

where $r = |\alpha| = \sqrt{a^2 + b^2}$ is the modulus, and $\theta = \arctan(b/a)$ (Figure A.4). θ is called the argument of α. Let us observe what happens when we multiply two complex numbers:

$$
\begin{aligned}
& r_1(\cos\theta_1 + i\sin\theta_1)r_2(\cos\theta_2 + i\sin\theta_2) \\
&= r_1r_2[(\cos\theta_1\cos\theta_2 - \sin\theta_1\sin\theta_2) + i(\cos\theta_1\sin\theta_2 + \cos\theta_2\sin\theta_1)] \\
&= r_1r_2[\cos(\theta_1 + \theta_2) + i\sin(\theta_1 + \theta_2)].
\end{aligned}
$$

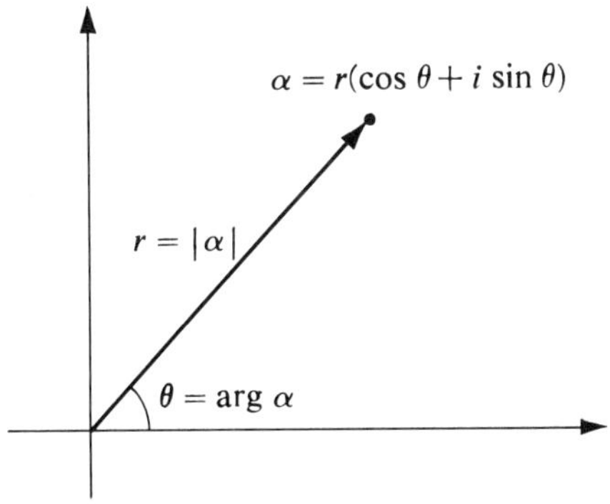

Figure A.4

Thus to multiply two complex numbers we multiply the moduli and add the arguments. This additive property of arguments should remind us of the similar property of the exponential, and for this reason we make the tentative definition

$$e^{i\theta} = \cos\theta + i\sin\theta \qquad \textit{(de Moivre's formula)}.$$

As we noted in the text, this definition makes sense because the Maclaurin series for both sides (assuming that such series are valid for complex numbers) agree term by term.

Thus a complex number α can very compactly be represented as $\alpha = re^{i\theta}$. If we place the value $\theta = \pi$ in de Moivre's formula, we obtain the special case

$$e^{\pi i} = -1, \quad \text{or} \quad e^{\pi i} + 1 = 0, \qquad \textit{(Euler's formula)}$$

relating the important mathematical constants 0, ± 1, e, π, and i.

The theory of infinite series can be developed for complex numbers in much the same way as for reals; absolute convergence is meaningful as long as we interpret an absolute values of a complex number to be its modulus. A power series of the form $\Sigma a_n z^n$, where z represents a complex variable, will then either

(a) converge only for $z = 0$,

(b) converge for all complex z with finite modulus,

(c) converge for all z such that $|z| < r$.

In the latter case, then, the power series converges in a circle of radius r centered at the origin.

In the study of differential equations in the complex plane, the location of singular points plays an important role because the power series for a solution, unless it terminates after a finite number of terms, will converge inside the largest circle containing no singular point of the equation inside, but at least one on its boundary. Restricting ourselves to the real case, as in this text, we are only able to see a small fraction of the total picture.

Select Bibliography

The following list of books is appended for those students who wish to read further on some particular topic. Each book either develops some topics in more detail, or discusses topics not included in the present text, or covers more advanced aspects of the theory. A short note after those titles which are not sufficiently specific describes their primary usefulness in this regard.

Boyce, W., and DiPrima R., *Elementary Differential Equations and Boundary Value Problems* (John Wiley & Sons, New York, 1969). A quite complete elementary text with good treatment of systems and an introduction to partial differential equations.

Brauer, F., Nohel, J. A., and Schneider, H., *Linear Mathematics* (W. A. Benjamin, New York, 1970). Introductory linear algebra and applications to systems.

Churchill, R., *Fourier Series and Boundary Value Problems* (McGraw-Hill Book Company, New York, 1951). Introductory partial differential equations, with emphasis on separation of variables and superposition of solutions.

Coddington, E., and Levinson, N., *Theory of Ordinary Differential Equations* (McGraw-Hill Book Company, New York, 1955). Advanced theory.

Erdélyi, A. (Ed.), *Tables of Integral Transforms*, Bateman Manuscript Project (McGraw-Hill Book Company, New York, 1954).

Goffman, C., *Real Functions* (Prindle, Weber & Schmidt, Boston, 1953).

Kreider, D., Kuller, R., Ostberg, D., and Perkins, F., *An Introduction to Linear Analysis* (Addison-Wesley Publishing Company, Reading, Mass., 1966). Linear ordinary and partial differential equations. No systems.

Olmsted, J., *Intermediate Analysis* (Appleton-Century-Crofts, New York, 1956).

Saaty, T., and Bram, J., *Nonlinear Mathematics* (McGraw-Hill Book Company, New York, 1964). Theory and some derivations of nonlinear equations.

Sansone, G., and Conti, R., *Nonlinear Differential Equations* (Pergamon Press (Macmillan), New York, 1964). Advanced theory of nonlinear equations.

Selby, S. (Ed.), *Standard Mathematical Tables* (The Chemical Rubber Company, Cleveland, Ohio, 1969).

Smith, C., Pike, R., and Murrill, P., *Formulation and Optimization of Mathematical Models* (International, Scranton, Penn., 1970). Applications and numerical methods.

Tenenbaum, M., and Pollard, H., *Ordinary Differential Equations* (Harper & Row Publishers, New York, 1963). An encyclopedia on the subject of elementary differential equations; includes elementary existence proofs.

Answers

Section 1.1

1. Second order, nonlinear.
3. Fourth order, second degree.
19. Yes.
21. No.

Section 1.2

1. $y^2 = 1 + ce^{2x}/(1 + x)^2$.
3. $y^2 = -3x^3/(cx^3 + 2)$.
5. $\sinh z + (1/t) = c$.
7. $y = 1 + c\exp(1 - x)^2$.
9. $x\cos x + 2y^2 = c$.
11. $\ln(y + 1) = -e^{x - x^2} + c$.
13. $-1/y + \ln y = \frac{1}{2}\ln(1 + x^2) + c$.
15. $y + y^3/3 = \frac{3}{2}\ln[(1 + x)/(1 - x)]$.
17. $x^4 + 4xy^3 = c$.
19. $x^2 + 2xy - y^2 = cx^2e^{-2x}$.
21. $3y^2 - 2x^2 + xy + 7x - 2y = c$.
23. $2y^2 + 3x^2 - 5xy + 8x + 3y = c$.
25. $4x^2 - 9y^2 + 7xy - 2x \quad 6y - c$.

Section 1.4

1. Exact; $x(x^2 - y^2) = c$.
3. Exact; $\frac{1}{2}x^2 + \frac{3}{2}x^2y^2 - (\frac{1}{2}y^2)\cos x = c$.
5. Exact; $y - \cos x \cosh y = c$.
7. Exact; $x\ln x + e^y = c$.
9. Exact; $y^3 + 3xy^2 + 12x^3y + x^4 = c$.
11. Exact; $x^2y - xy^2\cos x = c$.
13. Exact; $x^2y - (\frac{1}{3}y^3) - (\frac{1}{3}x^3) = c$.
15. Not exact.

Answers to 17–25; integrating factor given first, then solution.

17. y^{-1}; $x\ln xy = c$.
19. $y^{-2}x^{-1}$; $y + x/y = \ln cx$.
21. e^x; $e^xy\cos y = c$.
23. $\cos x^2$; $\sin x^2 = c(y^3 + 1)^{-1}$.
25. $(1 + x^2)^{-1}$; $\arctan x \ln y = c$.

Section 1.5

1. $y = e^{x^2}(\cos x + c)$.
3. $y = (x\ln x + c)/(x^3 + 3x^2)$.
5. $x^2y^3 = x\sin x + c$.
7. $(y - 1)e^y = \frac{1}{2}x^2\ln x - \frac{1}{4}x^2 + c$.

9. $xy = c + 1/(x^2 + 1)$.

11. $y = e^{2x} + ce^x$.

13. $y = c\exp(-\frac{1}{4}x) + \frac{1}{17}\cos x - \frac{4}{17}\sin x$.

15. $y = c\exp(-\frac{1}{4}x) + \frac{1}{5}e^x$.

17. $x^2 y \ln y = \sin x + c$.

19. $y \cosh x = e^x$.

21. $x^3 y^4 = \cos x + c$.

23. $y^{-3}e^x = \sin x + c$.

25. $xy^{-4} = 2e^{-2x} + ce^{-x}$.

27. $y^4 = \sin 3x + c$.

29. $e^x y^2 = 3\arcsin x + c$.

Section 1.6

1. $\dot{x}_3 + \frac{1}{k}x_3 = \frac{1}{k}[x_2(t) - 11]$, $x_3(0) = -22$, where $x_2(t)$ is the solution of equation (1.30).

3. The equation of motion of car 1 will be $x_1 = \frac{1}{2}a_1 t^2$; the equation of motion of car 2 will be $x_2 = \frac{1}{2}A(t-\tau)^2 - 11$ until time T, which is the positive root of the equation $(\frac{1}{2}a_1 - \frac{1}{2}A)T^2 + (A\tau - kA)T + kA\tau - \frac{1}{2}A\tau^2 = 0$; for $t \geqq T$ the motion of car 2 is described by

$$\dot{x}_2 + \frac{1}{k}x_2 = \frac{1}{k}(\tfrac{1}{2}a_1 t^2 - 11), x_2(T) = \tfrac{1}{2}AT^2.$$

5. $x_2(t) = \frac{1}{2}a_1 t^2 - a_1 kt - 11 + a_1 k^2(1 - e^{-t/k})$ for $0 \leqq t \leqq T$; $x_2(t) = -\frac{1}{2}a_1 T^2 - a_1 kT - 11 + a_1 Tt + k^2 a_1 e^{-t/k}(e^{T/k} - 1)$ for $t \geqq T$.

Section 1.7

1. $y = c_1 + c_2 e^{-2x} + e^x$.

3. $y = -\ln\cos(x + c_1) + c_2$.

5. $y = \frac{1}{3}x^3 + \frac{1}{2}x^2$.

7. $y = c_1 \tan c_1(x + c_2)$.

9. $\ln(1 + c_1 e^{-y}) = -c_1(x + c_2)$.

Section 1.8

1. $y = cx - \ln c$; none.

3. $y = cx - c\arccos c + \sqrt{1 - c^2}$; $y = \sin x$.

5. $y = cx + c^4$; none.

Section 2.1

1. $\ddot{\theta} + \eta_1\dot{\theta} + \eta_2\dot{\theta}^3 + (g/l)\sin\theta = 0$.

3. $\theta(t) \equiv 0$.

5. $\dot{Q}_1 - \dot{Q}_2 - \dot{Q}_3 = 0, R_1\dot{Q}_1 + R_2\dot{Q}_3 + (1/C)Q_1 = 0, LQ_2 - R_2\dot{Q}_3 = 0, Q_1(0) = Q_{01}, Q_2(0) = Q_{02}, \dot{Q}_2(0) = I_{02}, Q_3(0) = Q_{03}$.

Section 2.2

1. Dependent. 3. Independent. 5. Independent. 7. Independent.
9. Independent.

Section 2.3

1. $y_2 = (x-1)^{-1}$.

3. $y_2 = \cos x$.

5. $y_2 = xe^{2x}$.

7. $y_2 = x^{-2}\ln x$.

9. $y_2 = e^{-x}$.

Section 2.4

1. $L_1 + L_2 = (x + 3)D + (x^2 + 2)$; $L_1L_2 = 3xD^2 + (x^3 + 6)D + 4x^2$; $L_2L_1 = 3xD^2 + (x^3 + 9)D + 2x^2$.

3. $L_1 + L_2 = (\cos x + e^x)D - (\sin x + 1)$; $L_1L_2 = e^x \cos xD^2 + (e^x \cos x - e^x \sin x - \cos x)D + \sin x$; $L_2L_1 = e^x \cos xD^2 - (2e^x \sin x + \cos x)D + \sin x$.

5. $L_1 + L_2 = \dfrac{x^2+2}{x^2+1}D + \dfrac{1-x}{x}$; $L_1L_2 = \dfrac{1}{1+x^2}D^2 + \left[\dfrac{1}{x(1+x^2)} - 1\right]D - \left[\dfrac{1}{x} + \dfrac{1}{x^2(1+x^2)}\right]$;

$$L_2L_1 = \frac{1}{1+x^2}D^2 + \left[\frac{1}{x(1+x^2)} - \frac{2x}{(1+x^2)^2} - 1\right]D - \frac{1}{x}.$$

7. $L_1 + L_2 = (x^2 + 3x - 1)D - (x^3 - x^2 - 1)$; $L_1L_2 = (x^2 + 2x)(x - 1)D^2 - (x^5 + 2x^4 - x^3 - 3x + 1)D - (x^5 + 3x^4 + 7x^3)$; $L_2L_1 = (x^2 + 2x)(x - 1)D^2 - (x^5 + 2x^4 - x^3 - x + 3)D - (x^5 + x^3 - 2x^2 + 2x)$.

Section 3.1

21. $y = c_1e^x + c_2e^{2x} + c_3e^{3x}$.

23. $y = c_1 + c_2 \sin 2x + c_3 \cos 2x$.

25. $y = (c_1 + c_2x + c_3x^2)e^{-3x}$.

27. $y = c_1e^x + c_2e^{-2x} + c_3e^{-5x} + c_4e^{-6x}$.

29. $y = (c_1 + c_2x) \sin 2x + (c_3 + c_4x) \cos 2x$.

Section 3.2

1. $x = e^{-t}$.

3. $x = 2e^{2t} - e^{-2t}$.

5. $x = e^{2t} \cos 2t$.

7. $x = e^{-3t}(\cos 2t + 2 \sin 2t)$.

9. $x = t(1 - t)e^t$.

Section 3.4

1. $y_p = xe^x$.

3. $y_p = -\frac{1}{5}\cos 2x + \frac{2}{5}\sin 2x$.

5. $y_p = \frac{4}{5}xe^{-x}$.

7. $y_p = \frac{3}{10}x^2 \cos x$.

9. $y_p = \frac{3}{5}\cos x + \frac{1}{5}\sin x - \frac{1}{4}$.

11. $y_p = \frac{2}{3}x - \frac{7}{9} - e^{2x}(\frac{345}{439}\cos 3x + \frac{54}{439}\sin 3x)$.

13. $y_p = \frac{1}{12}x^4 + \frac{5}{6}x^3 + \frac{15}{2}x^2$.

15. $y_p = -xe^{2x} - \frac{1}{25}\sin 4x - \frac{1}{50}\cos 4x$.

Section 3.5

17. $x \sin x + (\cos x) \ln \cos x$.

19. $xe^{-x} \ln x$.

Section 3.7

1. $y = c_1x^{-1/2} + c_2x$.

3. $y = c_1x^3 + c_2x^{-3}$.

5. $y = c_1x^{-2} + c_2x^3 - \frac{1}{2}\ln x + \frac{1}{12}$.

7. $y = c_1x^{-1/3} + c_2x^{-1/3} \ln x$.

9. $y = c_1x^2 \cos(2 \ln x) + c_2x^2 \sin(2 \ln x)$.

11. $y = c_1x^{-1} + c_2x^{-1} \ln x + c_3x^{-1} \ln^2 x$.

13. $y = c_1x + c_2x^3 + c_3x^{-2}$.

15. $y = c_1 \cos(\ln x) + c_2 \sin(\ln x) + c_3x^2$.

Section 3.8

1. $\begin{pmatrix} 1 & -2e^t \\ 1 & e^t \end{pmatrix}$.

3. $\begin{pmatrix} 3e^{-t} & e^t \\ e^{-t} & e^t \end{pmatrix}$.

5. $\begin{pmatrix} 5e^{3t} & e^{4t} \\ 4e^{3t} & e^{4t} \end{pmatrix}$.

7. $\begin{pmatrix} e^{-t} & e^{-t} + te^{-t} \\ -e^{-t} & -te^{-t} \end{pmatrix}$.

9. $\begin{pmatrix} e^t & te^t \\ e^t & \frac{1}{2}e^t + te^t \end{pmatrix}$.

11. $\begin{pmatrix} -ie^{it} & ie^{-it} \\ e^{it} & e^{-it} \end{pmatrix}; \begin{pmatrix} \sin t & \cos t \\ \cos t & -\sin t \end{pmatrix}$.

13. $\begin{pmatrix} (-1+3i)e^{3it} & (-1-3i)e^{-3it} \\ e^{3it} & e^{-3it} \end{pmatrix}; \begin{pmatrix} -\sin 3t + 3\cos 3t & -\cos 3t - 3\sin 3t \\ \sin 3t & \cos 3t \end{pmatrix}$.

15. $\begin{pmatrix} (-1+i)e^{(2+i)t} & (-1-i)e^{(2-i)t} \\ e^{(2+i)t} & e^{(2-i)t} \end{pmatrix}; e^{2t}\begin{pmatrix} \cos t - \sin t & -\cos t - \sin t \\ \sin t & \cos t \end{pmatrix}$.

17. $\begin{pmatrix} e^{-2t} & 2e^{-3t} \\ e^{-2t} & e^{-3t} \end{pmatrix}$.

19. $\begin{pmatrix} (-1+i)e^{it} & (-1-i)e^{it} \\ -e^{it} & -e^{-it} \end{pmatrix}; \begin{pmatrix} \sin t - \cos t & \cos t + \sin t \\ \sin t & \cos t \end{pmatrix}$.

21. $\begin{pmatrix} e^{-t} \\ 3 \end{pmatrix}$.

23. $\begin{pmatrix} e^t \\ e^{-t} \end{pmatrix}$.

25. $\begin{pmatrix} t \\ t^2 \end{pmatrix}$.

Section 4.1

1. $6/s^3$.
3. $1/(s+1)^2$.
5. $(2s^2+1)/s^3$.
7. $\dfrac{1}{s-2} - \dfrac{1}{s^2+1} + \dfrac{1}{s^2}$.
9. $\dfrac{s}{4(s^2+4)} + \dfrac{2}{(s^2+4)} - \dfrac{s\cos 1}{s^2+1} - \dfrac{\sin 1}{s^2+1}$.

15. $4/3$.

17. $16/35$.

19. $\dfrac{1}{mc^{(n+1)/m}}\Gamma\left(\dfrac{n+1}{m}\right)$.

21. $(s^2 - \omega^2)/(s^2 + \omega^2)^2$; $2\omega s/(s^2 + \omega^2)^2$; $[(s - a)^2 - \omega^2/(s - a)^2 + \omega^2]^2$; $2\omega(s - a)/[(s - a)^2 + \omega^2]^2$.

Section 4.2

1. $-\frac{1}{6}e^t + \frac{1}{15}e^{-2t} + \frac{1}{10}e^{3t}$.
3. $\frac{6}{1001}e^{6t} - \frac{49}{936}e^{-7t} + \frac{75}{56}e^{-15t} - \frac{128}{99}e^{-16t}$.
5. $-1 + 7e^{t/2} - 6e^t + \frac{5}{2}te^t$.
7. $\frac{3}{10} - \frac{2}{5}e^t + \frac{1}{10}e^{3t}\cos t + \frac{1}{10}e^{3t}\sin t$.
9. $3\cos t - 3\cos\sqrt{2}t - \frac{3}{2}\sqrt{2}t\sin\sqrt{2}t$.
11. $\frac{3}{13}e^t - \frac{3}{13}e^{-2t}\cos 2t + \frac{2}{13}e^{-2t}\sin 2t$.
13. $-\frac{1}{2}\cos t - \frac{1}{4}\sin t + \left(\dfrac{6+\sqrt{3}}{24}\right)e^{\sqrt{3}t} + \left(\dfrac{6-\sqrt{3}}{24}\right)e^{-\sqrt{3}t}$.
15. $\frac{11}{45}e^{2t} + \frac{29}{20}e^{-3t} - \frac{25}{26}e^{-t} + \frac{1}{6}te^{-t}$.
17. $\frac{5}{4}e^{-t} - \frac{1}{4}e^t - \cos t + \frac{3}{2}\sin t$
19. $-\frac{1}{4}e^t + \frac{1}{2}te^t + \frac{1}{4}\cos\sqrt{3}t - \frac{1}{4}\sqrt{3}\sin\sqrt{3}t$.

Section 4.3

1. $e^t(\cos\sqrt{7}t + \sqrt{7}/7\sin\sqrt{7}t)$.
3. $-\frac{5}{2} - 3t - \frac{1}{2}e^{2t} + 4e^t$.
5. $t(3e^t - \frac{1}{2}) + 3e^t - \frac{3}{4}e^{2t} - 1$.
7. $\frac{103}{54}e^{-3t/2} - \frac{1}{2}e^{-t/2} + \frac{2}{3}t - \frac{8}{9}t + \frac{16}{27}$.
9. $-\frac{1}{4} - \frac{859}{500}e^{4t} + \frac{26}{125}e^{-t} - \frac{4}{25}te^{-t} - \frac{2}{5}t^te^{-t}$.
11. $-\frac{5}{32}e^{-2t} + \frac{5}{32}e^{2t} - \frac{5}{16}\sin 2t$.
13. $\frac{15}{16}e^{-5t} + \frac{1}{16}e^{-t} = \frac{3}{4}te^{-t}$.
15. $-\frac{3}{2} - \frac{3}{5}e^t - \frac{31}{10}e^{-t}\cos t - \frac{13}{10}e^{-t}\sin t$.
17. $y = 1 - 3\cos t, x = \sin t$.
19. $y = -5t, x = 2t^2 - 1$.

Section 4.5

1. $\chi_\alpha(t)(1 - e^{-(t-\alpha)} - (t - \alpha)e^{-(t-\alpha)})$.
3. $\chi_\alpha(t)\{\frac{1}{5} - \frac{1}{5}e^{-(t-\alpha)}[\cos 2(t - \alpha) + \sin 2(t - \alpha)]\}$.
5. $\chi_\alpha(t)\sin 2(t - \alpha)$.
7. $\frac{1}{5}\chi_\alpha(t)(e^{2(t-\alpha)} - e^{-3(t-\alpha)})$.
9. $\chi_\alpha(t)\sin 2(t - \alpha) - \chi_\beta(t)\sin 2(t - \beta)$.

Section 5.1

1. $x = 0$ (RSP).

3. $x = -1, \dfrac{1 \pm i\sqrt{31}}{2}$ (RSP).

5. $x = \pm\sqrt{2}$ (ISP).

7. $x = 0$ (RSP); $x = 1$ (RSP).

9. $x = \dfrac{-1 \pm i\sqrt{3}}{2}$ (RSP).

11. $x = 0$ (RSP).

13. $x = 2$ (RSP); $x = 3$ (ISP).

15. No real singular points.

17. $x = 0$ (ISP); $x = 1$ (RSP).

19. $x = 0$ (RSP).

Section 5.2

1. $y = a_0(1 + \frac{3}{4}x^2 + \frac{3}{16}x^4 + \frac{1}{64}x^6) + a_1\left[x + \sum_{k=1}^{\infty} \frac{5 \cdot 3 \cdot 1 \cdot (-1) \cdots (7 - 2k)k!}{2^k(2k+1)!} x^{2k+1}\right]$

3. $y = a_0\left[1 + \sum_{k=1}^{\infty} \frac{(-1)^k 2^k x^{2k}}{3^k \cdot 1 \cdot 3 \cdot 5 \cdots (2k-1)}\right] + a_1 \sum_{k=0}^{\infty} \frac{(-1)^k x^{2k+1}}{3^k k!}.$

5. $y = a_0\left[1 + \sum_{k=1}^{\infty} \frac{(-3)(-1) \cdot 3 \cdot 5 \cdots (2k-5)}{2^k k!} x^{2k}\right] + a_1 x(1 - \frac{2}{3}x^2).$

7. $y = a_0(1 - \frac{3}{2}x^2 - \frac{1}{6}x^4 - \frac{1}{30}x^6) + a_1\left[x + \sum_{k=1}^{\infty} \frac{(-5)(-3)(-1) \cdots (2k-7)}{2^k(2k+1)!} x^{2k+1}\right].$

9. $y = a_0 \sum_{k=0}^{\infty} \frac{(-1)^k(2k+1)!}{3^k 4^k (k!)^2} x^{2k} + a_1 \sum_{k=0}^{\infty} \frac{(-1)^k 4^k (k!)^2}{3^k(2k+1)!} x^{2k+1}.$

11. $y = a_0\left[1 + \sum_{k=1}^{\infty} \frac{(-1) \cdot 1 \cdot 3 \cdot 5 \cdots (2k-3)(2k+1)}{3^k k!} x^2\right] + a_1 x.$

13. $y = a_0\left[1 + \sum_{k=1}^{\infty} \frac{2^k(k+1)!(-1) \cdot 1 \cdot 3 \cdots (2k-3)}{(2k)!} x^{2k}\right] + a_1 x.$

15. $y = a_0 \sum_{k=0}^{\infty} \frac{(-1)^k 4^k x^{2k}}{(2k)!} + a_1 \sum_{k=0}^{\infty} \frac{(-1)^k 4^k x^{2k+1}}{(2k+1)!}.$

17. $y = a_0\left[1 + \sum_{k=1}^{\infty} \frac{3^k}{(3k)!}(x-3)^{3k}\right] + a_1\left[(x-3) + \sum_{k=1}^{\infty} \frac{2 \cdot 3^k}{(3k+1)!}(x-3)^{3k+1}\right].$

19. $y = a_0\left[1 + \sum_{k=1}^{\infty} \frac{(-1)^k 2^k (x-1)^{3k}}{3^k k! 2 \cdot 5 \cdot 8 \cdots (3k-1)}\right]$

$+ a_1\left[(x-1) + \sum_{k=1}^{\infty} \frac{(-1)^k 2^k (x-1)^{3k+1}}{4 \cdot 7 \cdot 10 \cdots (3k+1)3^k k!}\right].$

Section 5.4

1. $y_1 = \sum_{n=0}^{\infty} \frac{(-1)^n x^{n+1/2}}{2^n n!}; y_2 = \sum_{n=0}^{\infty} \frac{(-1)^n x^{n+1}}{3 \cdot 5 \cdot 7 \cdots (2n+1)}.$

3. $y_1 = 1 + \sum_{n=1}^{\infty} \frac{2^n x^n}{n!4 \cdot 7 \cdots (3n-2)}$; $y_2 = x^{2/3} + \sum_{n=1}^{\infty} \frac{2^n x^{n+2/3}}{n!5 \cdot 8 \cdots (3n+2)}$.

5. $y_1 = 1 + 4x + \frac{8}{3}(x^2)$; $y_2 = x^{1/2} + \sum_{n=0}^{\infty} \frac{(-1)^n(-3)(-1)\cdots(2n-5)}{2^n n!} x^{n+1/2}$.

7. $y_1 = x^{-3/2} + \sum_{n=1}^{\infty} \left(-\frac{7}{2}\right)^n \frac{(-3)(-1)1\cdots(2n-5)}{1 \cdot 3 \cdot 5 \cdots (2n-1)} x^{n-3/2}$; $y_2 = x^{-1} + \frac{7}{3}$.

9. $y_1 = 1$; $y_2 = x^{7/5} + \sum_{n=1}^{\infty} \frac{(-1)^n 3^n \cdot 7[2 \cdot 7 \cdot 12 \cdots (5n-3)]}{5^{2n}(5n+7)n!} x^{n+7/5}$.

11. $y_1 = 1 + x$; $y_2 = x^{2/3}\left[1 - \frac{1}{15}(x) - \frac{1}{90}(x^2) + \sum_{n=3}^{\infty} \frac{1 \cdot 4 \cdot 7 \cdots (3n-2)(-2)}{3^n(3n-1)(3n+2)n!} x^n\right]$.

13. $y_1 = \sum_{n=0}^{\infty} \frac{2^n}{3^n n!} x^{n-2/3}$; $y_2 = x + \sum_{n=1}^{\infty} \frac{2^n x^{n+1}}{8 \cdot 11 \cdot 14 \cdots (3n+5)}$.

15. $y_1 = 1 + \sum_{n=1}^{\infty} \frac{(-1)^n x^n}{n!(-1)5 \cdot 11 \cdots (6n-7)}$; $y_2 = x^{7/6}\left[1 + \sum_{n=1}^{\infty} \frac{(-7)^n x^n}{n!13 \cdot 19 \cdots (6n+7)}\right]$.

17. $y_1 = (x-1)^{1/2}\left[1 + \sum_{n=1}^{\infty} \frac{3 \cdot 5 \cdots (2n-1)(x-1)^n}{2^n n!9 \cdot 11 \cdots (2n+7)}\right]$;

$y_2 = (x-1)^{-3} + \frac{3}{5}(x-1)^{-2} + \frac{1}{5}(x-1)^{-1} + \frac{1}{15}$.

19. $y_1 = 1 + \sum_{n=1}^{\infty} \frac{(-4)^n 2^{n-1}}{n(2n-1)!}(x+7)^n$; $y_2 = \sum_{n=0}^{\infty} \frac{(-4)^n 2^n}{(2n+1)!}(x+7)^{n+1/2}$.

Section 6.1

1. Between 2.05 and 2.10 seconds.

3. $\tilde{\theta}_{max} \cong 1.0081$; $t \cong 1.70$ sec.

Section 6.3

1. (a) About 0.425 sec.
   (b) About 0.6375 radians.

3. (a) Slightly over 0.4 sec.
   (b) 0.7835 radians.

Section 6.5

1. (a) About 316 hours.
   (b) $x = 11388$, $y = 1170$.
   (c) 10,848 and 361,436.
   (d) 350 and 2177.

3. $xx'' + (\gamma x + \delta x^2 - x')x' - \alpha x^2(\gamma - \delta x) = 0$.
   $yy'' + (-\alpha y + \beta y^2 - y')y' - \alpha y^2(\gamma - \beta y) = 0$.

Section 7.1

1. $y_1 = 1$; $y_2 = y_1 \ln x + \sum_{n=1}^{\infty} \frac{(-1)^n x^n}{n!n}$.

3. $y_1 = 1$, $y_2 = \ln x$.

5. $y_1 = 1 + 3x + \frac{9}{4}x^2$; $y_2 = y_1 \ln x - \frac{9}{2} - \frac{27}{4}x + \sum_{n=3}^{\infty} \frac{2(-3)^n x^n}{2^n(n-2)(n-1)n}$.

7. $y_1 = x^{-2} + 2x^{-1} + \frac{1}{2}; y_2 = y_1 \ln x + 2 \sum_{k=1}^{\infty} \frac{(-1)^k x^{k-2}}{k!(k-2)(k-1)k}$.

9. $y_1 = \sum_{k=0}^{\infty} \frac{(-1)^k}{(k!)^2}\left(\frac{x}{2}\right)^{2k}; y_2 = y_1 \ln x - \sum_{k=1}^{\infty} \frac{(-1)^k H_k}{(k!)^2}\left(\frac{x}{2}\right)^{2k}$.

11. $y_1 = 1 + \sum_{k=1}^{\infty} \frac{(-1)^k 1^2 \cdot 3^2 \cdots (2k-3)^2}{6^k (k!)^2} x^k;$

$$y_2 = y_1 \ln x + \sum_{k=1}^{\infty} \frac{(-1)^k 1^2 \cdot 3^2 \cdots (2k-3)^2}{6(k!)^2}(-2 + 2H_{2k-2} - H_{k-1} - 2H_k)x^k.$$

13. $y_1 = x^3\left[1 + \sum_{n=0}^{\infty} \frac{5^n (n+1)(n+2)x^n}{2n!}\right];$

$$y_2 = y_1 \ln x + \sum_{n=1}^{\infty} \frac{5^n (n+1)(n+2)}{2n!}\left(\frac{1}{n+1} + \frac{1}{n+2} - \frac{3}{2} - H_{n-3}\right)x^{n+3}.$$

15. $y_1 = \sum_{k=0}^{\infty} \frac{x^k}{2^{3k}(k!)}; y_2 = y_1 \ln x + \sum_{k=1}^{\infty} \frac{(-H_k)}{2^{3k}(k!)^2} x^k$.

17. $y_1 = (x-1)^{\frac{1}{2}}\left[1 + \sum_{n=1}^{\infty} \frac{(-1)^n 1 \cdot 3 \cdots (2n-1)}{(n-1)!^2}(x-1)^n\right];$

$$y_2 = y_1 \ln(x-1) + 2(x-1)^{\frac{1}{2}} \sum_{n=1}^{\infty} \frac{(-1)^n 1 \cdot 3 \cdots (2n-1)}{(n-1)!^2}(H_{2n-1} - \tfrac{5}{2}H_{n-1})(x-1)^n.$$

19. $y_1 = \sum_{k=0}^{\infty} \frac{(-1)^k 2^k}{(k!)^2}(x+3)^k; y_2 = y_1 \ln(x+3) + \sum_{k=1}^{\infty} \frac{(-1)^{k+1} 2^{k+1} H_k}{(k!)^2}(x+3)^k$.

Section 7.2

1. $y_1 = a_0\left(1 + \sum_{k=1}^{\infty} \frac{(-1)^k (-1) \cdot 5 \cdot 11 \cdots (6k-7)x^{2k}}{2^k (2k)!}\right);$

$$y_2 = a_1\left(x + \sum_{k=1}^{\infty} \frac{(-1)^k 1 \cdot 4 \cdot 7 \cdots (3k-2)x^{2k+1}}{(2k+1)!}\right).$$

3. $y_1 = a_0 x^{-3/2}\left(1 + \sum_{k=1}^{\infty} \frac{(-1)^k x^k}{(-1) \cdot 1 \cdot 3 \cdots (2k-3)2^k k!}\right);$

$$y_2 = a_3 x^{3/2}\left(1 + \sum_{k=2}^{\infty} \frac{(-1)^{k-1} x^{k-1}}{2^{k-1}(k-1)! 5 \cdot 7 \cdot 9 \cdots (2k+1)}\right).$$

5. $y_1 = a_0(x^{-2} - x^{-1} + \frac{1}{2}); y_2 = a_3 x^{-2} \sum_{n=3}^{\infty} \frac{(-1)^{n-3}}{n!} x^n$.

7. $y_1 = a_0; y_2 = a_3 \sum_{k=1}^{\infty} \frac{(-1)^{k+1} x^{3k}}{3^{k-1} k!}$.

9. $y_1 = a_0(1 - \frac{5}{3}x + \frac{10}{9}x^2); y_2 = a_5 x^5\left[1 + \sum_{n=6}^{\infty} \frac{(-\frac{2}{3})^{n-5} 5!(n+4)!(n-3)!}{2 \cdot 9! n!(n-5)!} x^n\right]$.

11. $y_1 = a_0(1 - \frac{3}{2}x); y_2 = a_2\left[x^2 + \sum_{n=3}^{\infty} \frac{(-1)^{n-2}3^{n-2}x^n}{2^{n-3}n!}\right].$

13. $y_1 = a_0(x^{-2} - \frac{3}{4}x^{-1} + \frac{3}{16}); y_2 = a_5x^3\left[1 + \sum_{n=6}^{\infty} \frac{(-3)^{n-5}(n-4)(n-3)}{2^{n-4}\cdot 6\cdot 7\cdots n} x^{n-5}\right].$

15. $y_1 = a_0(1 + 3x + 28x^2 + 56x^3 + 70x^4 + 56x^5 + 28x^6); y_2 = a_7x^7(1 + \frac{1}{8}x).$

17. $y_1 = a_0\left[1 + \sum_{k=1}^{\infty} \frac{(-1)^k 3\cdot 7\cdots(4k-1)}{3^k(2k)!}(x-3)^{2k}\right];$

$$y_2 = a_1\left[(x-3) + \sum_{k=1}^{\infty} \frac{(-1)^k 5\cdot 9\cdots(4k+1)}{3^k(2k+1)!}(x-3)^{2k+1}\right].$$

19. $y_1 = a_0[(x+4)^{-5} - \frac{4}{3}(x+4)^{-4} + \frac{2}{3}(x+4)^{-3}];$

$$y_2 = a_4 \sum_{n=4}^{\infty} \frac{(-2)^{n-4}24(n-3)(x+4)^{n-5}}{n!}.$$

Section 7.3

1. $y_1 = \sum_{n=1}^{\infty} \frac{(-1^n)(-5)(-1)(3)\cdots(4n-9)}{2^n(n-1)!} x^{n-1};$

$$y_2 = y_1 \ln x + x^{-1}\left[1 + \sum_{n-1}^{\infty} (-1)^n(-5)(-1)\cdot 3\cdots(4n-9)\left(\frac{-1}{5} - 1 + \frac{1}{3} + \cdots + \frac{1}{4n-9} - H_{n-1}\right)x^n\right].$$

3. $y_1 = \frac{x}{2} + \sum_{\kappa=2}^{\infty} \frac{(-1)^\kappa}{\kappa!(\kappa-1)!}\left(\frac{x}{2}\right)^{2\kappa-1};$

$$y_2 = y_1 \ln x + \frac{1}{x} + \frac{x}{4} + \sum_{\kappa=2}^{\infty} \frac{(-1)^{\kappa+1}}{2^{2\kappa}\kappa!(\kappa-1)!}(H_\kappa + H_{\kappa-1})x^{2\kappa-1}.$$

5. $y_1 = \sum_{n=2}^{\infty} \frac{(-1)^{n+1}2^n}{(n-2)!} x^{n+1};$

$$y_2 = y_1 \ln x + x\left[1 + 2x + \sum_{n=2}^{\infty} \frac{(-1)^{n+1}2^n}{(n-2)!}(1 - H_{n-2})x^n\right].$$

7. $y_1 = \frac{4}{3}x^2 + \sum_{n=3}^{\infty} \frac{(-1)^n 1\cdot 4\cdots(3n-5)(n+1)(n+2)}{3^n(n-2)!} x^n;$

$$y_2 = y_1 \ln x + 1 - 2x + \frac{19}{9}x^2 + \sum_{n=3}^{\infty} \frac{(-1)^n 1\cdot 4\cdots(3n-5)(n+1)(n+2)}{3^n(n-2)!} \cdot\left[\frac{-3}{2} + 3 + \frac{3}{4} + \cdots + \frac{3}{3n-5} + \frac{1}{n+1} + \frac{1}{n+2} - \frac{1}{2} - H_{n-2}\right]x^n.$$

9. $y_1 = \sum_{n=1}^{\infty} \frac{x^{2n-3}}{2^{n-1}(n-1)!2n}; y_2 = y_1 \ln x + x^{-3}\left[1 + \sum_{n=1}^{\infty} \frac{(2nH_{n-2} + 1)x^{2n}}{2^n n!^2}\right].$

11. $y_1 = \frac{1}{8}x^{\frac{1}{2}} + \sum_{n=2}^{\infty} \frac{(-1)^n(-\frac{1}{2})(\frac{1}{2})\cdots(n-\frac{3}{2})}{4^n n!(n-1)!} x^{n-\frac{1}{2}}$;

$y_2 = y_1 \ln x + x^{-\frac{1}{2}}\left[1 - \frac{3}{8}x + \sum_{n=2}^{\infty} \frac{(-1)^n(-\frac{1}{2})(\frac{1}{2})\cdots(n-\frac{3}{2})}{4^n n!(n-1)!}\left(\frac{2}{3} + \cdots + \frac{2}{2n-3}\right.\right.$

$\left.\left. - (H_n + H_{n-1})\right)x^n\right]$.

13. $y_1 = 120x^3 + 300x^4 - 540x^5$; $y_2 = y_1 \ln x + 1 - 5x + 30x^2 - 4x^3 - 475x^4$

$+ 1675x^5 + \sum_{n=6}^{\infty} \frac{(-1)^{n+1}5!(n-6)!(n+1)x^n}{2(n-3)!}$.

15. $y_1 = \frac{9}{32}x^{3/2} + \sum_{n=3}^{\infty} \frac{(-1)^{n+1}3^n x^{n-\frac{1}{2}}}{4^n n!(n-2)!}$;

$y_2 = y_1 \ln x + x^{-\frac{1}{2}}\left[1 + \frac{3}{4}x + \frac{9}{64}x^2 + \sum_{n=3}^{\infty} \frac{(-3)^n}{4^n n!(n-2)!}(H_n + H_{n-2} + 1)x^n\right]$.

17. $y_1 = \sum_{n=3}^{\infty} \frac{2^{n-1}(n+1)}{(n-3)!}(x+1)^{n+2}$; $y_2 = y_1 \ln(x+1) + (x+1)^2\left[1 - 2(x+1)\right.$

$\left. + 6(x+1)^2 + \sum_{n=5}^{\infty} \frac{2^{n-1}(n+1)}{(n-3)!}\left(H_{n-3} - \frac{1}{2} - \frac{1}{n+1}\right)(x+1)^n\right]$.

19. $y_1 = -\frac{1}{4} + \sum_{n=5}^{\infty} \frac{(-1)(n-3)(n-3)(n-1)n}{2^n \cdot 6}(x+1)^{n-4}$;

$y_2 = y_1 \ln(x+1) + (x+1)^{-4} - \frac{1}{6}(x+1)^{-3} + \frac{1}{2}(x+1)^{-2} - \frac{1}{8}(x+1)^{-1} - \frac{47}{48}$

$+ \sum_{n=5}^{\infty} \frac{(-1)(n-3)(n-2)(n-1)n}{2^n \cdot 6}\left(\frac{1}{n-3} + \frac{1}{n-2}\right.$

$\left. + \frac{1}{n-1} + \frac{1}{n} + \frac{11}{6}\right)(x+1)^{n-4}$.

Section 7.5

1. $y = c_0(1 - \frac{3}{2}x^{-2}) + c_1(x^{-1} - \frac{1}{6}x^{-3})$.
3. $y = c_0 e^{1/x} + c_1 e^{2/x}$.
5. $y = c_0 \sin\frac{1}{x} + c_1 \cos\frac{1}{x}$.
7. $y = e^{2/x}\left(c_0 \cos\frac{3}{x} + c_1 \sin\frac{3}{x}\right)$.
9. $y = c_0(1 + 8x^{-3} + \frac{8}{5}x^{-6}) + c_1(x^{-1} + \frac{5}{2}x^{-4})$.

Index